PHYSICS AND ASTROPHYSICS OF
HADRONS AND HADRONIC MATTER

PHYSICS AND ASTROPHYSICS OF HADRONS AND HADRONIC MATTER

Editor

A. B. Santra

Narosa Publishing House

New Delhi Chennai Mumbai Kolkata

Physics and Astrophysics of Hadrons and Hadronic Matter
344 pgs | 16 tbs | 138 figs

Editor
A.B. Santra
Nuclear Physics Division
Bhabha Atomic Research Center
Mumbai, India

Copyright © 2008 Narosa Publishing House Pvt. Ltd.

N A R O S A   P U B L I S H I N G   H O U S E   P V T.   L T D.

22 Delhi Medical Association Road, Daryaganj, New Delhi 110 002
35-36 Greams Road, Thousand Lights, Chennai 600 006
306 Shiv Centre, D.B.C. Sector 17, K.U. Bazar P.O., Navi Mumbai 400 703
2F-2G Shivam Chambers, 53 Syed Amir Ali Avenue, Kolkata 700 019

www.narosa.com

Printed from the camera-ready copy provided by the Editor.

ISBN 978-81-7319-881-6

Published by N.K. Mehra for Narosa Publishing House Pvt. Ltd., 22 Delhi Medical Association Road, Daryaganj, New Delhi 110 002

Printed in India

Preface

Quantum Chromodynamics (QCD), the relativistic field theory of quarks and gluons, is the fundamental theory of strong interaction that governs the dynamics of hadronic phenomena. However, QCD is analytically intractable at the low energy regime. In such a situation, the search for building connections of QCD with low energy hadronic or nuclear many-body phenomenology has led to the development of Nuclear Effective Field Theory, which provides a systematic approach to hadron physics and few-nucleon systems at low energy.

Compact astrophysical systems, neutron stars in particular, are very dense objects of hadronic matter. One would like to know about the composition of hadronic matter and hadronic interactions at high densities prevalent inside the neutron stars. The neutron star thus serves as a laboratory for testing knowledge of hadron physics gathered in terrestrial laboratories, as well as studying the same at very high densities.

In the recent years, nuclear physicists are becoming more and more involved in research in both of these areas.

This volume is a well written document of the lecture series and seminars on the above topics presented at the International Workshop on "Physics and Astrophysics of Hadrons and Hadronic Matter" organized jointly by Nuclear Physics Division of Bhabha Atomic Research Centre (BARC) and Physics department of Visva Bharati during November 6 to 11, 2006 at the Physics Department of Visva Bharati, Santiniketan, India.

The generous financial support received from the Board of Research in Nuclear Sciences (BRNS) of Department of Atomic Energy, Government of India, Saha Institute of Nuclear Physics (SINP), Kolkata, Institute of Physics (IOP), Bhubaneswar, Visva Bharati, Santiniketan and Inter-University Centre for Astronomy and Astrophysics (IUCAA), Pune is gratefully acknowledged.

I am grateful to the authors for kindly writing up the lectures and promptly giving the manuscripts. I am indebted to Dr. R.K. Choudhury and Dr. S. Kailas of Nuclear Physics Division, BARC for their support and guidance. I thank Professor Somenath Chakrabarty and other faculty members, students and staff of the Physics Department of Visva Bharati for their kind help.

A. B. Santra
(BARC)

Contents

Physics and Astrophysics of Hadrons and Hardronic Matter
Editor: A. B. Santra

An Introduction to Chiral Perturbation Theory

Bastian Kubis

Helmholtz-Institut für Strahlen- und Kernphysik (Theorie),
Universität Bonn, Nussallee 14-16, D-53115 Bonn, Germany
email: kubis@itkp.uni-bonn.de

Abstract

A brief introduction to the low-energy effective field theory of the standard model, chiral perturbation theory, is presented.

1   Introduction

The phenomenology of the strong interactions at low energies, where the coupling constant of Quantum Chromodynamics (QCD) becomes large and renders perturbation theory useless, remains one of the major challenges of modern particle physics. Only two rigorous approaches to this part of the standard model are known: lattice QCD; and the effective field theory called chiral perturbation theory. Both also offer the only ways to provide a firmer foundation for nuclear physics, rooted in the standard model.

These lectures provide an introduction to chiral perturbation theory, organised as follows. In Sect. 2, we present a few fundamental ideas on effective field theories in general. Section 3 introduces the basic concept and construction of chiral Lagrangians; a first application is the determination of the ratios of light quark masses. In Sect. 4, chiral perturbation theory is extended to higher orders, in particular, the relation of the quark mass expansion of the pion mass to pion-pion scattering is discussed in some detail. Section 5 extends chiral perturbation theory to include nucleons; the complications in doing loop calculations are explained, and the quark mass expansion of the nucleon mass is discussed in relation to pion-nucleon scattering. A final outlook summarises some major omissions that could not be covered here.

Several useful pedagogical introductions and review articles on the subject have been consulted in the course of the preparation of these lectures, see in particular [1–7], many of them being much more comprehensive than the present article, which are therefore recommended for further reading.

2   Effective field theories

The large number of different areas in physics describe phenomena at very disparate scales of length, time, energy, or mass. It is a rather intuitive idea that, as long as one is only interested in a particular parameter range, scales much bigger or much smaller than the ones one is interested in should not influence the description of the system in question too strongly. Indeed, the two fundamental revolutions in physics in the early 20th century did not come about earlier because they involve scales far removed from our everyday experience: for velocities far smaller than the speed of light, $v \ll c$, relativity effects can safely be ignored; and for energies and time scales much larger than Planck's constant, $E \times t \gg \hbar$, quantum effects are rarely relevant.

One striking feature in particle physics is the extremely wide range of observed particle masses. Even ignoring neutrinos, the masses of the fermions comprise nearly six orders of magnitude, ranging from the electron mass $m_e = 0.511$ MeV to the top quark mass $m_t \approx 180$ GeV. On the other hand, we can very well calculate the properties of certain systems, say, the spectrum of the hydrogen atom, without having any precise knowledge of m_t at all. In this sense, very heavy particles do not seem to have a significant influence on the description of the system.

These rather informal ideas are explored more systematically in effective field theories. Let us, as an example, consider a theory with a set of "light" degrees of freedom l_i and heavy fields H_j, their respective masses well separated by a scale Λ,

$$m_{l_i} \ll \Lambda \lesssim M_{H_j} \ .$$

For energies well below Λ, the heavy particles can be integrated out of the generating functional, leaving behind an effective Lagrangian for the light degrees of freedom only,

$$\mathcal{L}(l_i, H_j) \xrightarrow{E \ll \Lambda} \mathcal{L}_{\text{eff}}(l_i) = \mathcal{L}_{d \leq 4} + \sum_{d > 4} \frac{1}{\Lambda^{d-4}} \sum_{i_d} g_{i_d} O_{i_d} \ , \tag{1}$$

i.e. the procedure potentially generates non-renormalisable operators of dimension larger than 4. In a so-called "decoupling" effective field theory, the effects of the heavy fields H_j enter $\mathcal{L}_{\text{eff}}(l_i)$ either as renormalisation effects of the effective coupling constants, or as new, higher-dimensional operators suppressed by inverse powers of the heavy mass M_{H_j}.

We want to briefly illustrate such decoupling effective theories with a classic example: electrodynamics far below the electron mass.

2.1 An example: light-by-light scattering

In Quantum Electrodynamics (QED), the only available mass scale is the mass of the electron m_e. If we consider QED at energies far below this mass scale, $\omega \ll m_e$, we should be able to write down an effective Lagrangian that contains only photons as dynamical degrees of freedom,

$$\mathcal{L}_{\text{QED}}[\psi, \bar{\psi}, A_\mu] \to \mathcal{L}_{\text{eff}}[A_\mu] \ .$$

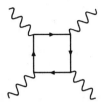

Figure 1: Feynman diagram for light-by-light scattering.

However, the electrons present in the full theory generate effective interactions for the photon fields, contributing e.g. to photon–photon scattering, see Fig. 1. However, we do not have to calculate the underlying loop diagrams explicitly in order to understand the structure of the effective theory; rather we can write it down directly in terms of the invariants $F_{\mu\nu}F^{\mu\nu} \propto \vec{E}^2 - \vec{B}^2$, $F_{\mu\nu}\tilde{F}^{\mu\nu} \propto \vec{E} \cdot \vec{B}$. Considering only terms with up to four photon fields, we find

$$\mathcal{L}_{\text{eff}} = \frac{1}{2}\left(\vec{E}^2 - \vec{B}^2\right) + \frac{e^4}{16\pi^2 m_e^4}\left[a\left(\vec{E}^2 - \vec{B}^2\right)^2 + b\left(\vec{E} \cdot \vec{B}\right)^2\right] + \dots , \qquad (2)$$

where the prefactor of the interaction term is taken from dimensional analysis, such that the coupling constants a and b are expected to be of order 1. These constants can only be calculated explicitly from the underlying theory, with the result $7a = b = 14/45$ [8].

The following points are to be noted from this brief example, which are typical for the construction of (low-energy) effective field theories:

1. We have constructed the interaction terms in the Lagrangian based on the *symmetries* (gauge invariance, Lorentz invariance) of the underlying theory, which ought to be shared by the effective one.

2. We could only guess the order of magnitude of the effective coupling constants a, b correctly, but their exact values are not determined by symmetry considerations alone; they have to be calculated explicitly from the dynamics of the fundamental theory.

3. By considering only the simplest invariant terms that can be constructed in terms of the field strength tensor $F_{\mu\nu}$ (and its adjoint) and no additional derivatives, we have implicitly performed a low-energy expansion of the amplitude, i.e. an expansion in powers of $(\omega/m_e)^{2n}$.

4. It is obvious from (2) that the calculation of cross sections etc. is much simpler and more efficiently done using $\mathcal{L}_{\text{eff}}$ than performing the calculation in full QED.

2.2 Weinberg's conjecture

The following statement by Weinberg lies at the very heart of the successful application of effective field theories, using an effective Lagrangian framework:

> Quantum Field Theory has no content besides unitarity,
> analyticity, cluster decomposition, and symmetries. [9]

This means that in order to calculate the S-matrix for any theory below some scale, simply use the most general effective Lagrangian consistent with these principles in terms of the appropriate asymptotic states. This is what we have done in the previous subsection for QED at very low energies; and we will now follow this principle in the construction of an effective theory for the strong interactions.

3 The strong interactions at low energies

3.1 The symmetries of Quantum Chromodynamics

The spectrum of states of strongly interacting particles displays some interesting features. Above a typical hadronic mass scale of about 1 GeV, there is a large number of states, both meson resonances and baryons. Only a very few (pseudoscalar) states, however, are significantly lighter than this mass scale: in particular the pions ($M_\pi \approx 140$ MeV), but also kaons ($M_K \approx 495$ MeV) and the eta ($M_\eta \approx 550$ MeV).

The widely accepted theory of the strong interactions is Quantum Chromodynamics, a theory formulated in terms of quark and gluon fields built on the principle of colour gauge invariance with the gauge group $SU(3)_c$. The running strong coupling constant leads to the phenomena of asymptotic freedom in the high-energy regime, but also to confinement of the quark and gluon degrees of freedom inside colour-neutral hadronic states. The large coupling prevents a use of perturbation theory at low energies, so there is no direct and obvious link between QCD and its *fundamental* degrees of freedom, and the *relevant* hadronic degrees of freedom as observed in the spectrum of mesons and baryons.

In order to construct a low-energy effective theory for the strong interactions, we have to investigate the symmetries of the QCD Lagrangian more closely. For this purpose, we decompose the quark fields into its chiral components according to

$$q = \frac{1}{2}(1 - \gamma_5)q + \frac{1}{2}(1 + \gamma_5)q = P_L q + P_R q = q_L + q_R . \tag{3}$$

Using this, we can write the QCD Lagrangian as

$$\begin{aligned} \mathcal{L}_{\text{QCD}} &= \mathcal{L}_{\text{QCD}}^0 - \mathcal{L}_{\text{QCD}}^m + \cdots , \\ \mathcal{L}_{\text{QCD}}^0 &= -\frac{1}{2}\text{Tr}\, G_{\mu\nu}^a G^{\mu\nu,a} + i\bar{q}_L \slashed{D} q_L + i\bar{q}_R \slashed{D} q_R , \end{aligned}$$

$$\mathcal{L}^m_{\text{QCD}} = \bar{q}_L \mathcal{M} q_R + \bar{q}_R \mathcal{M}^\dagger q_L , \tag{4}$$

where D_μ is the covariant derivative, $G^a_{\mu\nu}$ the gluon field strength tensor, q collects the light quark flavours $q^T = (u, d, s)$, and $\mathcal{M} = \text{diag}(m_u, m_d, m_s)$ is the quark mass matrix. The ellipse denotes the heavier quark flavours, gauge fixing terms etc. We note that, besides the obvious symmetries like Lorentz-invariance, $\text{SU}(3)_c$ gauge invariance, and the discrete symmetries P, C, T, $\mathcal{L}_{\text{QCD}}$ displays a *chiral* symmetry in the limit of vanishing quark masses (which is hence called "chiral limit"): $\mathcal{L}^0_{\text{QCD}}$ is invariant under chiral $U(3)_L \times U(3)_R$ flavour transformations,

$$(q_L, q_R) \longmapsto (L q_L, R q_R) , \quad L, R \in \text{U}(3)_{L,R} . \tag{5}$$

As the masses of the three light quarks are *small* on the typical hadronic scale,

$$m_{u,d,s} \ll 1 \text{ GeV} \approx \Lambda_{hadr} ,$$

there is hope that the real world is not too far from the chiral limit, such that one may invoke a perturbative expansion in the quark masses. The effective theory constructed in the following, based on this idea, is therefore called "chiral perturbation theory" (ChPT).

If we rewrite the symmetry group according to

$$\text{U}(3)_L \times \text{U}(3)_R = \text{SU}(3)_L \times \text{SU}(3)_R \times \text{U}(1)_V \times \text{U}(1)_A , \tag{6}$$

where we have introduced vector $V = L + R$ and axial vector $A = L - R$ transformations, and consider the Noether currents associated with this symmetry group, it turns out that the different parts of it are realised in very different ways in nature:

- The $\text{U}(1)_V$ current $V^{\mu,0} = \bar{q}\gamma_\mu q$, the quark number or baryon number current, is a conserved current in the standard model.

- The $\text{U}(1)_A$ current $A^{\mu,0} = \bar{q}\gamma_\mu\gamma_5 q$ is broken by quantum effects, the $\text{U}(1)_A$ anomaly, and is not a conserved current of the quantum theory.

As far as the chiral symmetry group $\text{SU}(3)_L \times \text{SU}(3)_R$ and its conserved currents

$$\begin{aligned} V^{\mu,a} &= \bar{q}\gamma^\mu \frac{\lambda^a}{2} q , & \partial_\mu V^{\mu,a} &= 0 , \\ A^{\mu,a} &= \bar{q}\gamma^\mu\gamma_5 \frac{\lambda^a}{2} q , & \partial_\mu A^{\mu,a} &= 0 , \end{aligned} \qquad a = 1, \ldots, 8$$

are concerned, they are certainly broken *explicitly* by the quark masses, but this is expected to be a small effect. Hence the main question is whether chiral symmetry is realised in nature in the Wigner–Weyl mode, i.e. the symmetry is manifest in the spectrum in terms of multiplets, or whether it is realised as the Goldstone mode, i.e. the symmetry is hidden or spontaneously broken.

5

Can chiral symmetry of the strong interactions be in the Wigner–Weyl mode? In this case, the conserved axial charges annihilating the vacuum,

$$Q_5^a |0\rangle = 0 , \quad Q_5^a = \int d^3x A^{0,a}(x) ,$$

would lead to parity doubling in the hadron spectrum. Phenomenologically, we find (approximate) $SU(3)_V$ multiplets, but no parity doubling is observed. Furthermore, unbroken chiral symmetry would lead to a vanishing difference of the vector–vector and axial–axial vacuum correlators, $\langle 0|VV|0\rangle - \langle 0|AA|0\rangle = 0$. This difference can be measured in hadronic tau decays $\tau \to \nu_\tau + n\,\pi$, leading to a non-vanishing result [10].

If chiral symmetry is, however, realised in the Goldstone mode, the Vafa–Witten theorem [11] asserts that the vector subgroup should remain unbroken, in accordance with the observation of hadronic multiplets, so the symmetry breaking pattern would be

$$SU(3)_L \times SU(3)_R \xrightarrow{\text{SSB}} SU(3)_V . \tag{7}$$

The axial charges then commute with the Hamiltonian, but do not leave the ground state invariant. As a consequence, massless excitations, so-called "Goldstone bosons" appear, which are non-interacting for vanishing energy. In the case at hand, the 8 Goldstone bosons should be pseudoscalars, which the lightest hadrons in the spectrum indeed are, namely $\pi^\pm$, π^0, $K^\pm$, K^0, $\bar{K}^0$, and η. The task is now to construct a low-energy theory for these Goldstone bosons. This is an example for a *non-decoupling* effective field theory: in contrast to the example of QED at energies below the electron mass described earlier, the transition from the full to the effective theory proceeds via a phase transition / via spontaneous symmetry breakdown, in the course of which new light degrees of freedom are generated.

3.2 Construction of the effective Lagrangian

We want to develop a general formalism [12] to construct the effective theory for the Goldstone bosons corresponding to a symmetry group G spontaneously broken to its subgroup H, hence $G \xrightarrow{\text{SSB}} H$, where $\dim(G) - \dim(H) = n$. We combine the Goldstone boson fields in a vector $\vec{\phi} = (\phi_1, \ldots, \phi_n)$, $\phi_i : M^4 \to \mathbb{R}$ (where M^4 denotes Minkowski space). The symmetry group G acts on $\vec{\phi}$ according to

$$g \in G : \quad \vec{\phi} \longmapsto \vec{\phi}' = \vec{f}(g, \vec{\phi}) ,$$

which has to obey the composition law

$$\vec{f}(g_1, \vec{f}(g_2, \vec{\phi})) = \vec{f}(g_1 g_2, \vec{\phi}) .$$

Consider the image of the origin $\vec{f}(g, \vec{0})$: elements leaving the origin invariant form a subgroup, the conserved subgroup H. Now we have

$$\forall g \in G \quad \forall h \in H \quad \vec{f}(gh, \vec{0}) = \vec{f}(g, \vec{0}) \ ,$$

therefore $\vec{f}$ maps the quotient space G/H onto the space of Goldstone boson fields. This mapping is invertible, as $\vec{f}(g_1, \vec{0}) = \vec{f}(g_2, \vec{0})$ implies $g_1 g_2^{-1} \in H$. Hence we conclude that the Goldstone bosons can be identified with elements of G/H. For $q_i \in G/H$, the action of G on G/H is then given by

$$g q_1 = q_2 h(g, q_1) \ , \quad h(g, q_1) \in H \ ,$$

therefore the coordinates of G/H transform nonlinearly under G.

In the case of QCD, we denote the group elements by $g \sim (g_R, g_L)$, $g_{R/L} \in \mathrm{SU}(3)_{R/L}$, with the composition law

$$g_1 g_2 = (g_{R_1}, g_{L_1})(g_{R_2}, g_{L_2}) = (g_{R_1} g_{R_2}, g_{L_1} g_{L_2}) \ .$$

The choice of a representative element inside each equivalence class is in principle arbitrary, the convention for $gH \in G/H$ is to rewrite $(g_R, g_L) = (\mathbb{E}, g_L g_R^{-1})(g_R, g_R)$ and to characterise each element of G/H, i.e. each Goldstone boson, uniquely by a unitary matrix

$$U = g_L g_R^{-1} = \exp\left(\frac{i\lambda_a \phi_a}{F'}\right) \ . \tag{8}$$

Here, the ϕ_a, $a = 1, \ldots, 8$ are the Goldstone boson fields, and F' is a dimensionful constant to be determined later. How does U transform under the chiral group? We find

$$(R, L)(\mathbb{E}, U) \cdot H = (\mathbb{E}, LUR^\dagger)(R, R) \cdot H = (\mathbb{E}, LUR^\dagger) \cdot H \ ,$$

therefore

$$U \xmapsto{G} LUR^\dagger \ . \tag{9}$$

3.3 The leading-order Lagrangian

Now we know how the Goldstone boson fields transform under chiral transformations, we can proceed to construct a Lagrangian in terms of the matrix U that is invariant under $\mathrm{SU}(3)_L \times \mathrm{SU}(3)_R$. As we want to construct a *low-energy* effective theory, the guiding principle is to use the power of momenta or derivatives to order the importance of various possible terms. "Low energies" here refer to a scale well below 1 GeV, i.e. an energy region where the Goldstone bosons are the only relevant degrees of freedom.

Lorentz invariance dictates that Lagrangian terms can only come in even powers of derivatives, hence $\mathcal{L}$ is of the form

$$\mathcal{L} = \mathcal{L}^{(0)} + \mathcal{L}^{(2)} + \mathcal{L}^{(4)} + \dots \ . \tag{10}$$

However, as U is unitary, therefore $UU^\dagger = \mathbb{E}$, $\mathcal{L}^{(0)}$ can only be a constant. Therefore, in accordance with the Goldstone theorem, the leading term in the Lagrangian is $\mathcal{L}^{(2)}$, which already involves derivatives. It can be shown to consist of one single term,

$$\mathcal{L}^{(2)} = \frac{F^2}{4} \langle \partial_\mu U \partial^\mu U^\dagger \rangle \ , \tag{11}$$

where F is another dimensionful constant, and

$$U = \exp\left(\frac{i\phi}{F'}\right) \ , \quad \phi = \sqrt{2} \begin{pmatrix} \frac{\phi_3}{\sqrt{2}} + \frac{\phi_8}{\sqrt{6}} & \pi^+ & K^+ \\ \pi^- & -\frac{\phi_3}{\sqrt{2}} + \frac{\phi_8}{\sqrt{6}} & K^0 \\ K^- & \bar{K}^0 & -\frac{2\phi_8}{\sqrt{6}} \end{pmatrix} . \tag{12}$$

Expanding U in powers of ϕ, $U = 1 + i\phi/F' - \phi^2/(2F'^2) + \dots$, we find the canonical kinetic terms

$$\mathcal{L}^{(2)} = \partial_\mu \pi^+ \partial^\mu \pi^- + \partial_\mu K^+ \partial^\mu K^- + \dots$$

exactly for $F' = F$. The invariance of (11) under $SU(3)_L \times SU(3)_R$ is easily verified:

$$\langle \partial_\mu U \partial^\mu U^\dagger \rangle \longmapsto \langle \partial_\mu U' \partial^\mu U'^\dagger \rangle = \langle L \partial_\mu U R^\dagger R \partial^\mu U^\dagger L^\dagger \rangle = \langle \partial_\mu U \partial^\mu U^\dagger \rangle \ .$$

We remark here that our derivation of (11) is somewhat heuristic. A more formal proof of the equivalence of QCD and its representation in terms of an effective Lagrangian, based on an analysis of the chiral Ward identities, is given in [13]. Furthermore, we have neglected anomalies in the above reasoning, which can be shown to enter only at next-to-leading order [14].

3.4 The constant F

In order to determine the constant F, we proceed to calculate the Noether currents V_a^μ, A_a^μ from $\mathcal{L}^{(2)}$:

$$V_a^\mu, A_a^\mu = R_a^\mu \pm L_a^\mu = i \frac{F^2}{4} \langle \lambda_a [\partial^\mu U, U^\dagger]_\mp \rangle \ . \tag{13}$$

Expanding the axial current in powers of ϕ, we find $A_a^\mu = -F \partial^\mu \phi_a + \mathcal{O}(\phi^3)$, such that we can calculate the matrix element of the axial current between a one-boson state and the vacuum,

$$\langle 0 | A_a^\mu | \phi_b(p) \rangle = ip^\mu \delta_{ab} F \ , \tag{14}$$

from which we conclude that F is the pion (meson) decay constant (in the chiral limit), which is measured in pion decay $\pi^+ \to \ell^+ \nu_\ell$, $F \approx F_\pi = 92.4$ MeV .

3.5 Explicit symmetry breaking: quark masses

So far, we have only considered the chiral limit $m_u = m_d = m_s = 0$. Accordingly, we have constructed a theory for massless Goldstone bosons, and indeed, $\mathcal{L}^{(2)}$ does not contain any mass terms. In nature, the quark masses are small, but certainly non-zero, therefore chiral symmetry is explicitly broken. In order to account for this fact, the quark masses have to be re-introduced perturbatively. For this purpose, we have to understand the transformation properties of the symmetry breaking term; then the (appropriately generalised) effective Lagrangian is still the right tool to systematically derive all symmetry relations of the theory.

From the QCD mass term $\mathcal{L}^m_{\text{QCD}} = \bar{q}_L \mathcal{M} q_R + \bar{q}_R \mathcal{M}^\dagger q_L$, we notice it *would* be invariant under chiral transformations if $\mathcal{M}$ transformed according to

$$\mathcal{M} \longmapsto \mathcal{M}' = L\mathcal{M}R^\dagger . \tag{15}$$

Assuming this, we now construct chirally invariant Lagrangian terms from U, derivatives thereon, plus the quark mass matrix $\mathcal{M}$; this procedure guarantees that chiral symmetry is broken in exactly the same way in the effective theory as it is in QCD.

At leading order, i.e. to linear order in the quark masses and without any further derivatives, we find exactly one term in the chiral Lagrangian, such that $\mathcal{L}^{(2)}$ is of the form

$$\mathcal{L}^{(2)} = \frac{F^2}{4} \langle \partial_\mu U \partial^\mu U^\dagger + 2B(\mathcal{M}U^\dagger + \mathcal{M}^\dagger U) \rangle . \tag{16}$$

Expanding once more in powers of ϕ, we can read off the mass terms and find

$$M^2_{\pi^\pm} = B(m_u + m_d) , \quad M^2_{K^\pm} = B(m_u + m_s) , \quad M^2_{K^0} = B(m_d + m_s) . \tag{17}$$

We recover the Gell-Mann–Oakes–Renner relation $M^2_{GB} \propto m_q$, which justifies the unified power counting for the expansion in numbers of derivatives as well as quark masses according to $m_q = \mathcal{O}(p^2)$. We furthermore find that the flavour-neutral states ϕ_3, ϕ_8 are mixed when isospin breaking due to a difference in the light quark masses is allowed for, $m_u - m_d \neq 0$:

$$\mathcal{L}^{(2)} \longrightarrow \frac{B}{2} \begin{pmatrix} \phi_3 \\ \phi_8 \end{pmatrix}^T \begin{pmatrix} m_u + m_d & \frac{1}{\sqrt{3}}(m_u - m_d) \\ \frac{1}{\sqrt{3}}(m_u - m_d) & \frac{1}{3}(m_u + m_d + 4m_s) \end{pmatrix} \begin{pmatrix} \phi_3 \\ \phi_8 \end{pmatrix} ,$$

which can be diagonalised by the rotation

$$\begin{pmatrix} \pi^0 \\ \eta \end{pmatrix} = \begin{pmatrix} \cos\epsilon & -\sin\epsilon \\ \sin\epsilon & \cos\epsilon \end{pmatrix} \begin{pmatrix} \phi_3 \\ \phi_8 \end{pmatrix} , \quad \epsilon = \frac{1}{2} \arctan\left(\frac{\sqrt{3}}{2} \frac{m_d - m_u}{m_s - \hat{m}} \right) ,$$

where $\hat{m} = (m_u + m_d)/2$. The mass eigenvalues receive corrections to the isospin limit, which are however of second order in $m_u - m_d$,

$$M_{\pi^0}^2 = B(m_u + m_d) - \mathcal{O}\big((m_u - m_d)^2\big) \ ,$$
$$M_\eta^2 = \frac{B}{3}(m_u + m_d + 4m_s) + \mathcal{O}\big((m_u - m_d)^2\big) \ . \tag{18}$$

In the isospin limit, we of course find $M_{\pi^\pm}^2 = M_{\pi^0}^2$, $M_{K^\pm}^2 = M_{K^0}^2$. Finally, we can deduce the Gell-Mann–Okubo mass formula (for the pseudoscalars)

$$4M_K^2 = 3M_\eta^2 + M_\pi^2 \ , \tag{19}$$

which is found to be fulfilled in nature to 7% accuracy.

3.6 Quark mass ratios

The unknown factor B in the Gell-Man–Oakes–Renner relations prevents a direct quark mass determination from pseudoscalar meson masses. However, we can form quark mass ratios in which B cancels:

$$\frac{m_u}{m_d} = \frac{M_{K^+}^2 - M_{K^0}^2 + M_{\pi^+}^2}{M_{K^0}^2 - M_{K^+}^2 + M_{\pi^+}^2} \approx 0.66 \ , \tag{20}$$

$$\frac{m_s}{m_d} = \frac{M_{K^0}^2 + M_{K^+}^2 - M_{\pi^+}^2}{M_{K^0}^2 - M_{K^+}^2 + M_{\pi^+}^2} \approx 22 \ . \tag{21}$$

In particular the result for m_u/m_d is remarkable: it is very different from 1, so why is there no large isospin violation observed in nature? The answer is threefold: first, in purely pionic physics, only $(m_d - m_u)^2$ occurs, hence strong isospin violation is of second order. Second, $(m_d - m_u)/m_s$ (as showing up e.g. in the $\pi^0\eta$ mixing angle ϵ) is small; and third, compared to the typical hadronic scale, $(m_d - m_u)/\Lambda_{\text{hadr}}$ is small, too.

We can calculate the (strong) pion mass difference

$$M_{\pi^0}^2 = M_{\pi^+}^2 \left\{ 1 - \frac{(m_d - m_u)^2}{8\hat{m}(m_s - \hat{m})} + \dots \right\} \ ,$$

and evaluate it numerically by plugging in the quark mass ratios to find

$$M_{\pi^+} - M_{\pi^0} \approx 0.1 \text{ MeV} \ , \tag{22}$$

while the experimental mass difference is $(M_{\pi^+} - M_{\pi^0})_{\text{exp}} \approx 4.6$ MeV. The difference, the by far larger effect, is due to the second source of isospin violation that we have neglected so far: electromagnetism.

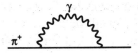

Figure 2: Photon loop diagram contributing to the pion self energy.

3.7 Electromagnetic effects

The coupling of $\mathcal{L}^{(2)}$ to external vector (v_μ) and axial vector (a_μ) currents is rather straightforward: we only have to replace the ordinary derivative by a covariant one according to

$$\partial_\mu U \to D_\mu U = \partial_\mu U - i[v_\mu, U] - i\{a_\mu, U\} \ . \tag{23}$$

If we insert the photon field for the vector current, $v_\mu = eA_\mu$, this will generate all the couplings necessary to calculate, say, the electromagnetic form factor of the pion, or pion Compton scattering.

However, including electromagnetism via minimal substitution alone does not generate the most general effects due to *virtual* photons. Consider, e.g., the contribution of a photon loop to the pion self-energy diagram, Fig. 2: for dimensional reasons, the contribution to the pion mass has to vanish in the chiral limit, while a non-vanishing term can be generated in certain models. Naively speaking, Fig. 2 neglects photon exchanges between the (charged) quarks *inside* the pion. We therefore have to generalise the chiral Lagrangian once more. We proceed [15] in analogy to the quark mass term, and now include the quark charge matrix as an additional element, $Q = e \operatorname{diag}(2, -1, -1)/3$. The part of the QCD Lagrangian coupling quarks to photons, decomposed into chiral components, takes the form

$$\mathcal{L}_{\text{QCD}}^{\text{em}} = -\bar{q}Q\slashed{A}q \ \longrightarrow \ -\bar{q}_L Q_L \slashed{A} q_L - \bar{q}_R Q_R \slashed{A} q_R \ . \tag{24}$$

If we postulate the following transformation law(s) for the spurion fields $Q_{L,R}$:

$$Q_L \longmapsto LQ_L L^\dagger \ , \quad Q_R \longmapsto RQ_R R^\dagger \ ,$$

then (24) is seen to be invariant under chiral transformations. We hence construct Lagrangian terms using $Q_{L,R}$, and set $Q_L = Q_R = Q$ in the end. The power counting is generalised to count $Q_{L,R} = \mathcal{O}(p)$. We find one single term at $\mathcal{O}(e^2) = \mathcal{O}(p^2)$:

$$\mathcal{L}_{\text{em}}^{(2)} \ = \ C\langle Q_L U Q_R U^\dagger \rangle \ . \tag{25}$$

which contributes to the masses of the charged mesons:

$$\left(M_{\pi^+}^2 - M_{\pi^0}^2\right)_{\text{em}} = \left(M_{K^+}^2 - M_{K^0}^2\right)_{\text{em}} = \frac{2Ce^2}{F^2} \ . \tag{26}$$

The equality of electromagnetic contributions to pion and kaon mass differences in the chiral limit is known as Dashen's theorem [16]. (25) has no contributions to neutral masses, or to $\pi^0 \eta$-mixing. With electromagnetic effects included, we find an improved quark mass ratio

$$\frac{m_u}{m_d} = \frac{M_{K^+}^2 - M_{K^0}^2 + 2M_{\pi^0}^2 - M_{\pi^+}^2}{M_{K^0}^2 - M_{K^+}^2 + M_{\pi^+}^2} = 0.55 , \tag{27}$$

which deviates significantly from (20).

3.8 $\pi\pi$ scattering to leading order

With the constants F, B (in products with quark masses), C fixed from phenomenology, the leading-order Lagrangian (16), (25) is completely determined, and we can go on and make predictions for other processes. A particularly important example is pion-pion scattering. For now, we revert to the isospin limit and set $m_u = m_d$, $e^2 = 0$, such that the scattering amplitude can be decomposed as

$$M(\pi^a \pi^b \to \pi^c \pi^d) = \delta^{ab}\delta^{cd} A(s,t,u) + \delta^{ac}\delta^{bd} A(t,u,s) + \delta^{ad}\delta^{bc} A(u,s,t) . \tag{28}$$

If we calculate the invariant amplitude $A(s,t,u)$ from $\mathcal{L}^{(2)}$, we find

$$A(s,t,u) = \frac{s - M_\pi^2}{F^2} , \tag{29}$$

a parameter-free prediction [17]. The isospin amplitudes are then given by

$$\begin{aligned}
T^{I=0} &= 3A(s,t,u) + A(t,u,s) + A(u,s,t) , \\
T^{I=1} &= A(t,u,s) - A(u,s,t) , \\
T^{I=2} &= A(t,u,s) + A(u,s,t) .
\end{aligned} \tag{30}$$

If we furthermore define the s-wave scattering lengths, proportional to the scattering amplitudes at threshold, $a_0^I = T^I(s = 4M_\pi^2, t = u = 0)/32\pi$, we find

$$a_0^0 = \frac{7M_\pi^2}{32\pi F_\pi^2} = 0.16 , \qquad a_0^2 = -\frac{M_\pi^2}{16\pi F_\pi^2} = -0.045 . \tag{31}$$

4 Chiral perturbation theory at higher orders

So far, we have only considered chiral Lagrangians at leading order, i.e. $\mathcal{O}(p^2)$. Are there good reasons to go beyond that level? First of all, although $\mathcal{O}(p^0)$ interactions are forbidden by chiral symmetry, all higher orders are allowed and therefore present in principle. They ought to be smaller at low energies, but for precision predictions, these corrections should be taken into account. Second, although ChPT is an effective field theory, it is still a quantum theory, i.e. we should also expect loop contributions. Remembering the $\pi\pi$ scattering amplitude (29), we notice that it is *real*, while unitarity requires the partial waves t_ℓ^I to obey

$$\operatorname{Im} t_\ell^I = \sqrt{1 - \frac{4M_\pi^2}{s}} \, |t_\ell^I|^2 \; . \tag{32}$$

The correct imaginary parts are only generated perturbatively by loops. But how do loop diagrams feature in the power counting scheme? What about divergences arising thereof, how does renormalisation work in such a theory?

4.1 Weinberg's power counting argument

Let us consider an arbitrary loop diagram based on the general effective Lagrangian $\mathcal{L}_{\mathrm{eff}} = \sum_d \mathcal{L}^{(d)}$, where d denotes the chiral power of the various terms. If we calculate a diagram with L loops, I internal lines, and V_d vertices of order d, the generic form of the corresponding amplitude $\mathcal{A}$ in terms of powers of momenta is

$$\mathcal{A} \propto \int (d^4 p)^L \frac{1}{(p^2)^I} \prod_d (p^d)^{V_d} \; . \tag{33}$$

Let $\mathcal{A}$ be of chiral dimension ν, then obviously $\nu = 4L - 2I + \sum_d dV_d$. We use the topological identity $L = I - \sum_d V_d + 1$ to eliminate I and find

$$\nu = \sum_d V_d (d - 2) + 2L + 2 \; . \tag{34}$$

The following points are to be noted about (34):

- The chiral Lagrangian starts with $\mathcal{L}^{(2)}$, i.e. $d \geq 2$, therefore the right-hand-side of (34) is a sum of non-negative terms. Consequently, for fixed ν, there is only a finite number of combinations L, V_d that can contribute.

- Each additional loop integration suppresses the amplitude by two orders in the momentum expansion.

As an example, let us consider $\pi\pi$ scattering. At lowest order p^2, only tree-level graphs composed of vertices of $\mathcal{L}^{(2)}$ contribute ($V_{d>2} = 0$, $L = 0$). The only graph is shown in Fig. 3(a). At $\mathcal{O}(p^4)$, there are two possibilities: either one-loop graphs composed only of lowest-order vertices ($V_{d>2} = 0$, $L = 1$), or tree graphs with exactly one insertion from $\mathcal{L}^{(4)}$ ($V_4 = 1$, $V_{d>4} = 0$, $L = 0$). Example graphs for both types are given in Fig. 3(b). Finally, at $\mathcal{O}(p^6)$, (34) allows for four different types of graphs: two-loop graphs with $\mathcal{L}^{(2)}$ vertices ($V_{d>2} = 0$, $L = 2$); one-loop graphs with one vertex from $\mathcal{L}^{(4)}$ ($V_4 = 1$, $V_{d>4} = 0$, $L = 1$); tree graphs with two insertions

13

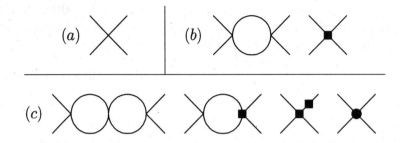

Figure 3: Feynman graphs contributing to $\pi\pi$ scattering at (a) $\mathcal{O}(p^2)$, (b) $\mathcal{O}(p^4)$, (c) $\mathcal{O}(p^6)$. The squares denote vertices from $\mathcal{L}^{(4)}$, the circle a vertex from $\mathcal{L}^{(6)}$.

from $\mathcal{L}^{(4)}$ ($V_4 = 2$, $V_{d>4} = 0$, $L = 0$); and tree graphs with one insertion from $\mathcal{L}^{(6)}$ ($V_4 = 0$, $V_6 = 1$, $V_{d>6} = 0$, $L = 0$). Typical examples for all these are displayed in Fig. 3(c).

4.2 Chiral symmetry breaking scale Λ_χ

Although we have now established a power counting scheme that determines the power of momenta p^n at which a certain diagram contributes, it is not yet clear compared to what *scale* these momenta are to be small. If we write the effective Lagrangian in the (slightly unconventional) form

$$\mathcal{L}_{\text{eff}} = \frac{F^2}{4}\left\{ \langle \partial_\mu U \partial^\mu U^\dagger \rangle + \frac{1}{\Lambda_\chi^2}\tilde{\mathcal{L}}^{(4)} + \frac{1}{\Lambda_\chi^4}\tilde{\mathcal{L}}^{(6)} + \ldots \right\} , \tag{35}$$

we need to know what the chiral symmetry breaking scale Λ_χ is. In other words, if we calculate higher-order corrections, what is their expected size $\propto p^2/\Lambda_\chi^2$, $\propto M_\pi^2/\Lambda_\chi^2$?

As a first argument, let us compare the two $\mathcal{O}(p^4)$ contributions to $\pi\pi$ scattering in Fig. 3(b) in a generic manner:

$$\propto \int \frac{d^4p}{(2\pi)^4} \frac{1}{(p^2)^2} \left(\frac{p^2}{F^2}\right)^2 \overset{\text{dim.reg}}{\propto} \frac{p^4}{(4\pi)^2 F^4} \log \mu ,$$

$$\propto \frac{F^2}{\Lambda_\chi^2} \frac{p^4}{F^4} \ell_i(\mu) .$$

The scale-dependent "low-energy constant" (LEC) $\ell_i(\mu)$ multiplying the tree-level graph compensates for the logarithmic scale dependence of the loop graph (as evaluated in dimensional regularisation). Naturally, the finite part of $\ell_i(\mu)$ should not be expected to be smaller than the shift induced by a change in the scale μ, therefore

$$\Lambda_\chi \approx 4\pi F \approx 1.2 \text{ GeV} . \tag{36}$$

14

As a second argument, we have to remember that we have constructed an effective theory for Goldstone bosons, which are the only dynamical degrees of freedom. The effective theory must fail once the energy reaches the resonance region, hence for $p^2/\Lambda_\chi^2 \approx p^2/M_{res}^2 \approx 1$. The resonance masses are channel-dependent, the lightest being $M_{res} = M_\rho = 770$ MeV, and typically $M_{res} \approx 1$ GeV, therefore this second estimate is roughly consistent with $\Lambda_\chi = 4\pi F$.

4.3 The chiral Lagrangian at higher orders

The number of independent terms and corresponding low-energy constants increases rapidly at higher orders. In fact, for chiral SU(n), $n = (2,3)$,[1]

$$
\begin{aligned}
\mathcal{L}^{(2)} &\quad \text{contains} \quad (2,2) \quad \text{constants} \quad (F, B), \\
\mathcal{L}^{(4)} &\quad \text{contains} \quad (7,10) \quad \text{constants} \quad [18, 19], \\
\mathcal{L}^{(6)} &\quad \text{contains} \quad (53,90) \quad \text{constants} \quad [20]
\end{aligned}
$$

(discounting so-called contact terms that depend on external fields only). As an example, we display $\mathcal{L}^{(4)}$ explicitly for chiral SU(3):

$$
\begin{aligned}
\mathcal{L}^{(4)} =\ & L_1 \langle D_\mu U^\dagger D^\mu U \rangle^2 + L_2 \langle D_\mu U^\dagger D_\nu U \rangle \langle D^\mu U^\dagger D^\nu U \rangle \\
& + L_3 \langle D_\mu U^\dagger D^\mu U D_\nu U^\dagger D^\nu U \rangle + L_4 \langle D_\mu U^\dagger D^\mu U \rangle \langle \chi^\dagger U + \chi U^\dagger \rangle \\
& + L_5 \langle D_\mu U^\dagger D^\mu U (\chi^\dagger U + \chi U^\dagger) \rangle + L_6 \langle \chi^\dagger U + \chi U^\dagger \rangle^2 \\
& + L_7 \langle \chi^\dagger U - \chi U^\dagger \rangle^2 + L_8 \langle \chi^\dagger U \chi^\dagger U + \chi U^\dagger \chi U^\dagger \rangle \\
& - i L_9 \langle F_R^{\mu\nu} D_\mu U D_\nu U^\dagger + F_L^{\mu\nu} D_\mu U^\dagger D_\nu U \rangle + L_{10} \langle U^\dagger F_R^{\mu\nu} U F_{L\mu\nu} \rangle \ . \quad (37)
\end{aligned}
$$

Here, $\chi = 2B(s+ip)$ collects (pseudo)scalar source terms, where s contains the quark mass matrix, $s = \mathcal{M} + \ldots$; vector and axial currents are combined as $r_\mu = v_\mu + a_\mu$, $l_\mu = v_\mu - a_\mu$, from which one can form field strength tensors, $F_R^{\mu\nu} = \partial^\mu r^\nu - \partial^\nu r^\mu - i[r^\mu, r^\nu]$, and similarly $F_L^{\mu\nu}$. We note that L_{1-3} multiply structures containing four derivatives; $L_{4,5}$ those with two derivatives and one quark mass term; the structures corresponding to L_{6-8} scale with the quark masses squared. $L_{9,10}$ only contribute to observables with external vector and axial vector sources. The seven similar constants in SU(2) that we will also partly use later are conventionally denoted by ℓ_i, $i = 1, \ldots, 7$.

4.4 The physics behind the low-energy constants

To better understand the role of the low-energy constants and the physics they incorporate, let us consider massive states (resonances) that are "integrated out" of

---

[1]In contrast to $\mathcal{L}^{(2)}$, which has the same form for both SU(2) and SU(3), the number of terms at higher orders is different in both theories because, although both have the same most general SU(N) Lagrangian, certain matrix-trace (Cayley-Hamilton) relations render some of the structures redundant, such that the minimal numbers of independent terms differ.

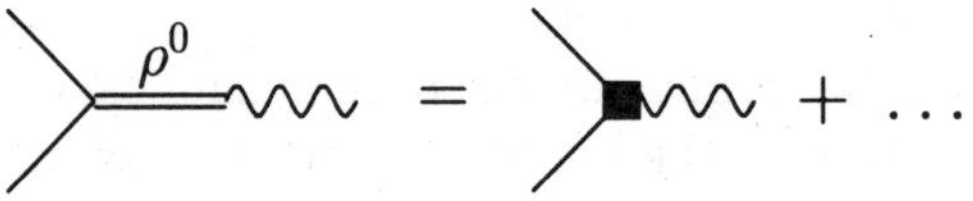

Figure 4: The contribution of the ρ-resonance to the pion vector form factor can, at small t, be represented by a point-like counterterm.

the theory, i.e. no dynamical degrees of freedom for energies below Λ_χ:

$$\mathcal{L}[U, \partial U, \ldots, H] \xrightarrow{\Lambda} \mathcal{L}_{\text{eff}}[U, \partial U, \ldots] . \tag{38}$$

A Lagrangian for the resonance fields, coupled to source terms, is of the form

$$\mathcal{L}[H] = \frac{1}{2} \left(\partial_\mu H \partial^\mu H - M_H^2 H^2 \right) + JH , \tag{39}$$

where J is a current formed of light degrees of freedom. In the path integral, we can use the fact the the heavy particle propagator $\Delta_H(x - y)$ is peaked for small separations $|x - y|$,

$$S_{\text{eff}}[J] = -\frac{1}{2} \int d^4x d^4y \, J(x) \Delta_H(x - y) J(y) \simeq \int d^4x \frac{1}{2M_H^2} J(x) J(y) + \ldots , \tag{40}$$

i.e. this generates local higher-order operators, with the couplings proportional to the inverse heavy masses. We conclude therefore that the effects of higher-mass states on the Goldstone boson interactions are hidden in the low-energy constants of higher-order terms in the chiral Lagrangian.

As an example, we consider the pion vector form factor $F_\pi^V(s)$, defined as

$$\langle \pi^a(p) \pi^b(p') | \bar{q} \frac{\tau^3}{2} \gamma_\mu q | 0 \rangle = i \epsilon^{a3b} (p' - p)_\mu F_\pi^V(s) , \quad s = (p + p')^2 . \tag{41}$$

In ChPT at $\mathcal{O}(p^4)$, there are loop diagrams contributing to $F_\pi^V(s)$ as well as a tree graph proportional to the low-energy constant $\bar{\ell}_6$. Expanding $F_\pi^V(s)$ for small s, we find for the radius term $\langle r^2 \rangle_\pi^V$

$$F_\pi^V(s) = 1 + \frac{1}{6} \langle r^2 \rangle_\pi^V s + \mathcal{O}(s^2) , \quad \langle r^2 \rangle_\pi^V = \frac{1}{(4\pi F_\pi)^2} (\bar{\ell}_6 - 1) . \tag{42}$$

Now consider the contribution of the ρ-resonance (see e.g. [21] for how to treat vector mesons in the context of chiral Lagrangians) to this form factor, as shown in Fig. 4.

Figure 5: Diagrams contributing to the pion self energy up to $\mathcal{O}(p^4)$.

Expanding the ρ propagator for $s \ll M_\rho^2$,

$$\frac{s}{M_\rho^2 - s} = \frac{s}{M_\rho^2}\left(1 + \frac{s}{M_\rho^2} + \cdots\right), \tag{43}$$

we find that identifying the leading term with the $\bar{\ell}_6$ contribution reproduces the empirical value for $\bar{\ell}_6$ nicely. This is a modern version of the time-honoured concept of "vector meson dominance": where allowed by quantum numbers, the numerical values of LECs are dominated by the contributions of vector resonances [22, 23].

4.5   The pion mass to $\mathcal{O}(p^4)$

As an example for a higher-order calculation, we consider the pion mass up to $\mathcal{O}(p^4)$. The necessary diagrams are shown in Fig. 5. We find for the pion propagator

$$\delta^{ab}\Delta(p) = i\int d^4x\, e^{ipx}\langle 0|T\phi^a(x)\phi^b(0)|0\rangle,$$

$$\Delta(p) = \frac{Z}{M_\pi^2 - p^2}\ (+2\ \text{loops}). \tag{44}$$

The physical pion mass is given, to this order, by

$$M_\pi^2 = M^2 + \frac{M^2}{2F^2}I + \frac{2\ell_3}{F^2}M^4, \tag{45}$$

where $M^2 = B(m_u + m_d)$, and the loop integral

$$I = \frac{1}{i}\int \frac{d^4l}{(2\pi)^4}\frac{1}{M^2 - l^2} \tag{46}$$

is actually divergent and has to be regularised. A "good" regularisation scheme, for our purposes, is dimensional regularisation, as it preserves all symmetries (which is much more difficult to achieve using a cutoff, say). In d dimensions, we find

$$I \to \frac{1}{i}\int \frac{d^dl}{(2\pi)^d}\frac{1}{M^2 - l^2} = \frac{M^{d-2}}{(4\pi)^{d/2}}\Gamma\left(1 - \frac{d}{2}\right), \tag{47}$$

which is finite for $d \neq 2,\, 4,\, 6, \ldots$, but still divergent for $d = 4$:

$$I \;\to\; \frac{M^2}{8\pi^2} \left\{ \frac{1}{d-4} + \log \frac{M}{\mu} + \ldots \right\} .$$

We now tune ℓ_3 such as to absorb this divergence (as well as the μ-dependence):

$$\ell_3 \;\to\; -\frac{1}{32\pi^2} \left(\frac{1}{d-4} + \log \frac{M}{\mu} + \frac{\bar{\ell}_3}{2} \right) , \qquad (48)$$

such that $\bar{\ell}_3$ contains the finite part of ℓ_3, and find

$$M_\pi^2 \;=\; M^2 - \frac{M^4}{32\pi^2 F^2} \bar{\ell}_3 + \mathcal{O}(M^6) . \qquad (49)$$

A few comments on renormalisation as we just saw it at work for the first time are in order. The required counterterm to cancel the divergence stems from $\mathcal{L}^{(4)}$; it is not sufficient to tune $\mathcal{L}^{(2)}$ parameters. This is typical for a non-renormalisable theory: going to to higher and higher orders, we need more and more counterterms. However, the fact that the theory is non-renormalisable does not mean it is non-calculable, the only disadvantage is the increasing number of LECs when calculating higher-order corrections. It is important to note furthermore that the LECs feature in different observables, and that their divergent parts and scale dependences are always the same. The renormalisation can be performed on the level of the generating functional in a manifestly chirally invariant way, and the β-functions of the ℓ_i [18], L_i [19] (and also of $\mathcal{L}^{(6)}$ LECs [24]) are known. The cancellation of divergences and scale dependence therefore serves as a powerful check on any specific calculations.

4.6 Quark mass expansion of the pion mass revisited

The expression (49) provides a correction to the Gell-Mann–Oakes–Renner relation: generically, it is of the form

$$M_\pi^2 = B(m_u + m_d) + A(m_u + m_d)^2 + \mathcal{O}(m_q^3) . \qquad (50)$$

Apart from naive order-of-magnitude expectations, how do we actually know that the leading term dominates? What if $\bar{\ell}_3$ turns out to be anomalously large? The consequences of this possibility were explored under the label of "generalised ChPT", which employs a different power counting scheme [25]. The essential question in order to determine the size of second-order corrections in the quark mass expansion of the pion mass is therefore: how can we learn something about $\bar{\ell}_3$?

18

4.7 $\pi\pi$ scattering at next-to-leading order

It turns out the process to study is, once more, $\pi\pi$ scattering. The $\pi\pi$ scattering lengths are known to next-to-next-to-leading order [26], the $\mathcal{O}(p^4)$ corrections for the $I = 0$ scattering length are of the form [18]

$$
\begin{aligned}
a_0^0 &= \frac{7M_\pi^2}{32\pi F_\pi^2}\left\{1 + \epsilon + \mathcal{O}(M_\pi^4)\right\} , \\
\epsilon &= \frac{5M_\pi^2}{84\pi^2 F_\pi^2}\left(\bar{\ell}_1 + 2\bar{\ell}_2 + \frac{3}{8}\bar{\ell}_3 + \frac{21}{10}\bar{\ell}_4 + \frac{21}{8}\right) .
\end{aligned}
\tag{51}
$$

There are two different types of LECs in the expression (51). $\bar{\ell}_1$, $\bar{\ell}_2$ come with structures containing four derivatives (like L_{1-3} in (37)), i.e. they survive in the chiral limit and can be determined from the momentum dependence of the $\pi\pi$ scattering amplitude, namely from d-waves. $\bar{\ell}_3$, $\bar{\ell}_4$ however are symmetry breaking terms that specify the quark mass dependence (comparable to $L_{4,5}$ in (37)), therefore they cannot be determined from $\pi\pi$ scattering alone.

One additional observable to consider is the scalar form factor of the pion $\Gamma(s)$, which is defined as

$$
\langle \pi^a(p)\pi^b(p') \,|\, \hat{m}(\bar{u}u + \bar{d}d) \,|\, 0\rangle = \delta^{ab}\Gamma(s) , \qquad s = (p + p')^2 .
\tag{52}
$$

At tree level, one has $\Gamma(s) = 2B\hat{m} = M_\pi^2 + \mathcal{O}(p^4)$ in accordance with the Feynman-Hellman theorem, $\Gamma(0) = \langle \pi \,|\, \hat{m}\,\bar{q}q \,|\, \pi\rangle = \hat{m}\,\partial M_\pi^2/\partial\hat{m}$. At next-to-leading order, one defines the scalar radius $\langle r^2\rangle_\pi^S$ according to

$$
\begin{aligned}
\Gamma(s) &= \Gamma(0)\left\{1 + \frac{1}{6}\langle r^2\rangle_\pi^S s + \mathcal{O}(s^2)\right\} , \\
\langle r^2\rangle_\pi^S &= \frac{3}{8\pi^2 F_\pi^2}\left(\bar{\ell}_4 - \frac{13}{12}\right) + \mathcal{O}(M_\pi^2) ,
\end{aligned}
\tag{53}
$$

therefore the scalar radius is directly linked to $\bar{\ell}_4$. Although the scalar form factor is not directly experimentally accessible, one can analyse $\Gamma(s)$ in dispersion theory and extract $\langle r^2\rangle_\pi^S$ that way.

If we plug in the low-energy theorems and eliminate $\bar{\ell}_1$, $\bar{\ell}_2$, $\bar{\ell}_4$ in favour of the d-wave scattering lengths $a_2^{I=0,2}$ and the scalar radius, we can rewrite ϵ in (51) as

$$
\epsilon = \frac{M_\pi^2}{3}\langle r^2\rangle_\pi^S + \frac{200\pi}{7}F_\pi^2 M_\pi^2\left(a_2^0 + 2a_2^2\right) - \frac{M_\pi^2}{672\pi^2 F_\pi^2}\left(15\bar{\ell}_3 - 353\right) .
\tag{54}
$$

Therefore all we need to do is to measure a_0^0 and extract $\bar{\ell}_3$ from the above relation. This will tell us how much of M_π^2 is due to the linear term in the quark mass expansion.

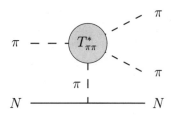

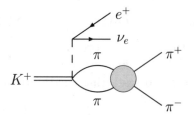

Figure 6: One-pion exchange contributions to $\pi N \rightarrow \pi\pi N$.

Figure 7: $\pi\pi$ rescattering in K_{e4} decays.

4.8 Experiments on $\pi\pi$ scattering

How can the $\pi\pi$ scattering lengths be measured? We wish to briefly comment on four different possibilities to extract information on the $\pi\pi$ interaction at low energies: pion reactions on nucleons, in particular $\pi N \rightarrow \pi\pi N$; so-called K_{e4} decays $K^+ \rightarrow \pi^+\pi^- e^+\nu_e$ [27, 28]; the cusp phenomenon in $K^+ \rightarrow \pi^+\pi^0\pi^0$ [29]; and the lifetime of pionium [30].

4.8.1 $\pi N \rightarrow \pi\pi N$

As shown schematically in Fig. 6, the process $\pi N \rightarrow \pi\pi N$ receives contributions from graphs containing the off-shell $\pi\pi$ amplitude $T^*_{\pi\pi}$, therefore it ought to be sensitive to $\pi\pi$ interactions. In order to isolate the one-pion-exchange contribution, one has to extrapolate to the pion pole at $t = M^2_\pi$, which is outside the physical region; this procedure is not without difficulties, and the data obtained thereof tend to be at relatively high energies, so it does not give direct access to scattering lengths without further theoretical input; see [31] for a ChPT-based analysis.

4.8.2 $K^+ \rightarrow \pi^+\pi^- e^+\nu_e$

Why is it possible to measure $\pi\pi$ scattering in such a kaon decay process? The first, naive explanation, is that the pions undergo final-state interactions, see Fig. 7, and they are the only strongly interacting particles in the final state, therefore they ought to be sensitive to this interaction. The more educated explanation is that K_{e4} decays can be described by form factors, which share the phases of $\pi\pi$ interaction due to Watson's final state theorem [32]. It can be shown [33] that the interference between s- and p-waves can be unambiguously extracted,

$$\delta^0_0(E_{\pi\pi}) - \delta^1_1(E_{\pi\pi}) \ ,$$

and as $E_{\pi\pi}$ is kinematically restricted to be smaller than M_K, these phases are measured close to threshold.

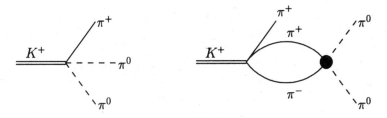

Figure 8: "Direct" and "rescattering" contributions to $K^+ \to \pi^+\pi^0\pi^0$.

4.8.3 Cusp in $K^+ \to \pi^+\pi^0\pi^0$

In a measurement of the kaon decay process $K^+ \to \pi^+\pi^0\pi^0$, the NA48/2 collaboration has detected a cusp phenomenon, i.e. a sudden change in slope, in the invariant mass spectrum of the $\pi^0\pi^0$ pair at $M_{\pi^0\pi^0} = 2M_{\pi^+}$ [29]. As suggested in Fig. 8, this phenomenon can be explained by an interference effect between "direct" tree graphs for $K^+ \to \pi^+\pi^0\pi^0$ and a $K^+ \to \pi^+\pi^+\pi^-$ decay, followed by $\pi^+\pi^- \to \pi^0\pi^0$ rescattering. The one-loop graph has a smooth part plus a part $v_{+-}(s)$, where

$$
v_{+-}(s) = \begin{cases} -\frac{1}{16\pi}\sqrt{\frac{4M_{\pi^\pm}^2}{s} - 1}\,, & s < 4M_{\pi^\pm}^2\,, \\[2mm] \frac{i}{16\pi}\sqrt{1 - \frac{4M_{\pi^\pm}^2}{s}}\,, & s > 4M_{\pi^\pm}^2\,. \end{cases} \tag{55}
$$

Below the $\pi^+\pi^-$ threshold, the loop graph is real and interferes directly with the tree contributions, while above threshold, it does not. Due to the square-root behaviour of $v_{+-}(s)$, a cusp is seen [34, 35]. The strength of this cusp is proportional to the scattering amplitude for $\pi^+\pi^- \to \pi^0\pi^0$ at threshold, hence a combination of $\pi\pi$ scattering lengths.

This behaviour is complicated at two-loop order as in contrast to K_{e4}, there are three strongly interacting particles in the final state [35–37]; in addition, virtual photons further modify the cusp structure. Nevertheless, the high statistics available in the experimental data in principle allow for a very precise determination of the scattering lengths, and the appropriate theoretical accuracy has to be provided.

4.8.4 Pionium lifetime

Pionium is a hadronic atom, a $\pi^+\pi^-$ system bound by electromagnetism. The energy levels of this system can in principle be calculated as in quantum mechanics for the hydrogen atom, however, they are perturbed by the strong interactions: the ground state is not stable, it decays according to

$$
A_{\pi^+\pi^-} \to \pi^0\pi^0,\, \gamma\gamma,\, \ldots\,.
$$

The decay width is given by the following (improved) Deser formula [38, 39] (further literature can be traced back from [40, 41]):

$$
\begin{aligned}
\Gamma &= \frac{2}{9}\alpha^3 p\,|\mathcal{A}(\pi^+\pi^- \to \pi^0\pi^0)_{\mathrm{thr}}|^2(1+\epsilon) \\
&= \frac{2}{9}\alpha^3 p\,|a_0^0 - a_0^2|^2(1+\delta)\,,
\end{aligned}
\tag{56}
$$

where p is the momentum of the π^0 in the decay in the centre-of-mass frame, and ϵ and δ are numerical correction factors accounting for isospin violation effects beyond leading order, $\delta = 0.058 \pm 0.012$ [42]. Taking information on the scattering lengths from elsewhere, one can predict the pionium lifetime as

$$
\tau = (2.9 \pm 0.1) \times 10^{-15} s\,.
\tag{57}
$$

Ultimately, one however wants to turn the argument around, measure the lifetime and extract $a_0^0 - a_0^2$. The corresponding experimental efforts are undertaken by the DIRAC collaboration, first results have been obtained [30].

4.8.5 Result on $\bar{\ell}_3$

For reasons of brevity, we just compare to the BNL-865 result for a_0^0 [27],

$$
a_0^0 = 0.216 \pm 0.013_{\mathrm{stat}} \pm 0.004_{\mathrm{syst}} \pm 0.005_{\mathrm{theo}}\,.
\tag{58}
$$

For an up-to-date compilation of the various experimental results, see e.g. [43] and references therein. From (58), one can extract a value for $\bar{\ell}_3 \approx 6 \pm 10$, which is compatible with the original estimate in [18] as well as lattice determinations (see [44] and the discussion in [43]). For the central value, the subleading correction for M_π^2 in (49) amounts to a mere 4%, therefore even from this seemingly rather loose bound, on can conclude that the leading term in the quark mass expansion of the pion mass dominates by far [45].

4.9 On the size of the corrections in a_0^0

If we remember the tree-level result for a_0^0 (31), $a_0^0(\mathrm{tree}) = 0.16$, the experimental result (58) seems somewhat surprising: higher-order corrections are of the order of 30%, rather than of the order of $M_\pi^2/(4\pi F_\pi)^2 \approx 2\%$. In order to understand this, we have to have another look at ϵ in (51). The $\bar{\ell}_i$ contain chiral logarithms, $\bar{\ell}_i \to -\log M_\pi^2$; collecting these together, ϵ contains logarithmic terms

$$
\epsilon = -\frac{9M_\pi^2}{32\pi^2 F_\pi^2}\log\frac{M_\pi^2}{\mu^2}
\tag{59}
$$

which, estimated at a scale $\mu = 1$ GeV, alone amount to 25%. We conclude that chiral logarithms potentially enhance higher-order corrections, and that the isoscalar s-wave $\pi\pi$ scattering length contains chiral logarithms with rather large coefficients, therefore corrections to the tree-level result are sizeable.

4.10 Quark mass ratios revisited

As another application of ChPT beyond leading order, we want to briefly revisit the ratios of the light quark masses. Forming dimensionless ratios, it turns out that one can write the $\mathcal{O}(p^4)$ corrections in the form [19]

$$\frac{M_K^2}{M_\pi^2} = \frac{m_s + \hat{m}}{m_u + m_d} \left\{ 1 + \Delta_M + \mathcal{O}(m_q^2) \right\} ,$$

$$\frac{(M_{K^0}^2 - M_{K^+}^2)_{\text{strong}}}{M_K^2 - M_\pi^2} = \frac{m_d - m_u}{m_s - \hat{m}} \left\{ 1 + \Delta_M + \mathcal{O}(m_q^2) \right\} ,$$

$$\text{where} \quad \Delta_M = \frac{8(M_K^2 - M_\pi^2)}{F_\pi^2}(2L_8 - L_5) + \text{chiral logs} . \tag{60}$$

The double ratio Q^2 is therefore particularly stable with respect to higher-order corrections,

$$Q^2 = \frac{m_s^2 - \hat{m}^2}{m_d^2 - m_u^2} = \frac{M_K^2}{M_\pi^2} \frac{M_K^2 - M_\pi^2}{(M_{K^0}^2 - M_{K^+}^2)_{\text{strong}}} \left\{ 1 + \mathcal{O}(m_q^2) \right\} . \tag{61}$$

(61) can be rewritten in the form of an ellipse equation for the quark mass ratios m_u/m_d, m_s/m_d (Leutwyler's ellipse [46]),

$$\left(\frac{m_u}{m_d}\right)^2 + \frac{1}{Q^2}\left(\frac{m_s}{m_d}\right)^2 = 1 . \tag{62}$$

We can use Dashen's theorem (26) to determine $(M_{K^0}^2 - M_{K^+}^2)_{\text{strong}}$ and therefore Q, with the result

$$Q_{\text{Dashen}} = 24.2 . \tag{63}$$

However, corrections to Dashen's theorem of $\mathcal{O}(e^2 m_q)$ are potentially large, different models yield a range [47]

$$1 \lesssim (M_{K^+}^2 - M_{K^0}^2)_{\text{em}}/(M_{\pi^+}^2 - M_{\pi^0}^2)_{\text{em}} \lesssim 2.5 , \tag{64}$$

inducing a rather large uncertainty in Q, $20.6 \lesssim Q \lesssim 24.2$. It would therefore be most desirable to obtain information on Q independent on meson mass relations. One such source will be introduced in the following subsection.

4.11 $\eta \to 3\pi$

The η meson has isospin $I = 0$, while three pions with angular momentum 0 cannot have $I = 0$, but only $I = 1$. $\eta \to 3\pi$ is therefore an isospin violating decay. Using the leading chiral Lagrangian $\mathcal{L}^{(2)}$ (including isospin breaking), the tree amplitude for $\eta \to \pi^+\pi^-\pi^0$ can be calculated to be

$$A(s,t,u) = \frac{B(m_u - m_d)}{3\sqrt{3}F_\pi^2} \left\{ 1 + \frac{3(s - s_0)}{M_\eta^2 - M_\pi^2} \right\} , \tag{65}$$

where $s = (p_\eta - p_{\pi^0})^2$, $t = (p_\eta - p_{\pi^+})^2$, $u = (p_\eta - p_{\pi^-})^2$, $3s_0 = s + t + u = M_\eta^2 + 2M_{\pi^+}^2 + M_{\pi^0}^2$. For $\eta \to 3\pi^0$, one finds the amplitude $A(s,t,u) + A(t,u,s) + A(u,s,t)$, with A as given in (65). We note that the electromagnetic term (25) does not contribute [48]. Terms of order $e^2 m_q$ were found to be very small [49], therefore $\eta \to 3\pi$ (potentially) allows for a much cleaner access to $m_u - m_d$ than the meson masses. We can rewrite $A(s,t,u)$ explicitly in terms of Q^2,

$$A(s,t,u) = \frac{1}{Q^2} \frac{M_K^2}{M_\pi^2} \left(M_\pi^2 - M_K^2\right) \frac{M(s,t,u)}{3\sqrt{3}F_\pi^2} ,$$

$$M(s,t,u) = \frac{3s - 4M_\pi^2}{M_\eta^2 - M_\pi^2} \quad \text{(at leading order)} . \tag{66}$$

Problems here arise from the fact that there are strong final-state interactions among the three pions: the one-loop corrections increase the width by a factor of 2.5,

$$\text{tree:} \quad \Gamma(\eta \to \pi^+\pi^-\pi^0) = 66 \text{ eV} ,$$

$$\text{one-loop:} \quad \Gamma(\eta \to \pi^+\pi^-\pi^0) = 160 \pm 50 \text{ eV} \quad [50] , \tag{67}$$

and even higher-order corrections are not negligible. Furthermore, there are partially contradictory experimental results (in particular on Dalitz plot parameters). This is therefore still a process of current interest, with strong ongoing experimental efforts for both final states [51].

The combined information on Q as deduced from various corrections to Dashen's theorem is shown in Fig. 9, together with two results obtained from studies of $\eta \to 3\pi$ [52, 53] (see also the discussion in [54]). Even with Q fixed, additional constraints are needed to find the position on the ellipse. Two possibilities are information on $\eta\eta'$ mixing, which, together with large-N_c arguments, leads to a determination of Δ_M in (60). Alternatively, the ratio $R = (m_s - \hat{m})/(m_d - m_u)$ can be extracted from baryon masses [55].

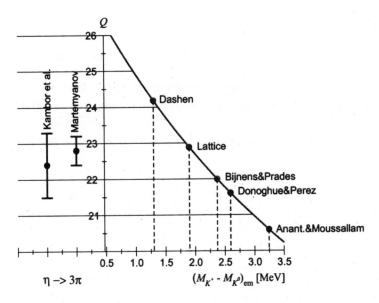

Figure 9: Various results on Q [47, 52, 53]. The figure is inspired by [46].

5 Chiral perturbation theory with baryons

So far, we have considered an effective field theory exclusively for (pseudo) Goldstone bosons. Perhaps the most important extension of this theory is the inclusion of nucleons or the baryon ground state octet in chiral SU(2) and SU(3), respectively. The problem, as we shall see later, is that the nucleon mass is a new, heavy mass scale that does not vanish in the chiral limit,

$$\lim_{m_q \to 0} m_N \approx m_N \ .$$

The idea for their incorporation in the theory is to view nucleons as (massive) matter fields coupled to pions and external sources. Their 3-momenta ought to remain small, of the order of M_π, in all processes. The number of baryons is therefore conserved, we consider no baryon–antibaryon creation/annihilation. In particular, in these lectures, we confine ourselves to processes with exactly one baryon.

For the construction of the meson-baryon Lagrangian, we proceed as before: we choose a suitable representation and transformation law for baryons under $SU(N)_L \times SU(N)_R$, and organise the effective Lagrangian according to an increasing number of momenta.

It turns out to be convenient to introduce a new field u for the Goldstone boson fields according to $u^2 = U$, which transforms as

$$u \longmapsto \sqrt{LUR^\dagger} = Lu\,K^\dagger(L,R,U) = K(L,R,U)\,u\,R^\dagger \tag{68}$$

Here, $K(L,R,U) \in \mathrm{SU}(N)$ is the so-called compensator field, that depends in a non-trivial way on L, R, and U. For $\mathrm{SU}(N)_V$ transformations $(L = R)$, (68) obviously reduces to $K(L,R,U) = L = R$.

A particularly convenient representation for nucleons and baryons is given by the following:

$$\psi = \begin{pmatrix} p \\ n \end{pmatrix} \longmapsto \psi' = K\psi \,,$$

$$B = \begin{pmatrix} \frac{\Sigma^0}{\sqrt{2}} + \frac{\Lambda}{\sqrt{6}} & \Sigma^+ & p \\ \Sigma^- & -\frac{\Sigma^0}{\sqrt{2}} + \frac{\Lambda}{\sqrt{6}} & n \\ \Xi^- & \Xi^0 & -\frac{2\Lambda}{\sqrt{6}} \end{pmatrix} \longmapsto B' = KBK^{-1} \,. \tag{69}$$

We introduce a covariant derivative $D^\mu = \partial^\mu + \Gamma^\mu$ with the chiral connection Γ^μ (vector)

$$\Gamma^\mu = \frac{1}{2}\big(u^\dagger(\partial^\mu - i\,r^\mu)u + u(\partial^\mu - i\,l^\mu)u^\dagger\big) \tag{70}$$

that transforms according to $\Gamma^\mu \mapsto K\Gamma^\mu K^\dagger - (\partial^\mu K)K^\dagger$, such that the covariant derivative has the expected transformation behaviour

$$D^\mu\psi \longmapsto KD^\mu\psi \,.$$

Furthermore, we shall use the chiral vielbein (axial vector)

$$u^\mu = i\big(u^\dagger(\partial^\mu - i\,r^\mu)u - u(\partial^\mu - i\,l^\mu)u^\dagger\big) \tag{71}$$

that transforms according to $u^\mu \mapsto Ku^\mu K^\dagger$. Finally, we can rewrite the (pseudo)scalar source term $\chi = 2B(s + i\,p) = 2B\mathcal{M} + \ldots$ as

$$\chi_+ = u^\dagger\chi u^\dagger + u\chi^\dagger u \,, \tag{72}$$

such that $\chi_+ \mapsto K\chi_+ K^\dagger$, and all the constitutive elements of the chiral Lagrangian transform with the compensator field K.

5.1 The leading-order chiral meson-baryon Lagrangian

We are now in the position to write down the leading-order meson-baryon Lagrangian, both for chiral SU(2) and SU(3):

$$\mathcal{L}_{\pi N}^{(1)} = \bar\psi\left(i\gamma_\mu D^\mu - m + \frac{g_A}{2}\gamma_\mu\gamma_5 u^\mu\right)\psi \,, \tag{73}$$

26

$$\mathcal{L}_{\phi B}^{(1)} = \langle \bar{B} \left(i\gamma_\mu D^\mu - m \right) B \rangle + \frac{D/F}{2} \langle \bar{B} \gamma_\mu \gamma_5 [u^\mu, B]_\pm \rangle . \tag{74}$$

While in meson ChPT, the Lagrangians come only in even powers of derivatives or momenta, odd powers of momenta are allowed in the meson-baryon sector due to the presence of spin (or, more general, Dirac structures):

$$\mathcal{L}_{\pi N} = \mathcal{L}_{\pi N}^{(1)} + \mathcal{L}_{\pi N}^{(2)} + \mathcal{L}_{\pi N}^{(3)} + \mathcal{L}_{\pi N}^{(4)} + \dots .$$

The new parameters of $\mathcal{L}_{\pi N}^{(1)}$ comprise m, the nucleon (baryon) mass in the chiral limit; g_A (in SU(2)), which, upon expansion of $u_\mu = 2a_\mu + \mathcal{O}(\pi)$, can be identified with the axial vector coupling that is known from neutron beta decay, $g_A = 1.26$; or D/F, the two axial vector couplings in SU(3), which can be determined from semileptonic hyperon decays and have to fulfil the SU(2) constraint $D + F = g_A$ ($D \approx 0.80, F \approx 0.46$).

5.2 Goldberger–Treiman relation

As a first consequence of (73), we want to derive the so-called Goldberger–Treiman relation. Setting external sources to zero, $v_\mu = a_\mu = 0$, and expanding in powers of the pion field, we find $u_\mu = -\partial_\mu \pi / F_\pi + \mathcal{O}(\pi^3)$ for the chiral vielbein, which results in the πNN vertex

$$\mathcal{L}_{\pi N}^{(1)} \rightarrow -\frac{g_A}{2F_\pi} \bar{\psi} \gamma^\mu \gamma_5 \partial_\mu \pi \psi . \tag{75}$$

The corresponding Feynman rule looks as follows:

$$\frac{g_A}{2F_\pi} \slashed{q} \gamma_5 \vec{\tau} \doteq V_{\pi NN} .$$

We can deduce a $N \rightarrow \pi N$ transition amplitude

$$T_{\pi NN} = -i\bar{u}(p')V_{\pi NN}u(p) = -i\frac{g_A m_N}{F_\pi} \bar{u}(p')\gamma_5 u(p)\vec{\tau} ,$$

which, compared to the canonical amplitude $-i\, g_{\pi N} \bar{u}(p')\gamma_5 u(p)\vec{\tau}$, yields the relation

$$g_{\pi N} = \frac{g_A m_N}{F_\pi} . \tag{76}$$

(76) is remarkable for relating weak (g_A) and strong (F_π) interaction quantities. Numerically it is rather well fulfilled in nature, $13.1 \dots 13.4 = 12.8$.

5.3 Weinberg's power counting for the one-baryon sector

We can derive a similar power counting formula as in (34) for the one-baryon sector. The chiral dimension ν of an arbitrary L-loop diagram with $V_d^{\pi\pi}$ meson-meson

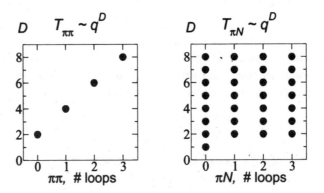

Figure 10: Schematic representation of loop contributions to chiral amplitudes in the Goldstone boson (left panel) and the meson-baryon (right panel) sector. The figure is inspired by [56].

vertices of order d and $V_{d'}^{\pi N}$ meson–baryon vertices of order d' is given by

$$\nu = 2L + 1 + \sum_d V_d^{\pi\pi}(d-2) + \sum_{d'} V_{d'}^{\pi N}(d'-1) \ . \tag{77}$$

Note that $d \geq 2$, $d' \geq 1$, such that the right hand side of (77) again consists of a sum of non-negative terms only. We conclude that at $\mathcal{O}(p^1)$, $\mathcal{O}(p^2)$, only tree diagrams contribute; tree plus one-loop diagrams enter at $\mathcal{O}(p^3)$, $\mathcal{O}(p^4)$; tree, one-loop, and two-loop diagrams can contribute to $\mathcal{O}(p^5)$, $\mathcal{O}(p^6)$, etc.

However, as it was noted in [56], in contrast to the meson sector, loop graphs do not necessarily obey the naive power counting rules as put forward in (77). The reason is that loop integrals cover all energy scales: while in the Goldstone boson sector, all mass scales are "small", such that naive power counting has to work in a mass-independent regularisation scheme (like dimensional regularisation), there is a new mass scale in the nucleon sector, the nucleon mass $m_N \approx \Lambda_\chi \approx 1$ GeV. The loop integration then also picks up momenta $p \sim m_N$.

Schematically, the situation is depicted in Fig. 10: in contrast to the Goldstone boson sector, higher-order loop graphs in the meson-baryon theory renormalise lower-order couplings at each order in the loop expansion. In the following, we shall discuss two possible remedies to restore the features of naive power counting for meson-baryon ChPT.

5.4   Heavy-baryon ChPT

The first remedy goes by the name of heavy-baryon chiral perturbation theory (HBChPT) [57, 58]. It is constructed in close analogy to heavy-quark effective field theory: we decompose the baryon momentum into a large part proportional to the

Figure 11: Nucleon self-energy graph. Full and dashed lines denote nucleon and pion, respectively.

nucleon velocity, plus a small residual momentum according to

$$p_\mu = m_N v_\mu + l_\mu \,, \quad v^2 = 1 \,, \quad v \cdot l \ll m_N \,. \tag{78}$$

In the heavy-baryon limit, the nucleon propagator is then of the form

$$\frac{1}{p^2 - m_N^2} \rightarrow \frac{1}{2m_N} \frac{1}{v \cdot l} + \mathcal{O}(1/m_N^2) \,.$$

This procedure eliminates the mass scale m_N from the propagator, which re-enters as a parametrical suppression factor. HBChPT is then a two-fold expansion in powers of $(p/\Lambda_\chi)^n$ and $(p/m_N)^n$, where $m_N \approx \Lambda_\chi$.

The nucleon field ψ is decomposed into velocity eigenstates according to

$$H_v(x) = e^{im_N v \cdot x} P_v^+ \psi(x) \,, \quad h_v(x) = e^{im_N v \cdot x} P_v^- \psi(x) \,, \tag{79}$$

where we have used the projectors $P_v^\pm = \frac{1}{2}(1 \pm \slashed{v})$ onto velocity eigenstates. H_v represents the "big" components of the spinor at low energies, while h_v are the "small" components. The exponential eliminates the large mass term from the time evolution of the field H_v. Written in terms of the new field H_v, the leading-order Lagrangian $\mathcal{L}_{\pi N}^{(1)}$ becomes

$$\mathcal{L}_{\pi N}^{(1)} = \bar{H}_v \big(iv \cdot D + g_A S \cdot u\big) H + \mathcal{O}(1/m_N) \,, \tag{80}$$

where we have introduced the Pauli-Lubanski spin vector $S_\mu = \frac{i}{2}\gamma_5 \sigma_{\mu\nu} v^\nu$. We note that the nucleon mass does not occur in (80), and the Dirac structure is massively simplified. $1/m_N$ corrections can be constructed systematically on the Lagrangian level in analogy to Foldy–Wouthuysen transformations [58].

5.5 Infrared regularisation

The second procedure is called infrared regularisation [59] (see also [60–62] for variants of this approach). It is an alternative way to regularise loop integrals and allows for a manifestly covariant way to calculate loops in baryon ChPT.

Let us consider the (relativistic) nucleon self-energy graph in Fig. 11. Evaluated at threshold in d dimensions, it yields the result

$$\frac{\Gamma(2 - \frac{d}{2})}{(4\pi)^{d/2}(d-3)} \frac{m_N^{d-3} + M_\pi^{d-3}}{m_N + M_\pi} . \tag{81}$$

We decompose (81) into two parts, the "regular" part $\propto m_N^{d-3}$, and the "infrared" part $\propto M_\pi^{d-3}$. The following properties of this decomposition can be shown to hold in general. The regular part scales with fractional powers of m_N, but has a regular expansion in M_π and momenta. It is this part that violates naive power counting, but as it can always be expanded as a polynomial, it can be absorbed by a redefinition of contact terms in the Lagrangian. The infrared part, on the other hand, scales with fractional powers of M_π; it contains all the "interesting" pieces of the loop diagram such as non-analytic structures and imaginary parts, and it obeys the naive power counting rules. It is therefore only the infrared part of the loop integral that we want to retain.

In [59], a very simple prescription to isolate the infrared part of the loop integral was given. With $a = M_\pi^2 - k^2 - i\epsilon$, $b = m_N^2 - (P-k)^2 - i\epsilon$, the scalar loop integral corresponding to Fig. 11 can be written as

$$H = \frac{1}{i} \int \frac{d^d k}{(2\pi)^d} \frac{1}{ab} = \int_0^1 dz \frac{1}{i} \int \frac{d^d k}{(2\pi)^d} \frac{1}{[(1-z)a + zb]^2} , \tag{82}$$

where we have introduced the Feynman parameter z. The Landau equations can be used to analyse the singularity structure of (82) in terms of values of z and the kinematic position where they occur:

$$z = \frac{M_\pi}{m_N + M_\pi} \quad \text{leading (pinch) singularity} \quad \leftrightarrow \quad P^2 = (m_N + M_\pi)^2 ,$$

$$z = 0 \quad \text{endpoint singularity} \quad \leftrightarrow \quad M_\pi^2 = 0 ,$$

$$z = 1 \quad \text{endpoint singularity} \quad \leftrightarrow \quad m_N^2 = 0 .$$

Clearly, $z = 1 \leftrightarrow m_N^2 = 0$ is not a "low-energy" singularity, hence one can obtain the infrared part I of H by avoiding the $z = 1$ endpoint singularity and extending the integration to $z = \infty$:

$$I = \int_0^\infty dz \frac{1}{i} \int \frac{d^d k}{(2\pi)^d} \frac{1}{[(1-z)a + zb]^2} . \tag{83}$$

The relation between the infrared regularisation scheme and the heavy-baryon expansion is shown schematically in Fig. 12: the infrared part I corresponds to a resummation of all $1/m_N$ corrections in a certain heavy-baryon diagram. This seems to be just the reverse of the expansion used to obtain the heavy-baryon propagator in the first place; however, we have interchanged summation and (highly irregular) loop integration, and the difference in the order of taking these limits is the regular part of the loop integral.

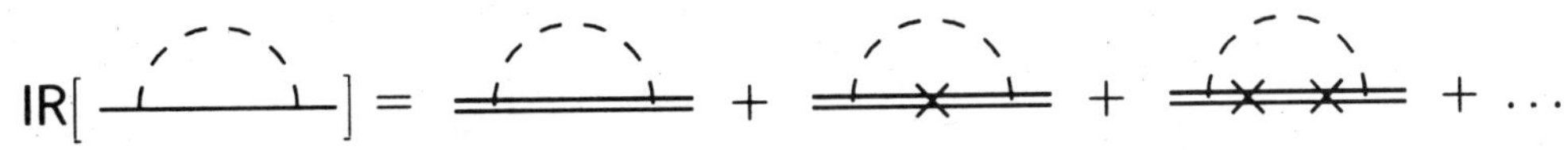

IR$\left[\underset{}{\overset{}{\frown}}\right]$ = $\underset{}{\overset{}{\frown}}$ + $\underset{}{\overset{}{\frown\times}}$ + $\underset{}{\overset{}{\times\frown\times}}$ + $\cdots$

Figure 12: Schematic relation of infrared regularisation and heavy-baryon expansion. The double line denotes the heavy-baryon propagator, crosses denote $1/m_N$ insertions.

Figure 13: Triangle graph contributing to the electromagnetic form factors of the nucleon.

In general, heavy-baryon loop integrals are easier to perform explicitly than those in infrared regularisation (although keeping track of all possible $1/m_N$ corrections is a considerable task in HBChPT when going to subleading loop orders). So does the difference between both procedures ever matter? Is the additional effort to calculate baryon ChPT in a manifestly Lorentz-invariant fashion worth it?

Sometimes, the difference does indeed matter, and in order to see this, we consider the electromagnetic form factors of the nucleon, defined by

$$\langle N(p')|J_\mu^{em}|N(p)\rangle = e\bar{u}(p')\left\{\gamma_\mu F_1^N(t) + \frac{i\sigma_{\mu\nu}q^\nu}{2m_N}F_2^N(t)\right\}u(p) . \tag{84}$$

An important contribution to the spectral function $\mathrm{Im}\, F_1^v(t)$ (where $F_1^v = F_1^p - F_1^n$) stems from the so-called triangle graph, Fig. 13. The "normal" threshold of the spectral function is at $t_{\mathrm{thr}} = 4M_\pi^2$; however there is an anomalous threshold very close by at $t_{\mathrm{anom}} = 4M_\pi^2 - M_\pi^4/m_N^2$. As both coincide in the heavy-baryon limit, $t_{\mathrm{anom}} = 4M_\pi^2 + \mathcal{O}(M_\pi^4)$, the analytic structure is distorted, as we can see comparing the contributions to the spectral function in infrared regularisation [63],

$$\mathrm{Im}\, F_1^v(t) \overset{\mathrm{IR}}{=} \frac{g_A^2}{192\pi F_\pi^2}(4m_N^2 - M_\pi^2)\left(1 - \frac{4M_\pi^2}{t}\right)^{3/2} + \cdots , \tag{85}$$

and in HBChPT [64],

$$\mathrm{Im}\, F_1^v(t) \overset{\mathrm{HB}}{=} \frac{g_A^2}{96\pi F_\pi^2}(5t - 8M_\pi^2)\left(1 - \frac{4M_\pi^2}{t}\right)^{1/2} + \cdots . \tag{86}$$

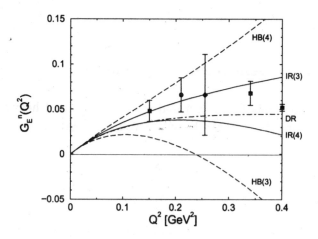

Figure 14: Neutron electric form factor at third and fourth order, in HBChPT and infrared regularisation, as a function of $Q^2 = -t$. Figure taken from [63].

While the infrared spectral function has the expected p-wave characteristic $\propto (1 - 4M_\pi^2/t)^{3/2}$, the threshold behaviour of the heavy-baryon spectral function is distorted.

Furthermore, the resummation of relativistic recoil effects in the nucleon propagator sometimes also helps to improve the phenomenological description of certain observables, as can be seen for the example of the neutron electric form factor

$$G_E^n(t) = F_1^n(t) + \frac{t}{4m_N^2} F_2^n(t) , \tag{87}$$

in Fig. 14. While convergence between leading and subleading loop order ($\mathcal{O}(p^3)$ and $\mathcal{O}(p^4)$, respectively) in HBChPT is rather poor, it is improved in infrared regularisation and, in addition, much closer to the data.

5.6 Quark mass dependence of the nucleon mass

As a further application of ChPT for nucleons, let us consider the quark mass expansion of the nucleon mass up to $\mathcal{O}(p^3)$. The result is

$$m_N = m - 4c_1 M_\pi^2 - \frac{3g_A^2 M_\pi^3}{32\pi F_\pi^2} + \mathcal{O}(M_\pi^4) . \tag{88}$$

The pion loop graph Fig. 11 yields a non-analytic term $\propto M_\pi^3 \propto \hat{m}^{3/2}$, but the leading correction term comes from a counterterm in $\mathcal{L}_{\pi N}^{(2)} = c_1 \bar{\psi} \langle \chi_+ \rangle \psi + \ldots$, with an unknown low-energy constant c_1. c_1 is closely related to the so-called πN σ-term.

We define the scalar form factor of the nucleon according to

$$\langle N(p')|\hat{m}(\bar{u}u + \bar{d}d)|N(p)\rangle = \sigma(t)\bar{u}(p')u(p) \ , \quad t = (p - p')^2 \ . \tag{89}$$

The σ-term is then given by

$$\sigma \equiv \sigma(0) = \frac{\hat{m}}{2m_N}\langle N|\bar{u}u + \bar{d}d|N\rangle \ . \tag{90}$$

σ to $\mathcal{O}(p^3)$ can be calculated from (88) using the Feynman–Hellmann theorem,

$$\sigma = \hat{m}\frac{\partial m_N}{\partial \hat{m}} = -4c_1 M_\pi^2 - \frac{9g_A^2 M_\pi^4}{32\pi F_\pi^2} + \mathcal{O}(M_\pi^4) \ , \tag{91}$$

so is is indeed given by c_1 at leading order. Furthermore, σ is related to the interesting question: how much do strange quarks contribute to nucleon properties? In this case, we consider the strangeness contribution to m_N, which is related to σ by

$$\sigma = \frac{\hat{m}}{2m_N}\frac{\langle N|\bar{u}u + \bar{d}d - 2\bar{s}s|N\rangle}{1 - y} \ . \quad y = \frac{2\langle N|\bar{s}s|N\rangle}{\langle N|\bar{u}u + \bar{d}d|N\rangle} \ . \tag{92}$$

Now $(m_s - \hat{m})(\bar{u}u + \bar{d}d - 2\bar{s}s)$ is the part of $\mathcal{L}_{\text{QCD}}$ that produces the SU(3) mass splittings and can therefore be related to differences in the baryon octet masses,

$$\sigma = \frac{\hat{\sigma}}{1 - y} \ , \quad \hat{\sigma} = \frac{\hat{m}}{m_s - \hat{m}}(m_\Xi + m_\Sigma - 2m_N) \simeq 26 \text{ MeV} \ . \tag{93}$$

Higher-order corrections lead to a modified value $\hat{\sigma} \to (36 \pm 7)$ MeV [65] (see also references therein for background, e.g. [66, 67]). But we conclude that, if we know σ, we know y and can therefore deduce information on the strangeness content of the nucleon.

We remember that we learnt about the quark mass dependence of M_π^2 from $\pi\pi$ scattering; it turns out that, here again, the quark mass dependence of m_N and the σ-term are closely related to πN scattering. πN scattering amplitudes can be separated into isospin even and odd parts,

$$T_{\pi N}^\pm = \frac{1}{2}\left[T(\pi^- p \to \pi^- p) \pm T(\pi^+ p \to \pi^+ p)\right] \ , \tag{94}$$

and we can decompose $T^\pm$ further into spin flip/non-flip amplitudes. Without going into all the details, let us consider the specific combinations $\bar{D}^\pm$ of πN amplitudes. In ChPT, the following relation can be proven:

$$\Sigma \equiv F_\pi^2 \bar{D}^+\left(s = u = m_N^2, t = 2M_\pi^2\right) = \sigma(2M_\pi^2) + \Delta_R \ . \tag{95}$$

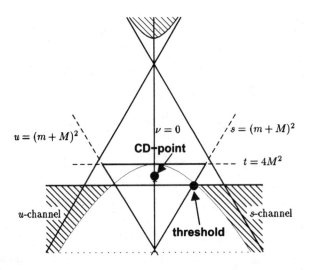

Figure 15: Mandelstam plane for πN scattering, with the position of the s-channel threshold and the Cheng–Dashen (CD) point. The figure is a modified version of one taken from [68].

The specific kinematic point $s = u = m_N^2$, $t = 2M_\pi^2$ at which $\bar{D}^+$ is to be evaluated is known as the Cheng–Dashen point. As shown in Fig. 15, it lies in the unphysical region (which is, in the s-channel, limited by $s \geq (m_N + M_\pi)^2$, $t \leq 0$). The remainder $\Delta_R = \mathcal{O}(M_\pi^4)$ in (95) is very small, $\Delta_R \lesssim 2$ MeV [69].

The procedure to learn about the strangeness content of the nucleon therefore consists of the following steps: (1) from πN scattering, deduce the amplitude $\bar{D}^+$ at the Cheng–Dashen point; (2) use (95) to calculate $\sigma(2M_\pi^2)$; (3) extrapolate $\sigma(t)$ to $t = 0$; (4) calculate y from σ using (93).

Step (1), the extrapolation into the unphysical region, is done using dispersion relations, with the result

$$\Sigma = 60 \pm 7 \text{ MeV} . \tag{96}$$

Step (2) is then safe due to the smallness of Δ_R. The crucial part is step (3),

$$\sigma(2M_\pi^2) = \sigma(0) + \Delta\sigma , \tag{97}$$

i.e. we have to understand the t-dependence of the scalar form factor $\sigma(t)$. A crude estimate would be to linearise the form factor,

$$\sigma(t) = \sigma(0)\left\{1 + \frac{1}{6}\langle r^2 \rangle_\sigma t + \ldots\right\} , \tag{98}$$

and assume $\langle r^2 \rangle_\sigma \simeq \langle r^2 \rangle_{\text{EM}} = 0.8 \text{ fm}^2$, leading to $\Delta\sigma \approx 3.5$ MeV. A complete one-loop calculation yields a value not too far from this, $\Delta\sigma = 4.6$ MeV. We would

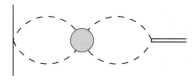

Figure 16: The scalar form factor of the nucleon beyond one loop. Full, dashed, and double lines denote nucleons, pions, and the scalar source, respectively.

deduce $\sigma \approx 55$ MeV and, consequently,

$$y = 1 - \frac{\hat{\sigma}}{\sigma} = 1 - \frac{35 \text{ MeV}}{55 \text{ MeV}} \approx 0.4 \ , \tag{99}$$

which appears far bigger than what one would naively expect. This was for some time known as the "σ-term puzzle": are 300 MeV of nucleon mass due to strange quarks?

The resolution lies in step (3), when one considers the scalar form factor beyond one loop, see Fig. 16: it involves isoscalar s-wave $\pi\pi$ rescattering, which we found earlier to be strong (see Sect. 4.9). A dispersive analysis yields $\langle r^2 \rangle_\sigma \approx 2\langle r^2 \rangle_{EM}$ and a large curvature term, with the result $\Delta\sigma \approx 15$ MeV [70]. The σ-term is then much smaller,

$$\sigma \approx 45 \text{ MeV} \ , \tag{100}$$

we now find $y \approx 0.2$ and a sizeable, but not outrageously large strangeness contribution to the nucleon mass,

$$\langle N | m_s \bar{s} s | N \rangle \simeq 130 \text{ MeV} \ . \tag{101}$$

5.7 πN scattering lengths

In order to improve the knowledge of the σ-term, it would be useful to have precise information on πN scattering in the threshold region (which is closest to the Cheng–Dashen point), or, more precisely, on the scattering lengths, defined by

$$a_\pm = \frac{1}{4\pi(1 + M_\pi/m_N)} T^\pm \left(s = (m_N + M_\pi)^2 \right) \ . \tag{102}$$

In ChPT at leading orders, these have the following expansion [71, 72]:

$$\begin{aligned}
a^- &= \frac{M_\pi}{8\pi(1 + M_\pi/m_N)F_\pi^2} + \mathcal{O}(M_\pi^3) \ , \\
a^+ &= 0 + \frac{M_\pi^2(-g_A^2 + 8m_N(-2c_1 + c_2 + c_3))}{16\pi m_N(1 + M_\pi/m_N)F_\pi^2} + \mathcal{O}(M_\pi^3) \ .
\end{aligned} \tag{103}$$

Numerically we have $a^- = 8.0 \times 10^{-2} M_\pi^{-1}$ plus small higher-order corrections. In contrast, a^+ vanishes at leading order p^1, receives contributions of several LECs at $\mathcal{O}(p^2)$, and can be seen to converge rather badly. This means that precisely the isoscalar scattering length that would be helpful in constraining the σ-term is barely known from ChPT.

Precise experimental information on the scattering lengths can be obtained from pionic hydrogen and pionic deuterium measurements, in analogy to the discussion of pionium lifetime for the $\pi\pi$ scattering lengths. The best values deduced thereof [73] (see also experimental references therein) are given by

$$a^- = (8.52 \pm 0.18) \times 10^{-2} M_\pi^{-1} \, , \quad a^+ = (0.15 \pm 0.22) \times 10^{-2} M_\pi^{-1} \, . \qquad (104)$$

a^+ is seen to be very small, and quite sensitive to isospin breaking corrections.

6 Outlook

Instead of a summary, we rather give an outlook on various aspects of ChPT that, due to lack of time and space, could not be covered in these lectures.

In the Goldstone boson sector, of course there are many more processes of high physical interest, e.g. various meson decays, further scattering problems etc. The whole sector of odd intrinsic parity ("chiral anomaly", Wess–Zumino–Witten term [14]) has not been touched upon, although it is responsible for such fundamental decays as $\pi^0 \to \gamma\gamma$ or $\eta \to \gamma\gamma$. A whole new set of Lagrangian terms is furthermore needed for weak matrix elements, e.g. for non-leptonic kaon decays. We have only in passing hinted at the links to dispersion theory, and relations of the chiral phenomenology to large-N_c arguments have been completely ignored.

In the single-baryon sector, some of the most noteworthy omissions include the highly interesting topic of isospin violation in πN scattering; the analysis of pion photo-/electroproduction $\gamma^{(*)} N \to \pi N$, which has proven to be one of the key successes in the development of baryon ChPT (see [74] and references therein); Compton scattering $\gamma N \to \gamma N$ and nucleon polarisabilities. An important extension of ChPT with baryons is the inclusion of explicit spin-3/2 (Δ) degrees of freedom ("small-scale expansion" [75]).

The topic of two-and-more-nucleon systems has been covered in the lectures by Chen, Hanhart [76], and Machleidt at this workshop. The connections to lattice QCD, which are another major point of research at present, concerning chiral extrapolations, ChPT in a finite volume and at non-zero lattice spacing, (partially) quenched ChPT have been touched upon in the lectures by Chen.

6.1 Acknowledgements

I would like to thank the organisers of "Physics and Astrophysics of Hadrons and Hadronic Matter" for a wonderful workshop, and in particular for their overwhelming hospitality during the week in Shantiniketan. I am grateful to Jürg Gasser and Ulf-G. Meißner for useful comments on the manuscript. This work was supported in parts by the DFG (SFB/TR 16) and the EU I3HP Project (RII3-CT-2004-506078).

References

[1] G. Ecker, Prog. Part. Nucl. Phys. **35** (1995) 1.

[2] A. V. Manohar, arXiv:hep-ph/9606222.

[3] U.-G. Meißner, arXiv:hep-ph/9711365.

[4] S. Scherer, Adv. Nucl. Phys. **27** (2003) 277.

[5] J. Gasser, Lect. Notes Phys. **629** (2004) 1.

[6] G. Colangelo, http://ltpth.web.psi.ch/zuoz2006.

[7] V. Bernard, U.-G. Meißner, arXiv:hep-ph/0611231.

[8] W. Heisenberg, H. Euler, Z. Phys. **98** (1936) 714.

[9] S. Weinberg, Physica A **96** (1979) 327.

[10] S. Schael *et al.* [ALEPH Collaboration], Phys. Rept. **421** (2005) 191.

[11] C. Vafa, E. Witten, Nucl. Phys. B **234** (1984) 173.

[12] S. R. Coleman, J. Wess, B. Zumino, Phys. Rev. **177** (1969) 2239; C. G. Callan, S. R. Coleman, J. Wess, B. Zumino, Phys. Rev. **177** (1969) 2247.

[13] H. Leutwyler, Annals Phys. **235** (1994) 165.

[14] J. Wess, B. Zumino, Phys. Lett. B **37** (1971) 95; E. Witten, Nucl. Phys. B **223** (1983) 422.

[15] R. Urech, Nucl. Phys. B **433** (1995) 234.

[16] R. F. Dashen, Phys. Rev. **183** (1969) 1245.

[17] S. Weinberg, Phys. Rev. Lett. **17** (1966) 616.

[18] J. Gasser, H. Leutwyler, Annals Phys. **158** (1984) 142.

[19] J. Gasser, H. Leutwyler, Nucl. Phys. B **250** (1985) 465.

[20] J. Bijnens, G. Colangelo, G. Ecker, JHEP **9902** (1999) 020.

[21] G. Ecker, J. Gasser, H. Leutwyler, A. Pich, E. de Rafael, Phys. Lett. B **223** (1989) 425.

[22] G. Ecker, J. Gasser, A. Pich, E. de Rafael, Nucl. Phys. B **321** (1989) 311.

[23] J. F. Donoghue, C. Ramirez, G. Valencia, Phys. Rev. D **39** (1989) 1947.

[24] J. Bijnens, G. Colangelo, G. Ecker, Annals Phys. **280** (2000) 100.

[25] M. Knecht, B. Moussallam, J. Stern, N. H. Fuchs, Nucl. Phys. B **457** (1995) 513.

[26] J. Bijnens, G. Colangelo, G. Ecker, J. Gasser, M. E. Sainio, Nucl. Phys. B **508** (1997) 263 [Erratum-ibid. B **517** (1998) 639].

[27] S. Pislak *et al.*, Phys. Rev. D **67** (2003) 072004.

[28] L. Masetti, arXiv:hep-ex/0610071.

[29] J. R. Batley *et al.* [NA48/2 Collaboration], Phys. Lett. B **633** (2006) 173.

[30] B. Adeva *et al.* [DIRAC Collaboration], Phys. Lett. B **619** (2005) 50.

[31] V. Bernard, N. Kaiser, U.-G. Meißner, Nucl. Phys. B **457** (1995) 147.

[32] K. M. Watson, Phys. Rev. **88** (1952) 1163.

[33] A. Pais, S. B. Treiman, Phys. Rev. **168** (1968) 1858.

[34] U.-G. Meißner, G. Müller, S. Steininger, Phys. Lett. B **406** (1997) 154 [Erratum-ibid. B **407** (1997) 454].

[35] N. Cabibbo, Phys. Rev. Lett. **93** (2004) 121801; N. Cabibbo, G. Isidori, JHEP **0503** (2005) 021.

[36] G. Colangelo, J. Gasser, B. Kubis, A. Rusetsky, Phys. Lett. B **638** (2006) 187.

[37] E. Gamiz, J. Prades, I. Scimemi, arXiv:hep-ph/0602023.

[38] S. Deser, M. L. Goldberger, K. Baumann, W. E. Thirring, Phys. Rev. **96** (1954) 774.

[39] A. Gall, J. Gasser, V. E. Lyubovitskij, A. Rusetsky, Phys. Lett. B **462** (1999) 335.

[40] J. Gasser, V. E. Lyubovitskij, A. Rusetsky, Phys. Lett. B **471** (1999) 244.

[41] H. Sazdjian, Phys. Lett. B **490** (2000) 203.

[42] J. Gasser, V. E. Lyubovitskij, A. Rusetsky, A. Gall, Phys. Rev. D **64** (2001) 016008.

[43] H. Leutwyler, arXiv:hep-ph/0612112.

[44] Ph. Boucaud *et al.* [ETM Collaboration], arXiv:hep-lat/0701012.

[45] G. Colangelo, J. Gasser, H. Leutwyler, Phys. Rev. Lett. **86** (2001) 5008.

[46] H. Leutwyler, Phys. Lett. B **378** (1996) 313.

[47] A. Duncan, E. Eichten, H. Thacker, Phys. Rev. Lett. **76** (1996) 3894; J. Bijnens, J. Prades, Nucl. Phys. B **490** (1997) 239; J. F. Donoghue, A. F. Perez, Phys. Rev. D **55** (1997) 7075; B. Ananthanarayan, B. Moussallam, JHEP **0406** (2004) 047.

[48] D. G. Sutherland, Phys. Lett. **23** (1966) 384.

[49] R. Baur, J. Kambor, D. Wyler, Nucl. Phys. B **460** (1996) 127.

[50] J. Gasser, H. Leutwyler, Nucl. Phys. B **250** (1985) 539.

[51] H. H. Adam *et al.* [WASA-at-COSY Collaboration], arXiv:nucl-ex/ 0411038; S. Giovannella *et al.* [KLOE Collaboration], arXiv:hep-ex/ 0505074; A. Starostin, Acta Phys. Slov. **56** (2005) 345.

[52] J. Kambor, C. Wiesendanger, D. Wyler, Nucl. Phys. B **465** (1996) 215.

[53] B. V. Martemyanov, V. S. Sopov, Phys. Rev. D **71** (2005) 017501.

[54] B. Borasoy, R. Nißler, Eur. Phys. J. A **26** (2005) 383.

[55] J. Gasser, H. Leutwyler, Phys. Rept. **87** (1982) 77.

[56] J. Gasser, M. E. Sainio, A. Švarc, Nucl. Phys. B **307** (1988) 779.

[57] E. Jenkins, A. V. Manohar, Phys. Lett. B **255** (1991) 558.

[58] V. Bernard, N. Kaiser, J. Kambor, U.-G. Meißner, Nucl. Phys. B **388** (1992) 315.

[59] T. Becher, H. Leutwyler, Eur. Phys. J. C **9** (1999) 643.

[60] P. J. Ellis, H. B. Tang, Phys. Rev. C **57** (1998) 3356.

[61] J. L. Goity, D. Lehmann, G. Prézeau, J. Saez, Phys. Lett. B **504** (2001) 21; D. Lehmann, G. Prézeau, Phys. Rev. D **65** (2002) 016001.

[62] M. R. Schindler, J. Gegelia, S. Scherer, Phys. Lett. B **586** (2004) 258.

[63] B. Kubis, U.-G. Meißner, Nucl. Phys. A **679** (2001) 698.

[64] V. Bernard, N. Kaiser, U.-G. Meißner, Nucl. Phys. A **611** (1996) 429.

[65] B. Borasoy, U.-G. Meißner, Annals Phys. **254** (1997) 192.

[66] J. Gasser, Annals Phys. **136** (1981) 62.

[67] V. Bernard, N. Kaiser, U.-G. Meißner, Z. Phys. C **60** (1993) 111.

[68] P. Büttiker, U.-G. Meißner, Nucl. Phys. A **668** (2000) 97.

[69] V. Bernard, N. Kaiser, U.-G. Meißner, Phys. Lett. B **389** (1996) 144.

[70] J. Gasser, H. Leutwyler, M. E. Sainio, Phys. Lett. B **253** (1991) 260.

[71] V. Bernard, N. Kaiser, U.-G. Meißner, Phys. Lett. B **309** (1993) 421.

[72] N. Fettes, U.-G. Meißner, Nucl. Phys. A **676** (2000) 311.

[73] U.-G. Meißner, U. Raha, A. Rusetsky, Phys. Lett. B **639** (2006) 478.

[74] V. Bernard, N. Kaiser, U.-G. Meißner, Int. J. Mod. Phys. E **4** (1995) 193.

[75] T. R. Hemmert, B. R. Holstein, J. Kambor, J. Phys. G **24** (1998) 1831.

[76] C. Hanhart, arXiv:nucl-th/0703028.

Physics and Astrophysics of Hadrons and Hardronic Matter
Editor: A. B. Santra

Nuclear Forces from Chiral Effective Field Theory

R. Machleidt

Department of Physics, University of Idaho
Moscow, ID 83844-0903, U.S.A.
email: machleid@uidaho.edu

Abstract

In this lecture series, I present the recent progress in our understanding of nuclear forces in terms of chiral effective field theory.

1 Introduction and Historical Perspective

The theory of nuclear forces has a long history (cf. Table 1). Based upon the seminal idea by Yukawa [1], first field-theoretic attempts to derive the nucleon-nucleon (NN) interaction focused on pion-exchange. While the one-pion exchange turned out to be very useful in explaining NN scattering data and the properties of the deuteron [2], multi-pion exchange was beset with serious ambiguities [3,4]. Thus, the "pion theories" of the 1950s are generally judged as failures—for reasons we understand today: pion dynamics is constrained by chiral symmetry, a crucial point that was unknown in the 1950s.

Historically, the experimental discovery of heavy mesons [5] in the early 1960s saved the situation. The one-boson-exchange (OBE) model [6,7] emerged which is still the most economical and quantitative phenomenology for describing the nuclear force [8,9]. The weak point of this model, however, is the scalar-isoscalar "sigma" or "epsilon" boson, for which the empirical evidence remains controversial. Since this boson is associated with the correlated (or resonant) exchange of two pions, a vast theoretical effort that occupied more than a decade was launched to derive the 2π-exchange contribution to the nuclear force, which creates the intermediate range attraction. For this, dispersion theory as well as field theory were invoked producing the Stony Brook [10], Paris [11,12], and Bonn [7,13] potentials.

The nuclear force problem appeared to be solved; however, with the discovery of quantum chromodynamics (QCD), all "meson theories" were relegated to models and the attempts to derive the nuclear force started all over again.

The problem with a derivation from QCD is that this theory is non-perturbative in the low-energy regime characteristic of nuclear physics, which makes direct solutions impossible. Therefore, during the first round of new attempts, QCD-inspired quark models [14] became popular. These models are able to reproduce qualitatively and, in some cases, semi-quantitatively the gross features of the nuclear force [15, 16]. However, on a critical note, it has been pointed out that these quark-based approaches are nothing but another set of models and, thus, do not represent any fundamental progress. Equally well, one may then stay with the simpler and much more quantitative meson models.

A major breakthrough occurred when the concept of an effective field theory (EFT) was introduced and applied to low-energy QCD. As outlined by Weinberg in a seminal paper [17], one has to write down the most general Lagrangian consistent with the assumed symmetry principles, particularly the (broken) chiral symmetry of QCD. At low energy, the effective degrees of freedom are pions and nucleons rather than quarks and gluons; heavy mesons and nucleon resonances are "integrated out". So, the circle of history is closing and we are back to Yukawa's meson theory, except that we have learned to add one important refinement to the theory: broken chiral symmetry is a crucial constraint that generates and controls the dynamics and establishes a clear connection with the underlying theory, QCD.

Table 1: Seven Decades of Struggle: The Theory of Nuclear Forces

1935	**Yukawa: Meson Theory**
1950's	*The "Pion Theories"* One-Pion Exchange: o.k. Multi-Pion Exchange: disaster
1960's	Many pions $\equiv$ multi-pion resonances: σ, ρ, ω, ... The One-Boson-Exchange Model: success
1970's	Refined meson models, including sophisticated 2π exchange contributions (Stony Brook, Paris, Bonn)
1980's	Nuclear physicists discover **QCD** Quark Cluster Models
1990's **and beyond**	Nuclear physicists discover **EFT** Weinberg, van Kolck **Back to Pion Theory!** *But, constrained by Chiral Symmetry: success*

Following the first initiative by Weinberg [18], pioneering work was performed by Ordóñez, Ray, and van Kolck [19, 20] who constructed a NN potential in coordinate space based upon chiral perturbation theory at next-to-next-to-leading order. The results were encouraging and many researchers became attracted to the new field [21–27]. As a consequence, nuclear EFT has developed into one of the most popular branches of modern nuclear physics [28, 29].

It is the purpose of these lectures to describe in some detail the recent progress in our understanding of nuclear forces in terms of nuclear EFT.

2 QCD and the Nuclear Force

Quantum chromodynamics (QCD) is the theory of strong interactions. It deals with quarks, gluons and their interactions and is part of the Standard Model of Particle Physics. QCD is a non-Abelian gauge field theory with color $SU(3)$ the underlying gauge group. The non-Abelian nature of the theory has dramatic consequences. While the interaction between colored objects is weak at short distances or high momentum transfer ("asymptotic freedom"); it is strong at long distances ($\gtrsim 1$ fm) or low energies, leading to the confinement of quarks into colorless objects, the hadrons. Consequently, QCD allows for a perturbative analysis at large energies, whereas it is highly non-perturbative in the low-energy regime. Nuclear physics resides at low energies and the force between nucleons is a residual QCD interaction. Therefore, in terms of quarks and gluons, the nuclear force is a very complicated problem.

3 Effective Field Theory for Low-Energy QCD

The way out of the dilemma of how to derive the nuclear force from QCD is provided by the effective field theory (EFT) concept. First, one needs to identify the relevant degrees of freedom. For the ground state and the low-energy excitation spectrum of an atomic nucleus as well as for conventional nuclear reactions, quarks and gluons are ineffective degrees of freedom, while nucleons and pions are the appropriate ones. Second; to make sure that this EFT is not just another phenomenology, the EFT must observe all relevant symmetries of the underlying theory. This requirement is based upon a 'folk theorem' by Weinberg [17]:

> If one writes down the most general possible Lagrangian, including *all* terms consistent with assumed symmetry principles, and then calculates matrix elements with this Lagrangian to any given order of perturbation theory, the result will simply be the most general possible S-matrix consistent with analyticity, perturbative unitarity, cluster decomposition, and the assumed symmetry principles.

Thus, the EFT program consists of the following steps:

1. Identify the degrees of freedom relevant at the resolution scale of (low-energy) nuclear physics: nucleons and pions.

2. Identify the relevant symmetries of low-energy QCD and investigate if and how they are broken.

3. Construct the most general Lagrangian consistent with those symmetries and the symmetry breaking.

4. Design an organizational scheme that can distinguish between more and less important contributions: a low-momentum expansion.

5. Guided by the expansion, calculate Feynman diagrams to the the desired accuracy for the problem under consideration.

We will now elaborate on these steps, one by one.

3.1 Symmetries of Low-Energy QCD

In this section, we will give a brief introduction into (low-energy) QCD, its symmetries and symmetry breaking. A more detailed introduction can be found in the excellent lecture series by Scherer and Schindler [30].

3.1.1 Chiral Symmetry

The QCD Lagrangian reads

$$\mathcal{L}_{\mathrm{QCD}} = \bar{q}(i\gamma^\mu \mathcal{D}_\mu - \mathcal{M})q - \frac{1}{4}\mathcal{G}_{\mu\nu,a}\mathcal{G}_a^{\mu\nu} \tag{1}$$

with the gauge-covariant derivative

$$\mathcal{D}_\mu = \partial_\mu + ig\frac{\lambda_a}{2}\mathcal{A}_{\mu,a} \tag{2}$$

and the gluon field strength tensor

$$\mathcal{G}_{\mu\nu,a} = \partial_\mu \mathcal{A}_{\nu,a} - \partial_\nu \mathcal{A}_{\mu,a} - gf_{abc}\mathcal{A}_{\mu,b}\mathcal{A}_{\nu,c}. \tag{3}$$

In the above, q denotes the quark fields and $\mathcal{M}$ the quark mass matrix. Further, g is the strong coupling constant and $\mathcal{A}_{\mu,a}$ are the gluon fields. The λ_a are the Gell-Mann matrices and the f_{abc} the structure constants of the $SU(3)_{\mathrm{color}}$ Lie algebra ($a, b, c = 1, \ldots, 8$); summation over repeated indices is always implied. The gluon-gluon term in the last equation arises from the non-Abelian nature of the gauge theory and is the reason for the peculiar features of the color force.

On a typical hadronic scale, i.e., on a scale of low-mass hadrons which are not Goldstone bosons, e.g., $m_\rho = 0.78$ GeV ≈ 1 GeV; the masses of the up (u), down (d), and—to a certain extend—strange (s) quarks are small [31]:

$$m_u = 2 \pm 1 \text{ MeV} \tag{4}$$
$$m_d = 5 \pm 2 \text{ MeV} \tag{5}$$
$$m_s = 95 \pm 25 \text{ MeV} \tag{6}$$

It is therefore of interest to discuss the QCD Lagrangian in the limit of vanishing quark masses:

$$\mathcal{L}_{\mathrm{QCD}}^0 = \bar{q}i\gamma^\mu \mathcal{D}_\mu q - \frac{1}{4}\mathcal{G}_{\mu\nu,a}\mathcal{G}_a^{\mu\nu}. \tag{7}$$

Defining right- and left-handed quark fields,

$$q_R = P_R q, \quad q_L = P_L q, \tag{8}$$

43

with

$$P_R = \frac{1}{2}(1 + \gamma_5), \quad P_L = \frac{1}{2}(1 - \gamma_5),$$ (9)

we can rewrite the Lagrangian as follows:

$$\mathcal{L}_{QCD}^0 = \bar{q}_R i \gamma^\mu \mathcal{D}_\mu q_R + \bar{q}_L i \gamma^\mu \mathcal{D}_\mu q_L - \frac{1}{4} \mathcal{G}_{\mu\nu,a} \mathcal{G}_a^{\mu\nu}.$$ (10)

Restricting ourselves now to up and down quarks, we see that $\mathcal{L}_{QCD}^0$ is invariant under the global unitary transformations

$$q_R = \begin{pmatrix} u_R \\ d_R \end{pmatrix} \longmapsto \exp\left(-i\Theta_i^R \frac{\tau_i}{2}\right) \begin{pmatrix} u_R \\ d_R \end{pmatrix}$$ (11)

and

$$q_L = \begin{pmatrix} u_L \\ d_L \end{pmatrix} \longmapsto \exp\left(-i\Theta_i^L \frac{\tau_i}{2}\right) \begin{pmatrix} u_L \\ d_L \end{pmatrix},$$ (12)

where τ_i ($i = 1, 2, 3$) are the generators of $SU(2)_{\text{flavor}}$, the usual Pauli spin matrices. *The right- and left-handed components of massless quarks do not mix.* This is $SU(2)_R \times SU(2)_R$ symmetry, also known as *chiral symmetry*. Noether's Theorem implies the existence of six conserved currents; three right-handed currents

$$R_i^\mu = \bar{q}_R \gamma^\mu \frac{\tau_i}{2} q_R \quad \text{with} \quad \partial_\mu R_i^\mu = 0$$ (13)

and three left-handed currents

$$L_i^\mu = \bar{q}_L \gamma^\mu \frac{\tau_i}{2} q_L \quad \text{with} \quad \partial_\mu L_i^\mu = 0.$$ (14)

It is useful to consider the following linear combinations; namely, three vector currents

$$V_i^\mu = R_i^\mu + L_i^\mu = \bar{q} \gamma^\mu \frac{\tau_i}{2} q \quad \text{with} \quad \partial_\mu V_i^\mu = 0$$ (15)

and three axial-vector currents

$$A_i^\mu = R_i^\mu - L_i^\mu = \bar{q} \gamma^\mu \gamma_5 \frac{\tau_i}{2} q \quad \text{with} \quad \partial_\mu A_i^\mu = 0,$$ (16)

which got their names from the fact that they transform as vectors and axial-vectors, respectively. Thus, the chiral $SU(2)_L \times SU(2)_R$ symmetry is equivalent to $SU(2)_V \times SU(2)_A$, where the vector and axial-vector transformations are given respectively by

$$q = \begin{pmatrix} u \\ d \end{pmatrix} \longmapsto \exp\left(-i\Theta_i^V \frac{\tau_i}{2}\right) \begin{pmatrix} u \\ d \end{pmatrix}$$ (17)

and

$$q = \begin{pmatrix} u \\ d \end{pmatrix} \longmapsto \exp\left(-i\Theta_i^A \gamma_5 \frac{\tau_i}{2}\right) \begin{pmatrix} u \\ d \end{pmatrix} . \tag{18}$$

Obviously, the vector transformations are isospin rotations and, therefore, invariance under vector transformations can be identified with isospin symmetry.

There are the six conserved charges,

$$Q_i^V = \int d^3x \, V_i^0 = \int d^3x \, q^\dagger(t, \vec{x}) \frac{\tau_i}{2} q(t, \vec{x}) \quad \text{with} \quad \frac{dQ_i^V}{dt} = 0 \tag{19}$$

and

$$Q_i^A = \int d^3x \, A_i^0 = \int d^3x \, q^\dagger(t, \vec{x}) \gamma_5 \frac{\tau_i}{2} q(t, \vec{x}) \quad \text{with} \quad \frac{dQ_i^A}{dt} = 0, \tag{20}$$

which are also generators of $SU(2)_V \times SU(2)_A$.

3.1.2 Explicit Symmetry Breaking

The mass term $-\bar{q}\mathcal{M}q$ in the QCD Lagrangian Eq. (1) breaks chiral symmetry explicitly. To better see this, let's rewrite $\mathcal{M}$,

$$\mathcal{M} = \begin{pmatrix} m_u & 0 \\ 0 & m_d \end{pmatrix} \tag{21}$$

$$= \frac{1}{2}(m_u + m_d) \begin{pmatrix} 1 & 0 \\ 0 & 1 \end{pmatrix} + \frac{1}{2}(m_u - m_d) \begin{pmatrix} 1 & 0 \\ 0 & -1 \end{pmatrix} \tag{22}$$

$$= \frac{1}{2}(m_u + m_d) I + \frac{1}{2}(m_u - m_d) \tau_3 . \tag{23}$$

The first term in the last equation in invariant under $SU(2)_V$ (isospin symmetry) and the second term vanishes for $m_u = m_d$. Thus, isospin is an exact symmetry if $m_u = m_d$. However, both terms in Eq. (23) break $SU(2)_A$. Since the up and down quark masses are small as compared to the typical hadronic mass scale of ≈ 1 GeV [cf. Eqs. (4) and (5)], the explicit chiral symmetry breaking due to non-vanishing quark masses is very small.

3.1.3 Spontaneous Symmetry Breaking

A (continuous) symmetry is said to be *spontaneously broken* if a symmetry of the Lagrangian is not realized in the ground state of the system. There is evidence that the chiral symmetry of the QCD Lagrangian is spontaneously broken—for dynamical reasons of nonperturbative origin which are not fully understood at this time. The most plausible evidence comes from the hadron spectrum. From chiral symmetry, one would naively expect the existence of degenerate hadron multiplets of opposite parity, i.e., for any hadron of positive parity one would expect a degenerate hadron state of negative parity and vice versa. However, these "parity doublets" are not observed in nature. For example, take the ρ-meson, a vector meson with negative parity (1^-) and mass 776 MeV. There does exist a 1^+ meson, the a_1, but it has a mass of 1230 MeV and, thus, cannot be perceived as degenerate with the ρ. On the other hand, the ρ meson comes in three charge states (equivalent to three isospin states), the $\rho^\pm$ and the ρ^0 with masses that differ by at most a few MeV. In summary, in the QCD ground state (the hadron spectrum) $SU(2)_V$ (isospin symmetry) is well observed, while $SU(2)_A$ (axial symmetry) is broken. Or, in other words, $SU(2)_V \times SU(2)_A$ is broken down to $SU(2)_V$.

A spontaneously broken global symmetry implies the existence of (massless) Goldstone bosons with the quantum numbers of the broken generators. The broken generators are the Q_i^A of Eq. (20) which are pseudoscalar. The Goldstone bosons are identified with the isospin triplet of the (pseudoscalar) pions, which explains why pions are so light. The pion masses are not exactly zero because the up and down quark masses are not exactly zero either (explicit symmetry breaking). Thus, pions are a truly remarkable species: they reflect spontaneous as well as explicit symmetry breaking.

3.2 Chiral Effective Lagrangians Involving Pions

The next step in our EFT program is to build the most general Lagrangian consistent with the (broken) symmetries discussed above. An elegant formalism for the construction of such Lagrangians was developed by Callan, Coleman, Wess, and Zumino (CCWZ) [32] who worked out the group-theoretical foundations of non-linear realizations of chiral symmetry. The Lagrangians given below are built upon the CCWZ formalism.

As discussed, the relevant degrees of freedom are pions (Goldstone bosons) and nucleons. Since the interactions of Goldstone bosons must vanish at zero momentum transfer and in the chiral limit ($m \to 0$), the low-energy expansion of the Lagrangian is arranged in powers of derivatives and pion masses. This is chiral perturbation theory (ChPT).

The Lagrangian consists of one part that deals with the interaction among pions, $\mathcal{L}_{\pi\pi}$, and another one that describes the interaction between pions and the nucleon, $\mathcal{L}_{\pi N}$:

$$\mathcal{L}_{\text{eff}} = \mathcal{L}_{\pi\pi} + \mathcal{L}_{\pi N} \tag{24}$$

with

$$\mathcal{L}_{\pi\pi} = \mathcal{L}_{\pi\pi}^{(2)} + \mathcal{L}_{\pi\pi}^{(4)} + \dots \tag{25}$$

and

$$\mathcal{L}_{\pi N} = \mathcal{L}_{\pi N}^{(1)} + \mathcal{L}_{\pi N}^{(2)} + \mathcal{L}_{\pi N}^{(3)} + \dots, \tag{26}$$

where the superscript refers to the number of derivatives or pion mass insertions (chiral dimension) and the ellipsis stands for terms of higher dimension.

The *leading order (LO)* $\pi\pi$ Lagrangian is given by [33]

$$\mathcal{L}_{\pi\pi}^{(2)} = \frac{f_\pi^2}{4} \operatorname{tr}\left[\partial^\mu U \partial_\mu U^\dagger + m_\pi^2 (U + U^\dagger) \right] \qquad (27)$$

and the LO relativistic πN Lagrangian reads [34]

$$\mathcal{L}_{\pi N}^{(1)} = \bar{\Psi} \left(i\gamma^\mu D_\mu - M_N + \frac{g_A}{2} \gamma^\mu \gamma_5 u_\mu \right) \Psi \qquad (28)$$

with

$$D_\mu = \partial_\mu + \Gamma_\mu \qquad (29)$$

$$\Gamma_\mu = \frac{1}{2}(\xi^\dagger \partial_\mu \xi + \xi \partial_\mu \xi^\dagger) = \frac{i}{4f_\pi^2} \boldsymbol{\tau} \cdot (\boldsymbol{\pi} \times \partial_\mu \boldsymbol{\pi}) + \dots \qquad (30)$$

$$u_\mu = i(\xi^\dagger \partial_\mu \xi - \xi \partial_\mu \xi^\dagger) = -\frac{1}{f_\pi} \boldsymbol{\tau} \cdot \partial_\mu \boldsymbol{\pi} + \dots \qquad (31)$$

$$U = \xi^2 = 1 + \frac{i}{f_\pi} \boldsymbol{\tau} \cdot \boldsymbol{\pi} - \frac{1}{2f_\pi^2} \pi^2 - \frac{i\alpha}{f_\pi^3}(\boldsymbol{\tau} \cdot \boldsymbol{\pi})^3 + \frac{8\alpha - 1}{8f_\pi^4} \pi^4 + \dots \qquad (32)$$

In Eq. (28) the chirally covariant derivative D_μ is applied which introduces the "gauge term" Γ_μ (also known as chiral connection), a vector current that leads to a coupling of pions with the nucleon. Besides this, the Lagrangian includes a coupling term which involves the axial vector u_μ. The $SU(2)$ matrix $U = \xi^2$ collects the Goldstone pion fields.

In the above equations, M_N denotes the nucleon mass, g_A the axial-vector coupling constant, and f_π the pion decay constant. Numerical values will be given later.

The coefficient α that appears in Eq. (32) is arbitrary. Therefore, diagrams with chiral vertices that involve three or four pions must always be grouped together such that the α-dependence drops out (cf. Fig. 4, below).

We apply the heavy baryon (HB) formulation of chiral perturbation theory [35] in which the relativistic πN Lagrangian is subjected to an expansion in terms of powers of $1/M_N$ (kind of a nonrelativistic expansion), the lowest order of which is

$$\begin{aligned} \widehat{\mathcal{L}}_{\pi N}^{(1)} &= \bar{N} \left(iD_0 - \frac{g_A}{2} \vec{\sigma} \cdot \vec{u} \right) N \\ &= \bar{N} \left[i\partial_0 - \frac{1}{4f_\pi^2} \boldsymbol{\tau} \cdot (\boldsymbol{\pi} \times \partial_0 \boldsymbol{\pi}) - \frac{g_A}{2f_\pi} \boldsymbol{\tau} \cdot (\vec{\sigma} \cdot \vec{\nabla})\boldsymbol{\pi} \right] N + \dots \qquad (33) \end{aligned}$$

In the relativistic formulation, the nucleon is represented by a four-component Dirac spinor field, Ψ, while in the HB version, the nucleon, N, is a Pauli spinor; in addition, all nucleon fields include Pauli spinors describing the isospin of the nucleon.

At dimension two, the relativistic πN Lagrangian reads

$$\mathcal{L}_{\pi N}^{(2)} = \sum_{i=1}^{4} c_i \bar{\Psi} O_i^{(2)} \Psi . \tag{34}$$

The various operators $O_i^{(2)}$ are given in Ref. [36]. The fundamental rule by which this Lagrangian—as well as all the other ones—are assembled is that they must contain *all* terms consistent with chiral symmetry and Lorentz invariance (apart from other trivial symmetries) at a given chiral dimension (here: order two). The parameters c_i are known as low-energy constants (LECs) and are determined empirically from fits to πN data.

The HB projected πN Lagrangian at order two is most conveniently broken up into two pieces,

$$\widehat{\mathcal{L}}_{\pi N}^{(2)} = \widehat{\mathcal{L}}_{\pi N,\,\text{fix}}^{(2)} + \widehat{\mathcal{L}}_{\pi N,\,\text{ct}}^{(2)} , \tag{35}$$

with

$$\widehat{\mathcal{L}}_{\pi N,\,\text{fix}}^{(2)} = \bar{N} \left[\frac{1}{2M_N} \vec{D} \cdot \vec{D} + i \frac{g_A}{4M_N} \{ \vec{\sigma} \cdot \vec{D}, u_0 \} \right] N \tag{36}$$

and

$$\widehat{\mathcal{L}}_{\pi N,\,\text{ct}}^{(2)} = \bar{N} \left[2 c_1 m_\pi^2 (U + U^\dagger) + \left(c_2 - \frac{g_A^2}{8M_N} \right) u_0^2 + c_3 u_\mu u^\mu \right.$$
$$\left. + \frac{i}{2} \left(c_4 + \frac{1}{4M_N} \right) \vec{\sigma} \cdot (\vec{u} \times \vec{u}) \right] N . \tag{37}$$

Note that $\widehat{\mathcal{L}}_{\pi N,\,\text{fix}}^{(2)}$ is created entirely from the HB expansion of the relativistic $\mathcal{L}_{\pi N}^{(1)}$ and thus has no free parameters ("fixed"), while $\widehat{\mathcal{L}}_{\pi N,\,\text{ct}}^{(2)}$ is dominated by the new πN contact terms proportional to the c_i parameters, besides some small $1/M_N$ corrections.

At dimension three, the relativistic πN Lagrangian can be formally written as

$$\mathcal{L}_{\pi N}^{(3)} = \sum_{i=1}^{23} d_i \bar{\Psi} O_i^{(3)} \Psi , \tag{38}$$

with the operators, $O_i^{(3)}$, listed in Refs. [36, 37]; not all 23 terms are of interest here. The new LECs that occur at this order are the d_i. Similar to the order two case, the HB projected Lagrangian at order three can be broken into two pieces,

$$\widehat{\mathcal{L}}_{\pi N}^{(3)} = \widehat{\mathcal{L}}_{\pi N,\,\text{fix}}^{(3)} + \widehat{\mathcal{L}}_{\pi N,\,\text{ct}}^{(3)} , \tag{39}$$

with $\widehat{\mathcal{L}}_{\pi N,\,\text{fix}}^{(3)}$ and $\widehat{\mathcal{L}}_{\pi N,\,\text{ct}}^{(3)}$ given in Refs. [36, 37].

3.3 Nucleon Contact Lagrangians

Nucleon contact interactions consist of four nucleon fields (four nucleon legs) and no meson fields. Such terms are needed to renormalize loop integrals, to make results reasonably independent of regulators, and to parametrize the unresolved short-distance contributions to the nuclear force. For more about contact terms, see Sec. 5.3.

Because of parity, nucleon contact interactions come only in even numbers of derivatives, thus,

$$\mathcal{L}_{NN} = \mathcal{L}_{NN}^{(0)} + \mathcal{L}_{NN}^{(2)} + \mathcal{L}_{NN}^{(4)} + \cdots \qquad (40)$$

The lowest order (or leading order) NN Lagrangian has no derivatives and reads [18]

$$\mathcal{L}_{NN}^{(0)} = -\frac{1}{2}C_S \bar{N}N\bar{N}N - \frac{1}{2}C_T \bar{N}\vec{\sigma}N\bar{N}\vec{\sigma}N , \qquad (41)$$

where N is the heavy baryon nucleon field. C_S and C_T are unknown constants which are determined by a fit to the NN data. The second order NN Lagrangian is given by [19]

$$
\begin{aligned}
\mathcal{L}_{NN}^{(2)} = &-C_1'[(\bar{N}\vec{\nabla}N)^2 + (\overline{\vec{\nabla}N}N)^2] - C_2'(\bar{N}\vec{\nabla}N) \cdot (\overline{\vec{\nabla}N}N) \\
&-C_3'\bar{N}N[\bar{N}\vec{\nabla}^2N + \overline{\vec{\nabla}^2N}N] \\
&-iC_4'[\bar{N}\vec{\nabla}N \cdot (\overline{\vec{\nabla}N} \times \vec{\sigma}N) + \overline{(\vec{\nabla}N)}N \cdot (\bar{N}\vec{\sigma} \times \vec{\nabla}N)] \\
&-iC_5'\bar{N}N(\overline{\vec{\nabla}N} \cdot \vec{\sigma} \times \vec{\nabla}N) - iC_6'(\bar{N}\vec{\sigma}N) \cdot (\overline{\vec{\nabla}N} \times \vec{\nabla}N) \\
&-(C_7'\delta_{ik}\delta_{jl} + C_8'\delta_{il}\delta_{kj} + C_9'\delta_{ij}\delta_{kl}) \\
&\times [\bar{N}\sigma_k\partial_i N\bar{N}\sigma_l\partial_j N + \overline{\partial_i N}\sigma_k N\overline{\partial_j N}\sigma_l N] \\
&-(C_{10}'\delta_{ik}\delta_{jl} + C_{11}'\delta_{il}\delta_{kj} + C_{12}'\delta_{ij}\delta_{kl})\bar{N}\sigma_k\partial_i N\overline{\partial_j N}\sigma_l N \\
&-(\frac{1}{2}C_{13}'(\delta_{ik}\delta_{jl} + \delta_{il}\delta_{kj}) \\
&+C_{14}'\delta_{ij}\delta_{kl})[\overline{\partial_i N}\sigma_k\partial_j N + \overline{\partial_j N}\sigma_k\partial_i N]\bar{N}\sigma_l N .
\end{aligned}
\qquad (42)
$$

Similar to C_S and C_T, the C_i' are unknown constants which are fixed in a fit to the NN data. Obviously, these contact Lagrangians blow up quite a bit with increasing order, which why we do not give $\mathcal{L}_{NN}^{(4)}$ explicitly here.

4 Nuclear Forces from EFT: Overview

In the beginning of Sec. 3, we spelled out the steps we have to take to accomplish our EFT program for the derivation of nuclear forces. So far, we discussed steps one to three. What is left are steps four (low-momentum expansion) and five (Feynman diagrams). In this section, we will say more about the expansion we are using and give an overview of the Feynman diagrams that arise order by order.

4.1 Chiral Perturbation Theory and Power Counting

In ChPT, we analyze contributions in terms of powers of small momenta over the large scale: $(Q/\Lambda_\chi)^\nu$, where Q stands for a momentum (nucleon three-momentum or pion four-momentum) or a pion mass and $\Lambda_\chi \approx 1$ GeV is the chiral symmetry breaking scale (hadronic scale). Determining the power ν at which a given diagram contributes has become known as power counting. For a non-iterative contribution involving A nucleons, the power ν is given by

$$\nu = -2 + 2A - 2C + 2L + \sum_i \Delta_i \,, \tag{43}$$

with

$$\Delta_i \equiv d_i + \frac{n_i}{2} - 2 \,, \tag{44}$$

where C denotes the number of separately connected pieces and L the number of loops in the diagram; d_i is the number of derivatives or pion-mass insertions and n_i the number of nucleon fields involved in vertex i; the sum runs over all vertices contained in the diagram under consideration. Note that for an irreducible NN diagram ($A = 2$), the above formula reduces to

$$\nu = 2L + \sum_i \Delta_i \tag{45}$$

The power ν is bounded from below; e.g., for $A = 2$, $\nu \geq 0$. This fact is crucial for the power expansion to be of any use.

4.2 The Hierarchy of Nuclear Forces

Chiral perturbation theory and power counting imply that nuclear forces emerge as a hierarchy ruled by the power ν, Fig. 1.

The NN amplitude is determined by two classes of contributions: contact terms and pion-exchange diagrams. There are two contacts of order Q^0 $[\mathcal{O}(Q^0)]$ represented by the four-nucleon graph with a small-dot vertex shown in the first row of Fig. 1. The corresponding graph in the second row, four nucleon legs and a solid square, represents the seven contact terms of $\mathcal{O}(Q^2)$. Finally, at $\mathcal{O}(Q^4)$, we have 15 contact contributions represented by a four-nucleon graph with a solid diamond.

Now, turning to the pion contributions: At leading order [LO, $\mathcal{O}(Q^0)$, $\nu = 0$], there is only the well-known static one-pion exchange (1PE), second diagram in the first row of Fig. 1. Two-pion exchange (2PE) starts at next-to-leading order (NLO, $\nu = 2$) and all diagrams of this leading-order two-pion exchange are shown. Further 2PE contributions occur in any higher order. Of this sub-leading 2PE, we

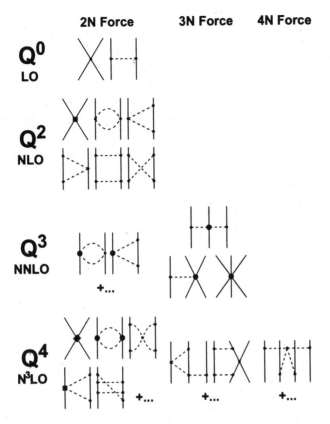

Figure 1: Hierarchy of nuclear forces in ChPT. Solid lines represent nucleons and dashed lines pions. Further explanations are given in the text.

show only two representative diagrams at next-to-next-to-leading order (NNLO) and three diagrams at next-to-next-to-next-to-leading order (N^{3}LO).

Finally, there is also three-pion exchange, which shows up for the first time at N^{3}LO (two loops; one representative 3π diagram is included in Fig. 1). At this order, the 3π contribution is negligible [38].

One important advantage of ChPT is that it makes specific predictions also for many-body forces. For a given order of ChPT, two-nucleon forces (2NF), three-nucleon forces (3NF), . . . are generated on the same footing (cf. Fig. 1). At LO, there are no 3NF, and at NLO, all 3NF terms cancel [18, 39]. However, at NNLO and higher orders, well-defined, nonvanishing 3NF occur [39, 40]. Since 3NF show up for the first time at NNLO, they are weak. Four-nucleon forces (4NF) occur first at N^{3}LO and, therefore, they are even weaker.

5 Two-Nucleon Forces

In this section, we will elaborate in detail on the two-nucleon force contributions of which we have given a rough overview in the previous section.

5.1 Pion-Exchange Contributions in ChPT

The effective pion Lagrangians presented in Sec. 3.2 are the crucial ingredients for the evaluation of the pion-exchange contributions to the NN interaction. We will derive these contributions now order by order.

We will state our results in terms of contributions to the momentum-space NN amplitude in the center-of-mass system (CMS), which takes the general form

$$
\begin{aligned}
V(\vec{p}\,',\vec{p}) = \ & V_C + \boldsymbol{\tau}_1 \cdot \boldsymbol{\tau}_2 W_C \\
& + \left[V_S + \boldsymbol{\tau}_1 \cdot \boldsymbol{\tau}_2 W_S \right] \vec{\sigma}_1 \cdot \vec{\sigma}_2 \\
& + \left[V_{LS} + \boldsymbol{\tau}_1 \cdot \boldsymbol{\tau}_2 W_{LS} \right] \left(-i\vec{S} \cdot (\vec{q} \times \vec{k}) \right) \\
& + \left[V_T + \boldsymbol{\tau}_1 \cdot \boldsymbol{\tau}_2 W_T \right] \vec{\sigma}_1 \cdot \vec{q}\, \vec{\sigma}_2 \cdot \vec{q} \\
& + \left[V_{\sigma L} + \boldsymbol{\tau}_1 \cdot \boldsymbol{\tau}_2 W_{\sigma L} \right] \vec{\sigma}_1 \cdot (\vec{q} \times \vec{k})\, \vec{\sigma}_2 \cdot (\vec{q} \times \vec{k}),
\end{aligned} \tag{46}
$$

where $\vec{p}\,'$ and $\vec{p}$ denote the final and initial nucleon momenta in the CMS, respectively; moreover,

$$
\begin{aligned}
\vec{q} &\equiv \vec{p}\,' - \vec{p} & \text{is the momentum transfer,} \\
\vec{k} &\equiv \tfrac{1}{2}(\vec{p}\,' + \vec{p}) & \text{the average momentum,} \\
\vec{S} &\equiv \tfrac{1}{2}(\vec{\sigma}_1 + \vec{\sigma}_2) & \text{the total spin,}
\end{aligned} \tag{47}
$$

and $\vec{\sigma}_{1,2}$ and $\boldsymbol{\tau}_{1,2}$ are the spin and isospin operators, respectively, of nucleon 1 and 2. For on-energy-shell scattering, V_α and W_α ($\alpha = C, S, LS, T, \sigma L$) can be expressed as functions of q and k (with $q \equiv |\vec{q}|$ and $k \equiv |\vec{k}|$), only.

Our formalism is similar to the one used by the Munich group [22, 41, 42] except for two differences: all our momentum space amplitudes differ by an over-all factor of (-1) and our spin-orbit potentials, V_{LS} and W_{LS}, differ by an additional factor of (-2). Our conventions are more in tune with what is commonly used in nuclear physics.

In all expressions given below, we will state only the *nonpolynomial* contributions to the NN amplitude. Note, however, that dimensional regularization typically generates also polynomial terms. These polynomials are absorbed by the contact interactions to be discussed in a later section and, therefore, they are of no interest here.

Q^2
(NLO)

Q^3
(N²LO)

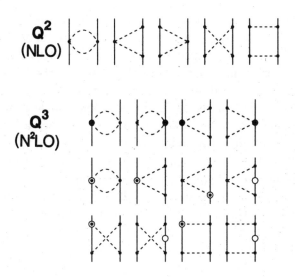

Figure 2: Two-pion exchange contributions to the NN interaction at order two and three in small momenta. Solid lines represent nucleons and dashed lines pions. Small dots denote vertices from the leading order πN Lagrangian $\widehat{\mathcal{L}}_{\pi N}^{(1)}$, Eq. (33). Large solid dots are vertices proportional to the LECs c_i from the second order Lagrangian $\widehat{\mathcal{L}}_{\pi N,\,\text{ct}}^{(2)}$, Eq. (37). Symbols with an open circles are relativistic $1/M_N$ corrections contained in the second order Lagrangian $\widehat{\mathcal{L}}_{\pi N}^{(2)}$, Eqs. (35). Only a few representative examples of $1/M_N$ corrections are shown and not all.

5.1.1   Zeroth Order (LO)

At order zero $[\nu = 0,\ \mathcal{O}(Q^0)$, lowest order, leading order, LO], there is only the well-known static one-pion exchange, second diagram in the first row of Fig. 1 which is given by:

$$V_{1\pi}(\vec{p}\,',\vec{p}) = -\frac{g_A^2}{4f_\pi^2}\,\boldsymbol{\tau}_1 \cdot \boldsymbol{\tau}_2\,\frac{\vec{\sigma}_1 \cdot \vec{q}\,\vec{\sigma}_2 \cdot \vec{q}}{q^2 + m_\pi^2}\,. \tag{48}$$

At first order $[\nu = 1,\ \mathcal{O}(Q)]$, there are no pion-exchange contributions (and also no contact terms).

5.1.2   Second Order (NLO)

Non-vanishing higher-order graphs start at second order ($\nu = 2$, next-to-leading order, NLO). The most efficient way to evaluate these loop diagrams is to use covariant perturbation theory and dimensional regularization. This is the method applied by the Munich group [22, 41, 42]. One starts with the relativistic versions of the πN Lagrangians (cf. Sec. 3.2) and sets up four-dimensional (covariant) loop integrals. Relativistic vertices and nucleon propagators are then expanded in powers of $1/M_N$. The divergences that occur in conjunction with the four-dimensional loop integrals are treated by means of dimensional regularization, a prescription which is consistent with chiral symmetry and power counting. The results derived in this way are the same obtained when starting right away with the HB versions of the πN Lagrangians. However, as it turns out, the method used by the Munich group is more efficient in dealing with the rather tedious calculations.

Two-pion exchange occurs first at second order, also know as leading-order 2π exchange. The graphs are shown in the first row of Fig. 2. Since a loop creates already $\nu = 2$, the vertices involved at this order can only be from the leading/lowest order Lagrangian $\widehat{\mathcal{L}}_{\pi N}^{(1)}$, Eq. (33), i. e., they carry only one derivative. These vertices are denoted by small dots in Fig. 2. Concerning the box diagram, we should note that we include only the non-iterative part of this diagram which is obtained by subtracting the iterated 1PE contribution Eq. (65) or Eq. (66), below, but using $M_N^2/E_p \approx M_N^2/E_{p''} \approx M_N$ at this order (NLO). Summarizing all contributions from irreducible two-pion exchange at second order, one obtains [22]:

$$
W_C = -\frac{L(q)}{384\pi^2 f_\pi^4} \left[4m_\pi^2 (5g_A^4 - 4g_A^2 - 1) + q^2(23g_A^4 - 10g_A^2 - 1) \right.
$$
$$
\left. + \frac{48g_A^4 m_\pi^4}{w^2} \right], \tag{49}
$$

$$
V_T = -\frac{1}{q^2} V_S = -\frac{3g_A^4 L(q)}{64\pi^2 f_\pi^4}, \tag{50}
$$

where

$$
L(q) \equiv \frac{w}{q} \ln \frac{w+q}{2m_\pi} \tag{51}
$$

and

$$
w \equiv \sqrt{4m_\pi^2 + q^2}. \tag{52}
$$

5.1.3 Third Order (NNLO)

The two-pion exchange diagrams of order three ($\nu = 3$, next-to-next-to-leading order, NNLO) are very similar to the ones of order two, except that they contain one insertion from $\widehat{\mathcal{L}}_{\pi N}^{(2)}$, Eq. (35). The resulting contributions are typically either proportional to one of the low-energy constants c_i or they contain a factor $1/M_N$. Notice that relativistic $1/M_N$ corrections can occur for vertices and nucleon propagators. In Fig. 2, we show in row 2 the diagrams with vertices proportional to c_i (large solid dot), Eq. (37), and in row 3 and 4 a few representative graphs with a $1/M_N$ correction (symbols with an open circle). The number of $1/M_N$ correction graphs is large and not all are shown in the figure. Again, the box diagram is corrected for a contribution from the iterated 1PE. If the iterative 2PE of Eq. (65) is used, the expansion of the factor $M_N^2/E_p = M_N - p^2/2M_N + \ldots$ is applied and the term proportional to $(-p^2/2M_N)$ is subtracted from the third order box diagram contribution. Then, one obtains for the full third order contribution [22]:

$$
V_C = \frac{3g_A^2}{16\pi f_\pi^4} \left\{ \frac{g_A^2 m_\pi^5}{16 M_N w^2} - \left[2m_\pi^2 (2c_1 - c_3) - q^2 \left(c_3 + \frac{3g_A^2}{16 M_N} \right) \right] \right. \\
\left. \times \widetilde{w}^2 A(q) \right\},
\tag{53}
$$

$$
W_C = \frac{g_A^2}{128\pi M_N f_\pi^4} \left\{ 3g_A^2 m_\pi^5 w^{-2} \right. \\
\left. - \left[4m_\pi^2 + 2q^2 - g_A^2 (4m_\pi^2 + 3q^2) \right] \widetilde{w}^2 A(q) \right\},
\tag{54}
$$

$$
V_T = -\frac{1}{q^2} V_S = \frac{9 g_A^4 \widetilde{w}^2 A(q)}{512 \pi M_N f_\pi^4},
\tag{55}
$$

$$
W_T = -\frac{1}{q^2} W_S \\
= -\frac{g_A^2 A(q)}{32\pi f_\pi^4} \left[\left(c_4 + \frac{1}{4M_N} \right) w^2 - \frac{g_A^2}{8M_N} (10 m_\pi^2 + 3q^2) \right],
\tag{56}
$$

$$
V_{LS} = \frac{3 g_A^4 \widetilde{w}^2 A(q)}{32\pi M_N f_\pi^4},
\tag{57}
$$

$$
W_{LS} = \frac{g_A^2 (1 - g_A^2)}{32\pi M_N f_\pi^4} w^2 A(q),
\tag{58}
$$

with

$$
A(q) \equiv \frac{1}{2q} \arctan \frac{q}{2m_\pi}
\tag{59}
$$

and

$$
\widetilde{w} \equiv \sqrt{2m_\pi^2 + q^2}.
\tag{60}
$$

As discussed in Sec. 5.1.5, below, we prefer the iterative 2PE defined in Eq. (66), which leads to a different NNLO term for the iterative 2PE. This changes the $1/M_N$ terms in the above potentials. The changes are obtained by adding to Eqs. (53)-(56) the following terms:

$$
V_C = -\frac{3 g_A^4}{256 \pi f_\pi^4 M_N} (m_\pi \omega^2 + \widetilde{\omega}^4 A(q))
\tag{61}
$$

$$
W_C = \frac{g_A^4}{128\pi f_\pi^4 M_N} (m_\pi \omega^2 + \widetilde{\omega}^4 A(q))
\tag{62}
$$

$$
V_T = -\frac{1}{q^2} V_S = \frac{3 g_A^4}{512 \pi f_\pi^4 M_N} (m_\pi + \omega^2 A(q))
\tag{63}
$$

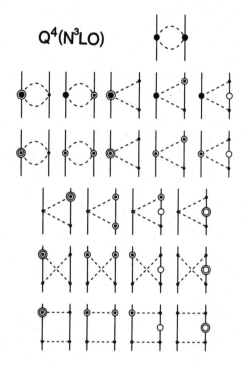

Figure 3: One-loop 2π-exchange contributions to the NN interaction at order four. Basic notation as in Fig. 2. Symbols with a large solid dot and an open circle denote $1/M_N$ corrections of vertices proportional to c_i. Symbols with two open circles mark relativistic $1/M_N^2$ corrections. Both corrections are part of the third order Lagrangian $\widehat{\mathcal{L}}_{\pi N}^{(3)}$, Eq. (39). Representative examples for all types of one-loop graphs that occur at this order are shown.

$$W_T = -\frac{1}{q^2}W_S = -\frac{g_A^4}{256\pi f_\pi^4 M_N}(m_\pi + \omega^2 A(q)) \qquad (64)$$

5.1.4 Fourth Order (N³LO)

This order, which may also be denoted by next-to-next-to-next-to-leading order (N³LO), is very involved. Three-pion exchange (3PE) occurs for the first time at this order. The 3PE contribution at N³LO has been calculated by the Munich group and found to be negligible [38]. Therefore, we will ignore it.

The 2PE contributions at N³LO can be subdivided into two groups, one-loop graphs, Fig. 3, and two-loop diagrams, Fig. 4. Since these contributions are very complicated, we have moved them to Appendix A.

Q⁴
(N³LO)

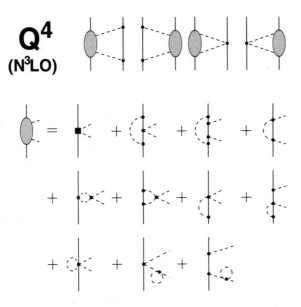

Figure 4: Two-loop 2π-exchange contributions at order four. Basic notation as in Fig. 2. The oval stands for all one-loop πN graphs some of which are shown in the lower part of the figure. The solid square represents vertices proportional to the LECs d_i which are introduced by the third order Lagrangian $\mathcal{L}_{\pi N}^{(3)}$, Eq. (38). More explanations are given in the text.

5.1.5 Iterated One-Pion-Exchange

Besides all the irreducible 2PE contributions presented above, there is also the reducible 2PE which is generated from iterated 1PE. This "iterative 2PE" is the only 2PE contribution which produces an imaginary part. Thus, one wishes to formulate this contribution such that relativistic elastic unitarity is satisfied. There are several ways to achieve this.

Kaiser $et\ al.$ [22] define the iterative 2PE contribution as follows,

$$V_{2\pi,it}^{(KBW)}(\vec{p}\,',\vec{p}) = \frac{M_N^2}{E_p} \int \frac{d^3p''}{(2\pi)^3} \frac{V_{1\pi}(\vec{p}\,',\vec{p}\,'')\,V_{1\pi}(\vec{p}\,'',\vec{p})}{p^2 - p''^2 + i\epsilon} \tag{65}$$

with $V_{1\pi}$ given in Eq. (48).

Since we adopt the relativistic scheme developed by Blankenbecler and Sugar [43] (BbS) (see beginning of Sec. 5.4), we prefer the following formulation which is consistent with the BbS approach (and, of course, with relativistic elastic unitarity):

$$V_{2\pi,it}^{(EM)}(\vec{p}\,',\vec{p}) = \int \frac{d^3p''}{(2\pi)^3} \frac{M_N^2}{E_{p''}} \frac{V_{1\pi}(\vec{p}\,',\vec{p}\,'')\,V_{1\pi}(\vec{p}\,'',\vec{p})}{p^2 - p''^2 + i\epsilon}\,. \tag{66}$$

The iterative 2PE contribution has to be subtracted from the covariant box diagram, order by order. For this, the expansion $M_N^2/E_p = M_N - p^2/2M_N + \ldots$ is applied in Eq. (65) and $M_N^2/E_{p''} = M_N - p''^2/2M_N + \ldots$ in Eq. (66). At NLO, both choices for the iterative 2PE collapse to the same, while at NNLO there are obvious differences.

5.2 *NN* Scattering in Peripheral Partial Waves Using the Perturbative Amplitude

After the tedious mathematics of the previous section, it is time for more tangible affairs. The obvious question to address now is: How does the derived NN amplitude compare to empirical information? Since our derivation includes only one- and two-pion exchanges, we are dealing here with the long- and intermediate-range part of the NN interaction. This part of the nuclear force is probed in the peripheral partial waves of NN scattering. Thus, in this section, we will calculate the phase shifts that result from the NN amplitudes presented in the previous section and compare them to the empirical phase shifts as well as to the predictions from conventional meson theory. Besides the irreducible two-pion exchanges derived above, we must also include 1PE and iterated 1PE.

In this section [44], which is restricted to just peripheral waves, we will always consider neutron-proton (np) scattering and take the charge-dependence of 1PE due to pion-mass splitting into account, since it is appreciable. With the definition

$$V_{1\pi}(m_\pi) \equiv -\frac{g_A^2}{4f_\pi^2}\frac{\vec{\sigma}_1\cdot\vec{q}\,\vec{\sigma}_2\cdot\vec{q}}{q^2+m_\pi^2}\,, \tag{67}$$

the charge-dependent 1PE for np scattering is

$$V_{1\pi}^{(np)}(\vec{p}\,',\vec{p}) = -V_\pi(m_{\pi^0}) + (-1)^{I+1}\,2\,V_\pi(m_{\pi^\pm})\,, \tag{68}$$

where I denotes the isospin of the two-nucleon system. We use $m_{\pi^0} = 134.9766$ MeV, $m_{\pi^\pm} = 139.5702$ MeV [31], and

$$M_N = \frac{2M_pM_n}{M_p+M_n} = 938.9182\ \text{MeV}\,. \tag{69}$$

Also in the iterative 2PE, we apply the charge-dependent 1PE, i.e., in Eq. (66) we replace $V_{1\pi}$ with $V_{1\pi}^{(np)}$.

The perturbative relativistic T-matrix for np scattering in peripheral waves is

$$T(\vec{p}\,',\vec{p}) = V_{1\pi}^{(np)}(\vec{p}\,',\vec{p}) + V_{2\pi,it}^{(\text{EM},np)}(\vec{p}\,',\vec{p}) + V_{2\pi,irr}(\vec{p}\,',\vec{p})\,, \tag{70}$$

where $V_{2\pi,irr}$ refers to any or all of the irreducible 2PE contributions presented in Sec. 5.1, depending on the order at which the calculation is conducted. In the

Table 2: Low-energy constants, LECs, used for a NN potential at N^3LO, Sec. 5.4.4, and in the calculation of the peripheral NN phase shifts shown in Fig. 5 (column "NN periph. Fig. 5"). The c_i belong to the dimension-two πN Lagrangian, Eq. (37), and are in units of GeV^{-1}, while the $\bar{d}_i$ are associated with the dimension-three Lagrangian, Eq. (38), and are in units of GeV^{-2}. The column "πN empirical" shows determinations from πN data.

LEC	NN potential at N^3LO	NN periph. Fig. 5	πN empirical
c_1	-0.81	-0.81	-0.81 ± 0.15[a]
c_2	2.80	3.28	3.28 ± 0.23[b]
c_3	-3.20	-3.40	-4.69 ± 1.34[a]
c_4	5.40	3.40	3.40 ± 0.04[a]
$\bar{d}_1 + \bar{d}_2$	3.06	3.06	3.06 ± 0.21[b]
$\bar{d}_3$	-3.27	-3.27	-3.27 ± 0.73[b]
$\bar{d}_5$	0.45	0.45	0.45 ± 0.42[b]
$\bar{d}_{14} - \bar{d}_{15}$	-5.65	-5.65	-5.65 ± 0.41[b]

[a]Table 1, Fit 1 of Ref. [48].
[b]Table 2, Fit 1 of Ref. [37].

calculation of the irreducible 2PE, we use the average pion mass $m_\pi = 138.039$ MeV and, thus, neglect the charge-dependence due to pion-mass splitting. The charge-dependence that emerges from irreducible 2π exchange was investigated in Ref. [45] and found to be negligible for partial waves with $L \geq 3$.

For the T-matrix given in Eq. (70), we calculate phase shifts for partial waves with $L \geq 3$ and $T_{lab} \leq 300$ MeV. At order four in small momenta, partial waves with $L \geq 3$ do not receive any contributions from contact interactions and, thus, the non-polynomial pion contributions uniquely predict the F and higher partial waves. We use $f_\pi = 92.4$ MeV [31] and $g_A = 1.29$. Via the Goldberger-Treiman relation, $g_A = g_{\pi NN} \, f_\pi / M_N$, our value for g_A is consistent with $g_{\pi NN}^2/4\pi = 13.63 \pm 0.20$ which is obtained from πN and NN analysis [46, 47].

The LECs used in this calculation are shown in Table 2, column "NN periph. Fig. 5". Note that many determinations of the LECs, c_i and $\bar{d}_i$, can be found in the literature. The most reliable way to determine the LECs from empirical πN information is to extract them from the πN amplitude inside the Mandelstam triangle (unphysical region) which can be constructed with the help of dispersion relations from empirical πN data. This method was used by Büttiker and Meißner [48]. Unfortunately, the values for c_2 and all $\bar{d}_i$ parameters obtained in Ref. [48] carry uncertainties, so large that the values cannot provide any guidance. Therefore, in Table 2, only c_1, c_3, and c_4 are from Ref. [48], while the other LECs are taken

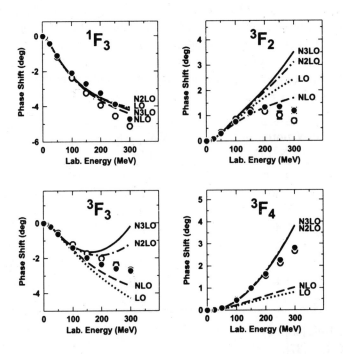

Figure 5: F-wave phase shifts of neutron-proton scattering for laboratory kinetic energies below 300 MeV. We show the predictions from chiral pion exchange at leading order (LO), next-to-leading order (NLO), next-to-next-to-leading order (N2LO), and next-to-next-to-next-to-leading order (N3LO). The solid dots and open circles are the results from the Nijmegen multi-energy np phase shift analysis [49] and the VPI single-energy np analysis SM99 [50], respectively.

from Ref. [37] where the πN amplitude in the physical region was considered. To establish a link between πN and NN, we apply the values from the above determinations in our calculations of the NN peripheral phase shifts. In general, we use the mean values; the only exception is c_3, where we choose a value that is, in terms of magnitude, about one standard deviation below the one from Ref. [48]. With the exception of c_3, phase shift predictions do not depend sensitively on variations of the LECs within the quoted uncertainties.

In Fig. 5, we show the phase-shift predictions for neutron-proton scattering in F waves for laboratory kinetic energies below 300 MeV (for G and H waves, see Ref. [26]). The orders displayed are defined as follows:

- Leading order (LO) is just 1PE, Eq. (68).

- Next-to-leading order (NLO) is 1PE, Eq. (68), plus iterated 1PE, Eq. (66), plus the contributions of Sec. 5.1.2 (order two), Eqs. (49) and (50).

- Next-to-next-to-leading order (denoted by N2LO in the figures) consists of NLO plus the contributions of Sec. 5.1.3 (order three), Eqs. (53)-(58) and (61)-(64).

- Next-to-next-to-next-to-leading order (denoted by N3LO in the figures) consists of N2LO plus the contributions of Sec. 5.1.4 (order four), Eqs. (99)-(112) and (115)-(124).

It is clearly seen in Fig. 5 that the leading order 2π exchange (NLO) is a rather small contribution, insufficient to explain the empirical facts. In contrast, the next order (N2LO) is very large, several times NLO. This is due to the $\pi\pi NN$ contact interactions proportional to the LECs c_i that are introduced by the second order Lagrangian $\mathcal{L}_{\pi N}^{(2)}$, Eq. (34). These contacts are supposed to simulate the contributions from intermediate Δ-isobars and correlated 2π exchange which are known to be large (see, e. g., Ref. [13]).

At N3LO a clearly identifiable trend towards convergence emerges. Obviously, $^1F_3$ and $^3F_4$ appear fully converged. However, in $^3F_2$ and $^3F_3$, N3LO differs noticeably from NNLO, but the difference is much smaller than the one between NNLO and NLO. This is what we perceive as a trend towards convergence.

In Fig. 6, we conduct a comparison between the predictions from chiral one- and two-pion exchange at N3LO and the corresponding predictions from conventional meson theory (curve 'Bonn'). As representative for conventional meson theory, we choose the Bonn meson-exchange model for the NN interaction [13], since it contains a comprehensive and thoughtfully constructed model for 2π exchange. This 2π model includes box and crossed box diagrams with NN, $N\Delta$, and $\Delta\Delta$ intermediate states as well as direct $\pi\pi$ interaction in S- and P-waves (of the $\pi\pi$ system) consistent with empirical information from πN and $\pi\pi$ scattering. Besides this the Bonn model also includes (repulsive) ω-meson exchange and irreducible diagrams of π and ρ exchange (which are also repulsive). However, note that in the phase shift predictions displayed in Fig. 6, the "Bonn" curve includes only the 1π and 2π contributions from the Bonn model; the short-range contributions are left out since the purpose of the figure is to compare different models/theories for $\pi + 2\pi$. In all waves shown we see, in general, good agreement between N3LO and Bonn. In $^3F_2$ and $^3F_3$ above 150 MeV and in $^3F_4$ above 250 MeV the chiral model at N3LO is more attractive than the Bonn 2π model. Note, however, that the Bonn model is relativistic and, thus, includes relativistic corrections up to infinite orders. Thus, one may speculate that higher orders in ChPT may create some repulsion, moving the Bonn and the chiral predictions even closer together [51].

The 2π exchange contribution to the NN interaction can also be derived from *empirical* πN and $\pi\pi$ input using dispersion theory, which is based upon unitarity, causality (analyticity), and crossing symmetry. The amplitude $N\bar{N} \to \pi\pi$ is constructed from $\pi N \to \pi N$ and $\pi N \to \pi\pi N$ data using crossing properties and analytic continuation; this amplitude is then 'squared' to yield the $N\bar{N}$ amplitude

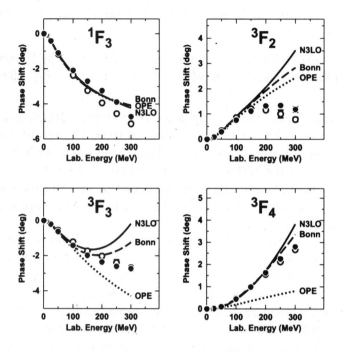

Figure 6: *F*-wave phase shifts of neutron-proton scattering for laboratory kinetic energies below 300 MeV. We show the results from one-pion-exchange (OPE), and one- plus two-pion exchange as predicted by ChPT at next-to-next-to-next-to-leading order (N3LO) and by the Bonn Full Model [13] (Bonn). Note that the "Bonn" curve does not include the repulsive ω and $\pi\rho$ exchanges of the full model. Empirical phase shifts (solid dots and open circles) as in Fig. 5.

which is related to NN by crossing symmetry [52]. The Paris group [11, 12] pursued this path and calculated NN phase shifts in peripheral partial waves. Naively, the dispersion-theoretic approach is the ideal one, since it is based exclusively on empirical information. Unfortunately, in practice, quite a few uncertainties enter into the approach. First, there are ambiguities in the analytic continuation and, second, the dispersion integrals have to be cut off at a certain momentum to ensure reasonable results. In Ref. [13], a thorough comparison was conducted between the predictions by the Bonn model and the Paris approach and it was demonstrated that the Bonn predictions always lie comfortably within the range of uncertainty of the dispersion-theoretic results. Therefore, there is no need to perform a separate comparison of our chiral N3LO predictions with dispersion theory, since it would not add anything that we cannot conclude from Fig. 6.

Finally, we need to compare the predictions with the empirical phase shifts. In F waves the N^{3}LO predictions above 200 MeV are, in general, too attractive. Note, however, that this is also true for the predictions by the Bonn $\pi + 2\pi$ model. In the *full* Bonn model, besides $\pi + 2\pi$, (repulsive) ω and $\pi\rho$ exchanges are included which move the predictions right on top of the data. The exchange of a ω meson or combined $\pi\rho$ exchange are 3π exchanges. Three-pion exchange occurs first at chiral order four. It has be investigated by Kaiser [38] and found to be negligible, at this order. However, 3π exchange at order five appears to be sizable [53] and may have impact on F waves. Besides this, there is the usual short-range phenomenology. In ChPT, this short-range interaction is parametrized in terms of four-nucleon contact terms (since heavy mesons do not have a place in that theory). Contact terms of order four (N^{3}LO) do not contribute to F-waves, but order six does. In summary, the remaining small discrepancies between the N3LO predictions and the empirical phase shifts may be straightened out in fifth or sixth order of ChPT.

5.3 NN Contact Potentials

In conventional meson theory, the short-range nuclear force is described by the exchange of heavy mesons, notably the $\omega(782)$. The qualitative short-distance behavior of the NN potential is obtained by Fourier transform of the propagator of a heavy meson,

$$\int d^3q \frac{e^{i\vec{q}\cdot\vec{r}}}{m_\omega^2 + \vec{q}^2} \sim \frac{e^{-m_\omega r}}{r}. \tag{71}$$

ChPT is an expansion in small momenta Q, too small to resolve structures like a $\rho(770)$ or $\omega(782)$ meson, because $Q \ll \Lambda_\chi \approx m_{\rho,\omega}$. But the latter relation allows us to expand the propagator of a heavy meson into a power series,

$$\frac{1}{m_\omega^2 + Q^2} \approx \frac{1}{m_\omega^2}\left(1 - \frac{Q^2}{m_\omega^2} + \frac{Q^4}{m_\omega^4} - + \ldots\right), \tag{72}$$

where the ω is representative for any heavy meson of interest. The above expansion suggests that it should be possible to describe the short distance part of the nuclear force simply in terms of powers of Q/m_ω, which fits in well with our over-all power scheme since $Q/m_\omega \approx Q/\Lambda_\chi$.

A second purpose of contact terms is renormalization. Dimensional regularization of the loop integrals of pion-exchanges (cf. Sec. 5.1) typically generates polynomial terms with coefficients that are, in part, infinite or scale dependent. Contact terms pick up infinities and remove scale dependence.

The partial-wave decomposition of a power Q^ν has an interesting property. First note that Q can only be either the momentum transfer between the two interacting nucleons q or the average momentum k [cf. Eq. (47) for their definitions]. In any case, for even ν,

$$Q^\nu = f_{\frac{\nu}{2}}(\cos\theta), \tag{73}$$

where f_m stands for a polynomial of degree m and θ is the CMS scattering angle.

The partial-wave decomposition of Q^ν for a state of orbital-angular momentum L involves the integral

$$I_L^{(\nu)} = \int_{-1}^{+1} Q^\nu P_L(\cos\theta) d\cos\theta = \int_{-1}^{+1} f_{\frac{\nu}{2}}(\cos\theta) P_L(\cos\theta) d\cos\theta, \qquad (74)$$

where P_L is a Legendre polynomial. Due to the orthogonality of the P_L,

$$I_L^{(\nu)} = 0 \quad \text{for} \quad L > \frac{\nu}{2}. \qquad (75)$$

Consequently, contact terms of order zero contribute only in S-waves, while order-two terms contribute up to P-waves, order-four terms up to D-waves, etc..

We will now present, one by one, the various orders of NN contact terms together with their partial-wave decomposition [54]. Note that, due to parity, only even powers of Q are allowed.

5.3.1   Zeroth Order

The contact potential at order zero reads:

$$V^{(0)}(\vec{p}',\vec{p}) = C_S + C_T\, \vec\sigma_1 \cdot \vec\sigma_2 \qquad (76)$$

Partial wave decomposition yields:

$$\begin{aligned}
V^{(0)}(^1S_0) &= \tilde{C}_{1S_0} = 4\pi\,(C_S - 3\,C_T) \\
V^{(0)}(^3S_1) &= \tilde{C}_{3S_1} = 4\pi\,(C_S + C_T)
\end{aligned} \qquad (77)$$

5.3.2   Second Order

The contact potential contribution of order two is given by:

$$\begin{aligned}
V^{(2)}(\vec{p}',\vec{p}) =\ & C_1 q^2 + C_2 k^2 \\
+\ & \left(C_3 q^2 + C_4 k^2\right) \vec\sigma_1 \cdot \vec\sigma_2 \\
+\ & C_5 \left(-i\vec{S}\cdot(\vec{q}\times\vec{k})\right) \\
+\ & C_6(\vec\sigma_1\cdot\vec{q})(\vec\sigma_2\cdot\vec{q}) \\
+\ & C_7(\vec\sigma_1\cdot\vec{k})(\vec\sigma_2\cdot\vec{k})
\end{aligned} \qquad (78)$$

Second order partial wave contributions:

$$\begin{aligned}
V^{(2)}(^1S_0) &= C_{1S_0}(p^2 + p'^2) \\
&= 4\pi\left(C_1 + \frac{1}{4}C_2 - 3C_3 - \frac{3}{4}C_4 - C_6 - \frac{1}{4}C_7\right)(p^2 + p'^2) \\
V^{(2)}(^3P_0) &= C_{3P_0}\, pp'
\end{aligned}$$

$$= 4\pi \left(-\frac{2}{3}C_1 + \frac{1}{6}C_2 - \frac{2}{3}C_3 + \frac{1}{6}C_4 - \frac{2}{3}C_5 + 2C_6 - \frac{1}{2}C_7 \right) pp'$$

$$V^{(2)}(^1P_1) = C_{1P_1} \, pp'$$

$$= 4\pi \left(-\frac{2}{3}C_1 + \frac{1}{6}C_2 + 2C_3 - \frac{1}{2}C_4 + \frac{2}{3}C_6 - \frac{1}{6}C_7 \right) pp'$$

$$V^{(2)}(^3P_1) = C_{3P_1} \, pp'$$

$$= 4\pi \left(-\frac{2}{3}C_1 + \frac{1}{6}C_2 - \frac{2}{3}C_3 + \frac{1}{6}C_4 - \frac{1}{3}C_5 - \frac{4}{3}C_6 + \frac{1}{3}C_7 \right) pp'$$

$$V^{(2)}(^3S_1) = C_{3S_1}(p^2 + p'^2)$$

$$= 4\pi \left(C_1 + \frac{1}{4}C_2 + C_3 + \frac{1}{4}C_4 + \frac{1}{3}C_6 + \frac{1}{12}C_7 \right) (p^2 + p'^2)$$

$$V^{(2)}(^3S_1 - {}^3D_1) = C_{3S_1 - {}^3D_1}p^2$$

$$= 4\pi \left(-\frac{2\sqrt{2}}{3}C_6 - \frac{2\sqrt{2}}{12}C_7 \right) p^2$$

$$V^{(2)}(^3P_2) = C_{3P_2} \, pp'$$

$$= 4\pi \left(-\frac{2}{3}C_1 + \frac{1}{6}C_2 - \frac{2}{3}C_3 + \frac{1}{6}C_4 + \frac{1}{3}C_5 \right) pp' \qquad (79)$$

5.3.3 Fourth Order

The contact potential contribution of order four reads:

$$
\begin{aligned}
V^{(4)}(\vec{p}\,', \vec{p}) = {}& D_1 q^4 + D_2 k^4 + D_3 q^2 k^2 + D_4(\vec{q} \times \vec{k})^2 \\
+ {}& \left(D_5 q^4 + D_6 k^4 + D_7 q^2 k^2 + D_8(\vec{q} \times \vec{k})^2 \right) \vec{\sigma}_1 \cdot \vec{\sigma}_2 \\
+ {}& \left(D_9 q^2 + D_{10} k^2 \right) \left(-i\vec{S} \cdot (\vec{q} \times \vec{k}) \right) \\
+ {}& \left(D_{11} q^2 + D_{12} k^2 \right) (\vec{\sigma}_1 \cdot \vec{q})(\vec{\sigma}_2 \cdot \vec{q}) \\
+ {}& \left(D_{13} q^2 + D_{14} k^2 \right) (\vec{\sigma}_1 \cdot \vec{k})(\vec{\sigma}_2 \cdot \vec{k}) \\
+ {}& D_{15} \left(\vec{\sigma}_1 \cdot (\vec{q} \times \vec{k}) \, \vec{\sigma}_2 \cdot (\vec{q} \times \vec{k}) \right) \qquad (80)
\end{aligned}
$$

The rather lengthy partial-wave expressions of this order have been relegated to Appendix B.

5.4 Constructing a Chiral NN Potential

5.4.1 Conceptual Questions

The two-nucleon system is non-perturbative as evidenced by the presence of a shallow bound state (the deuteron) and large scattering lengths. Weinberg [18] showed that the strong enhancement of the scattering amplitude arises from purely nucleonic intermediate states. He therefore suggested to use perturbation theory to calculate the NN potential and to apply this potential in a scattering equation to obtain the NN amplitude. We adopt this prescription.

Since the irreducible diagrams that make up the potential are calculated using covariant perturbation theory (cf. Sec. 5.1), it is consistent to start from the covariant Bethe-Salpeter (BS) equation [55] describing two-nucleon scattering. In operator notation, the BS equation reads

$$T = \mathcal{V} + \mathcal{V}\mathcal{G}T \tag{81}$$

with T the invariant amplitude for the two-nucleon scattering process, $\mathcal{V}$ the sum of all connected two-particle irreducible diagrams, and $\mathcal{G}$ the relativistic two-nucleon propagator. The BS equation is equivalent to a set of two equations

$$\begin{align}
T &= V + V\,g\,T \tag{82} \\
V &= \mathcal{V} + \mathcal{V}\,(\mathcal{G} - g)\,V \tag{83} \\
&= \mathcal{V} + \mathcal{V}_{1\pi}\,(\mathcal{G} - g)\,\mathcal{V}_{1\pi} + \dots, \tag{84}
\end{align}$$

where g is a covariant three-dimensional propagator which preserves relativistic elastic unitarity. We choose the propagator g proposed by Blankenbecler and Sugar (BbS) [43] (for more details on relativistic three-dimensional reductions of the BS equation, see Ref. [7]). The ellipsis in Eq. (84) stands for terms of irreducible 3π and higher pion exchanges which we neglect.

Note that when we speak of covariance in conjunction with (heavy baryon) ChPT, we are not referring to manifest covariance. Relativity and relativistic off-shell effects are accounted for in terms of a Q/M_N expansion up to the given order. Thus, Eq. (84) is evaluated in the following way,

$$V \approx \mathcal{V}(\text{on-shell}) + \mathcal{V}_{1\pi}\,\mathcal{G}\,\mathcal{V}_{1\pi} - V_{1\pi}\,g\,V_{1\pi}, \tag{85}$$

where the pion-exchange content of $\mathcal{V}(\text{on-shell})$ is $V_{1\pi} + V_{2\pi}'$ with $V_{1\pi}$ the on-shell 1PE given in Eq. (48) and $V_{2\pi}'$ the irreducible 2π exchanges calculated in Sec. 5.1, *but without the box*. $\mathcal{V}_{1\pi}$ denotes the relativistic (off-shell) 1PE. Notice that the term $(\mathcal{V}_{1\pi}\,\mathcal{G}\,\mathcal{V}_{1\pi} - V_{1\pi}\,g\,V_{1\pi})$ represents what has been called "the (irreducible part of the) box diagram contribution" in Sec. 5.1 where it was evaluated at various orders.

The full chiral NN potential V is given by irreducible pion exchanges V_π and contact terms V_{ct},

$$V = V_\pi + V_{\text{ct}} \tag{86}$$

with

$$V_\pi = V_{1\pi} + V_{2\pi} + \dots, \tag{87}$$

where the ellipsis denotes irreducible 3π and higher pion exchanges which are omitted. Two-pion exchange contributions appear in various orders

$$V_{2\pi} = V_{2\pi}^{(2)} + V_{2\pi}^{(3)} + V_{2\pi}^{(4)} + \cdots \tag{88}$$

as calculated in Sec. 5.1. Contact terms come in even orders,

$$V_{\rm ct} = V_{\rm ct}^{(0)} + V_{\rm ct}^{(2)} + V_{\rm ct}^{(4)} + \cdots \tag{89}$$

and were presented in Sec. 5.3. The potential V is calculated at a given order. For example, the potential at NNLO includes 2PE up to $V_{2\pi}^{(3)}$ and contacts up to $V_{\rm ct}^{(2)}$. At N^{3}LO, contributions up to $V_{2\pi}^{(4)}$ and $V_{\rm ct}^{(4)}$ are included.

The potential V satisfies the relativistic BbS equation, Eq. (82). Defining

$$\widehat{V}(\vec{p}\,',\vec{p}) \equiv \frac{1}{(2\pi)^3} \sqrt{\frac{M_N}{E_{p'}}}\, V(\vec{p}\,',\vec{p}) \sqrt{\frac{M_N}{E_p}} \tag{90}$$

and

$$\widehat{T}(\vec{p}\,',\vec{p}) \equiv \frac{1}{(2\pi)^3} \sqrt{\frac{M_N}{E_{p'}}}\, T(\vec{p}\,',\vec{p}) \sqrt{\frac{M_N}{E_p}} \tag{91}$$

with $E_p \equiv \sqrt{M_N^2 + \vec{p}^2}$ (the factor $1/(2\pi)^3$ is added for convenience), the BbS equation collapses into the usual, nonrelativistic Lippmann-Schwinger (LS) equation,

$$\widehat{T}(\vec{p}\,',\vec{p}) = \widehat{V}(\vec{p}\,',\vec{p}) + \int d^3p''\, \widehat{V}(\vec{p}\,',\vec{p}\,'') \frac{M}{p^2 - p''^2 + i\epsilon}\, \widehat{T}(\vec{p}\,'',\vec{p})\,. \tag{92}$$

Since $\widehat{V}$ satisfies Eq. (92), it can be used like a usual nonrelativistic potential, and $\widehat{T}$ is the conventional nonrelativistic T-matrix.

Iteration of $\widehat{V}$ in the LS equation requires cutting $\widehat{V}$ off for high momenta to avoid infinities, This is consistent with the fact that ChPT is a low-momentum expansion which is valid only for momenta $Q \ll \Lambda_\chi \approx 1$ GeV. Thus, we multiply $\widehat{V}$ with a regulator function

$$\widehat{V}(\vec{p}\,',\vec{p}) \longmapsto \widehat{V}(\vec{p}\,',\vec{p})\, e^{-(p'/\Lambda)^{2n}}\, e^{-(p/\Lambda)^{2n}} \tag{93}$$

$$\approx \widehat{V}(\vec{p}\,',\vec{p}) \left\{ 1 - \left[\left(\frac{p'}{\Lambda}\right)^{2n} + \left(\frac{p}{\Lambda}\right)^{2n} \right] + \cdots \right\} \tag{94}$$

with the 'cutoff parameter' Λ around 0.5 GeV. Equation (94) provides an indication of the fact that the exponential cutoff does not necessarily affect the given order at which the calculation is conducted. For sufficiently large n, the regulator introduces contributions that are beyond the given order. Assuming a good rate of convergence of the chiral expansion, such orders are small as compared to the given order and,

Table 3: χ^2/datum for the reproduction of the 1999 np database [56] by families of np potentials at NLO and NNLO constructed by the Juelich group [57].

T_{lab} bin (MeV)	# of np data	— Juelich np potentials —	
		NLO	NNLO
0–100	1058	4–5	1.4–1.9
100–190	501	77–121	12–32
190–290	843	140–220	25–69
0–290	2402	67–105	12–27

thus, do not affect the accuracy at the given order. In our calculations we use, of course, the full exponential, Eq. (93), and not the expansion. On a similar note, we also do not expand the square-root factors in Eqs. (90-91) because they are kinematical factors which guarantee relativistic elastic unitarity.

5.4.2 What Order?

Since in nuclear EFT we are dealing with a perturbative expansion, at some point, we have to raise the question, to what order of ChPT we have to go to obtain the precision we need. To discuss this issue on firm grounds, we show in Table 3 the χ^2/datum for the fit of the world np data below 290 MeV for a family of np potentials at NLO and NNLO. The NLO potentials produce the very large χ^2/datum between 67 and 105, and the NNLO are between 12 and 27. The rate of improvement from one order to the other is very encouraging, but the quality of the reproduction of the np data at NLO and NNLO is obviously insufficient for reliable predictions.

Based upon these facts, it has been pointed out in 2002 by Entem and Machleidt [25, 26] that one has to proceed to N³LO. Consequently, the first N³LO potential was published in 2003 [27].

At N³LO, there are 24 contact terms (24 parameters) which contribute to the partial waves with $L \leq 2$ (cf. Sec. 5.3). In Table 4, column 'Q^4/N³LO', we show how these terms/parameters are distributed over the various partial waves. For comparison, we also show the number of parameters used in the Nijmegen partial wave analysis (PWA93) [49] and in the high-precision CD-Bonn potential [9]. The table reveals that, for S and P waves, the number of parameters used in high-precision phenomenology and in EFT at N³LO are about the same. Thus, the EFT approach provides retroactively a justification for what the phenomenologists of the 1990's were doing. At NLO and NNLO, the number of parameters is substantially smaller than for PWA93 and CD-Bonn, which explains why these orders are insufficient for a quantitative potential. This fact is also clearly reflected in Fig. 7 where phase shifts are shown for potentials constructed at NLO, NNLO, and N³LO.

Table 4: Number of parameters needed for fitting the np data in phase-shift analysis and by a high-precision NN potential *versus* the number of NN contact terms of EFT based potentials at different orders.

| | Nijmegen partial-wave analysis [49] | CD-Bonn high-precision potential [9] | — Contact Potentials — | | |
			Q^0 LO	Q^2 NLO/NNLO	Q^4 N^{3}LO
$^1S_0$	3	4	1	2	4
$^3S_1$	3	4	1	2	4
$^3S_1$-$^3D_1$	2	2	0	1	3
$^1P_1$	3	3	0	1	2
$^3P_0$	3	2	0	1	2
$^3P_1$	2	2	0	1	2
$^3P_2$	3	3	0	1	2
$^3P_2$-$^3F_2$	2	1	0	0	1
$^1D_2$	2	3	0	0	1
$^3D_1$	2	1	0	0	1
$^3D_2$	2	2	0	0	1
$^3D_3$	1	2	0	0	1
$^3D_3$-$^3G_3$	1	0	0	0	0
$^1F_3$	1	1	0	0	0
$^3F_2$	1	2	0	0	0
$^3F_3$	1	2	0	0	0
$^3F_4$	2	1	0	0	0
$^3F_4$-$^3H_4$	0	0	0	0	0
$^1G_4$	1	0	0	0	0
$^3G_3$	0	1	0	0	0
$^3G_4$	0	1	0	0	0
$^3G_5$	0	1	0	0	0
Total	35	38	2	9	24

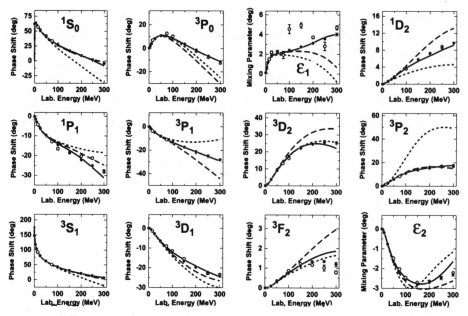

Figure : Phase parameters for np scattering as calculated from NN potentials at different orders of ChPT. The dotted line is NLO [57], the dashed NNLO [57], and the solid N³LO [27]. Partial waves with total angular momentum $J \leq 2$ are displayed. Solid dots represent the Nijmegen multienergy np phase shift analysis [49] and open circles are the GWU/VPI single-energy np analysis SM99 [50].

5.4.3 Charge-Dependence

For an accurate fit of the low-energy pp and np data, charge-dependence is important. We include charge-dependence up to next-to-leading order of the isospin-violation scheme (NLØ, in the notation of Ref. [58]). Thus, we include the pion mass difference in 1PE and the Coulomb potential in pp scattering, which takes care of the LØ contributions. At order NLØ, we have the pion mass difference in 2PE at NLO, $\pi\gamma$ exchange [59], and two charge-dependent contact interactions of order Q^0 which make possible an accurate fit of the three different $^1S_0$ scattering lengths, a_{pp}, a_{nn}, and a_{np}.

5.4.4 A Quantitative NN Potential at N³LO

NN Scattering. The fitting procedure starts with the peripheral partial waves because they depend on fewer parameters. Partial waves with $L \geq 3$ are exclusively determined by 1PE and 2PE because the N³LO contacts contribute to $L \leq 2$ only. 1PE and 2PE at N³LO depend on the axial-vector coupling constant, g_A (we use $g_A = 1.29$), the pion decay constant, $f_\pi = 92.4$ MeV, and eight low-energy constants (LECs) that appear in the dimension-two and dimension-three πN Lagrangians,

Table 5: χ^2/datum for the reproduction of the 1999 **np database** [56] by various np potentials. (Numbers in parentheses are the values of cutoff parameters in units of MeV used in the regulators of the chiral potentials.)

T_{lab} bin (MeV)	# of **np** data	Idaho N^{3}LO [27] (500–600)	Juelich N^{3}LO [60] (600/700–450/500)	Argonne V_{18} [61]
0–100	1058	1.0–1.1	1.0–1.1	0.95
100–190	501	1.1–1.2	1.3–1.8	1.10
190–290	843	1.2–1.4	2.8–20.0	1.11
0–290	2402	1.1–1.3	1.7–7.9	1.04

Table 6: χ^2/datum for the reproduction of the 1999 **pp database** [56] by various pp potentials. Notation as in Fig. 5.

T_{lab} bin (MeV)	# of **np** data	Idaho N^{3}LO [27] (500–600)	Juelich N^{3}LO [60] (600/700–450/500)	Argonne V_{18} [61]
0–100	795	1.0–1.7	1.0–3.8	1.0
100–190	411	1.5–1.9	3.5–11.6	1.3
190–290	851	1.9–2.7	4.3–44.4	1.8
0–290	2057	1.5–2.1	2.9–22.3	1.4

Eqs. (37) and (38). In the fitting process, we varied three of them, namely, c_2, c_3, and c_4. We found that the other LECs are not very effective in the NN system and, therefore, we kept them at the values determined from πN (cf. Table 2). The most influential constant is c_3, which has to be chosen on the low side (slightly more than one standard deviation below its πN determination) for an optimal fit of the NN data. As compared to a calculation that strictly uses the πN values for c_2 and c_4, our choices for these two LECs lower the $^3F_2$ and $^1F_3$ phase shifts bringing them into closer agreement with the phase shift analysis. The other F waves and the higher partial waves are essentially unaffected by our variations of c_2 and c_4. Overall, the fit of all $J \geq 3$ waves is very good.

We turn now to the lower partial waves. Here, the most important fit parameters are the ones associated with the 24 contact terms that contribute to the partial waves with $L \leq 2$. In addition, we have two charge-dependent contacts which are used to fit the three different $^1S_0$ scattering lengths, a_{pp}, a_{nn}, and a_{np}.

In the optimization procedure, we fit first phase shifts, and then we refine the fit by minimizing the χ^2 obtained from a direct comparison with the data. The

71

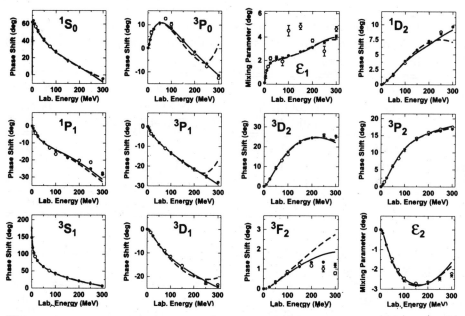

Figure 8: Neutron-proton phase parameters as described by two potentials at N³LO. The solid curve is calculated from the Idaho N³LO potential [27] while the dashed curve is from the Juelich [60] one. Solid dots and open circles as in Fig. 7.

χ^2/datum for the fit of the np data below 290 MeV is shown in Table 5, and the corresponding one for pp is given in Table 6. These tables show that at N³LO a χ^2/datum comparable to the high-precision Argonne V_{18} [61] potential can, indeed, be achieved. The "Idaho" N³LO potential [27] produces a χ^2/datum = 1.1 for the world np data below 290 MeV which compares well with the χ^2/datum = 1.04 by the Argonne potential. In 2005, also the Juelich group produced several N³LO NN potentials [60], the best of which fits the np data with a χ^2/datum = 1.7 and the worse with a χ^2/datum = 7.9 (see Table 5). While 7.9 is clearly unacceptable for any meaningful application, a χ^2/datum of 1.7 is reasonable, although it does not meet the precision standard that few-nucleon physicists established in the 1990's.

Turning to pp, the χ^2 for pp data are typically larger than for np because of the higher precision of pp data. Thus, the Argonne V_{18} produces a χ^2/datum = 1.4 for the world pp data below 290 MeV and the best Idaho N³LO pp potential obtains 1.5. The fit by the best Juelich N³LO pp potential results in a χ^2/datum = 2.9 which, again, is not quite consistent with the precision standards established in the 1990's. The worst Juelich N³LO pp potential produces a χ^2/datum of 22.3 and is incompatible with reliable predictions.

Phase shifts of np scattering from the best Idaho (solid line) and Juelich (dashed line) N³LO np potentials are shown in Figure 8. The phase shifts confirm what the corresponding χ^2 have already revealed.

Table 7: Deuteron properties as predicted by various NN potentials are compared to empirical information. (Deuteron binding energy B_d, asymptotic S state A_S, asymptotic D/S state η, deuteron radius r_d, quadrupole moment Q, D-state probability P_D; the calculated r_d and Q are without meson-exchange current contributions and relativistic corrections.)

	Idaho N³LO [27] (500)	Juelich N³LO [60] (550/600)	CD-Bonn[9]	AV18[61]	Empirical[a]
B_d (MeV)	2.224575	2.218279	2.224575	2.224575	2.224575(9)
A_S (fm$^{-1/2}$)	0.8843	0.8820	0.8846	0.8850	0.8846(9)
η	0.0256	0.0254	0.0256	0.0250	0.0256(4)
r_d (fm)	1.975	1.977	1.966	1.967	1.97535(85)
Q (fm^2)	0.275	0.266	0.270	0.270	0.2859(3)
P_D (%)	4.51	3.28	4.85	5.76	

[a] See Table XVIII of Ref. [9] for references; the empirical value for r_d is from Ref. [62].

The Deuteron. The reproduction of the deuteron parameters is shown in Table 7. We present results for two N³LO potentials, namely, Idaho [27] and Juelich [60]. Remarkable are the predictions by the chiral potentials for the deuteron radius which are in good agreement with the latest empirical value obtained by the isotope-shift method [62]. All NN potentials of the past (Table 7 includes two representative examples, namely, CD-Bonn [9] and AV18 [61]) fail to reproduce this very precise new value for the deuteron radius.

In Fig. 9, we display the deuteron wave functions derived from the N³LO potentials and compare them with wave functions based upon conventional NN potentials from the recent past. Characteristic differences are noticeable; in particular, the chiral wave functions are shifted towards larger r which explains the larger deuteron radius.

6 Many-Nucleon Forces

As noted before, an important advantage of the EFT approach to nuclear forces is that it creates two- and many-nucleon forces on an equal footing.

6.1 Three-Nucleon Forces

The first non-vanishing 3NF terms occur at NNLO and are shown in Fig. 10 (cf. also Fig. 1, row 'Q^3/NNLO', column '3N Force'). There are three diagrams: the 2PE,

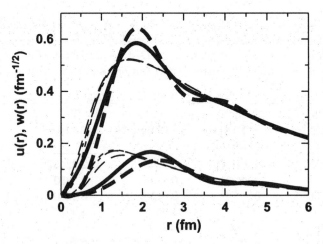

Figure 9: Deuteron wave functions: the family of larger curves are S-waves, the smaller ones D-waves. The thick lines represent the wave functions derived from chiral NN potentials at order N^{3}LO (thick solid: Idaho [27], thick dashed: Juelich [60]). The thin dashed, dash-dotted, and dotted lines refer to the wave functions of the CD-Bonn [9], Nijm-I [8], and AV18 [61] potentials, respectively.

1PE, and 3N-contact interactions [39, 40]. The 2PE 3N-potential is given by

$$V_{2PE}^{3NF} = \left(\frac{g_A}{2f_\pi}\right)^2 \frac{1}{2} \sum_{i \neq j \neq k} \frac{(\vec{\sigma}_i \cdot \vec{q}_i)(\vec{\sigma}_j \cdot \vec{q}_j)}{(q_i^2 + m_\pi^2)(q_j^2 + m_\pi^2)} F_{ijk}^{\alpha\beta} \tau_i^\alpha \tau_j^\beta \tag{95}$$

with $\vec{q}_i \equiv \vec{p}_i{}' - \vec{p}_i$, where $\vec{p}_i$ and $\vec{p}_i{}'$ are the initial and final momenta of nucleon i, respectively, and

$$F_{ijk}^{\alpha\beta} = \delta^{\alpha\beta} \left[-\frac{4c_1 m_\pi^2}{f_\pi^2} + \frac{2c_3}{f_\pi^2} \vec{q}_i \cdot \vec{q}_j \right] + \frac{c_4}{f_\pi^2} \sum_\gamma \epsilon^{\alpha\beta\gamma} \tau_k^\gamma \vec{\sigma}_k \cdot [\vec{q}_i \times \vec{q}_j] . \tag{96}$$

The vertex involved in this 3NF term is the two-derivative $\pi\pi NN$ vertex (solid square in Fig. 10) which we encountered already in the 2PE contribution to the NN potential at NNLO. Thus, there are no new parameters and the contribution is fixed by the LECs used in NN. The 1PE contribution is

$$V_{1PE}^{3NF} = D \frac{g_A}{8f_\pi^2} \sum_{i \neq j \neq k} \frac{\vec{\sigma}_j \cdot \vec{q}_j}{q_j^2 + m_\pi^2} (\boldsymbol{\tau}_i \cdot \boldsymbol{\tau}_j)(\vec{\sigma}_i \cdot \vec{q}_j) \tag{97}$$

and, finally, the 3N contact term reads

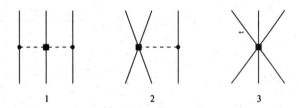

Figure 10: The three-nucleon force at NNLO (from Ref. [40]).

$$V_{ct}^{3NF} = E \frac{1}{2} \sum_{j \neq k} \tau_j \cdot \tau_k . \qquad (98)$$

The last two 3NF terms involve two new vertices (that do not appear in the 2N problem), namely, the $\pi NNNN$ vertex with parameter D and a $6N$ vertex with parameters E. To pin them down, one needs two observables that involve at least three nucleons. In Ref. [40], the triton binding energy and the nd doublet scattering length $^2a_{nd}$ were used. Alternatively, one may also choose the binding energies of ^{3}H and ^{4}He [63]. Once D and E are fixed, the results for other 3N, 4N, ... observables are predictions. In Refs. [63,64], the first calculations of the structure of light nuclei (^{6}Li and ^{7}Li) were reported. Recently, the structure of nuclei with $A = 10 - 13$ nucleons has been calculated using the *ab initio* no-core shell model and applying chiral two and three-nucleon forces [65]. The results are very encouraging. Concerning the famous 'A_y puzzle', the above 3NF terms yield some improvement of the predicted nd A_y, however, the problem is not solved [40].

Note that the 3NF expressions given in Eqs. (95)-(98) above are the ones that occur at NNLO, and all calculations to date have included only those. Since we have to proceed to N^{3}LO for sufficient accuracy of the 2NF, then consistency requires that we also consider the 3NF at N^{3}LO. The 3NF at N^{3}LO is very involved as can be seen from Fig. 11, but it does not depend on any new parameters. It is presently under construction [66]. So, for the moment, we can only hope that the A_y puzzle may be solved by a complete calculation at N^{3}LO.

6.2 Four-Nucleon Forces

In ChPT, four-nucleon forces (4NF) appear for the first time at N^{3}LO ($\nu = 4$). Thus, N^{3}LO is the leading order for 4NF. Assuming a good rate of convergence, a contribution of order $(Q/\Lambda_\chi)^4$ is expected to be rather small. Thus, ChPT predicts 4NF to be essentially insignificant, consistent with experience. Still, nothing is fully proven in physics unless we have performed explicit calculations. Very recently, the

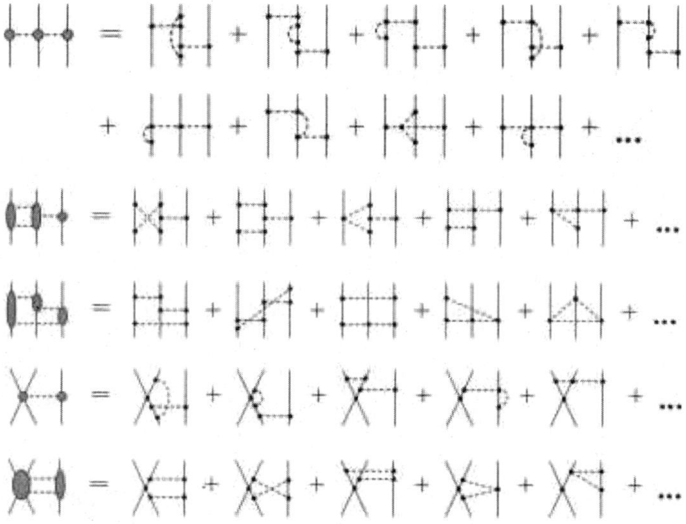

Figure 11: Three-nucleon force contributions at N^{3}LO (from Ref. [66]).

first such calculation has been performed: The chiral 4NF, Fig. 12, has been applied in a calculation of the ^{4}He binding energy and found to contribute a few 100 keV [68]. It should be noted that this preliminary calculation involves many approximations, but it certainly provides the right order of magnitude of the result, which is indeed very small as compared to the full ^{4}He binding energy of 28.3 MeV.

7 Conclusions

The theory of nuclear forces has made great progress since the turn of the millennium. Nucleon-nucleon potentials have been developed that are based on proper theory (EFT for low-energy QCD) and are of high-precision, at the same time. Moreover, the theory generates two- and many-body forces on an equal footing and provides a theoretical explanation for the empirically known fact that 2NF $\gg$ 3NF $\gg$ 4NF ....

At N^{3}LO [26, 27], the accuracy can be achieved that is necessary and sufficient for reliable microscopic nuclear structure predictions. First calculations applying the N^{3}LO NN potential [27] in the conventional shell model [69, 70], the *ab initio* no-core shell model [71–73], the coupled cluster formalism [74–78], and the unitary-model-operator approach [79] have produced promising results.

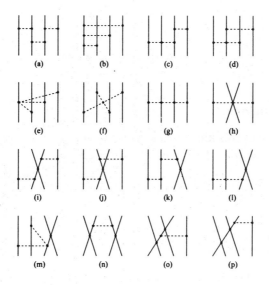

Figure 12: The four-nucleon force at N³LO (from Ref. [67]).

The 3NF at NNLO is known [39, 40] and has been applied in few-nucleon reactions [40, 80, 81] as well as the structure of light nuclei [63–65]. However, the famous 'A_y puzzle' of nucleon-deuteron scattering is not resolved by the 3NF at NNLO. Thus, one important outstanding issue is the 3NF at N³LO, which is under construction [66].

Another open question that needs to be settled is whether Weinberg power counting, which is applied in all current NN potentials, is consistent. This controversial issue is presently being debated in the literature [82, 83].

Acknowledgement

It is a pleasure to thank the organizers of this workshop, particularly, Ananda Santra, for their warm hospitality. I gratefully acknowledge numerous discussions with my collaborator D. R. Entem. This work was supported in part by the U.S. National Science Foundation under Grant No. PHY-0099444.

A Fourth Order Two-Pion Exchange Contributions

The fourth order 2PE contributions consist of two classes: the one-loop (Fig. 3) and the two-loop diagrams (Fig. 4).

A.1 One-loop diagrams

This large pool of diagrams can be analyzed in a systematic way by introducing the following well-defined subdivisions.

A.1.1 c_i^2 contributions.

The only contribution of this kind comes from the football diagram with both vertices proportional to c_i (first row of Fig. 3). One obtains [41]:

$$V_C = \frac{3L(q)}{16\pi^2 f_\pi^4}\left[\left(\frac{c_2}{6}w^2 + c_3\widetilde{w}^2 - 4c_1 m_\pi^2\right)^2 + \frac{c_2^2}{45}w^4\right], \tag{99}$$

$$W_T = -\frac{1}{q^2}W_S$$

$$= \frac{c_4^2 w^2 L(q)}{96\pi^2 f_\pi^4}. \tag{100}$$

A.1.2 c_i/M_N contributions.

This class consists of diagrams with one vertex proportional to c_i and one $1/M_N$ correction. A few graphs that are representative for this class are shown in the second row of Fig. 3. Symbols with a large solid dot and an open circle denote $1/M_N$ corrections of vertices proportional to c_i. They are part of $\widehat{\mathcal{L}}_{\pi N}^{(3)}$, Eq. (39). The result for this group of diagrams is [41]:

$$V_C = -\frac{g_A^2 L(q)}{32\pi^2 M_N f_\pi^4}\big[(c_2 - 6c_3)q^4 + 4(6c_1 + c_2 - 3c_3)q^2 m_\pi^2$$

$$+ 6(c_2 - 2c_3)m_\pi^4 + 24(2c_1 + c_3)m_\pi^6 w^{-2}\big], \tag{101}$$

$$W_C = -\frac{c_4 q^2 L(q)}{192\pi^2 M_N f_\pi^4}\big[g_A^2(8m_\pi^2 + 5q^2) + w^2\big], \tag{102}$$

$$W_T = -\frac{1}{q^2}W_S$$

$$= -\frac{c_4 L(q)}{192\pi^2 M_N f_\pi^4}\big[g_A^2(16m_\pi^2 + 7q^2) - w^2\big], \tag{103}$$

$$V_{LS} = \frac{c_2\, g_A^2}{8\pi^2 M_N f_\pi^4}w^2 L(q), \tag{104}$$

$$W_{LS} = -\frac{c_4 L(q)}{48\pi^2 M_N f_\pi^4}\big[g_A^2(8m_\pi^2 + 5q^2) + w^2\big]. \tag{105}$$

A.1.3 $1/M_N^2$ corrections.

These are relativistic $1/M_N^2$ corrections of the leading order 2π exchange diagrams. Typical examples for this large class are shown in row 3–6 of Fig. 3. This time, there is no correction from the iterated 1PE, Eq. (65) or Eq. (66), since the expansion of the factor M_N^2/E_p does not create a term proportional to $1/M_N^2$. The total result for this class is [42],

$$V_C = -\frac{g_A^4}{32\pi^2 M_N^2 f_\pi^4}\left[L(q)\left(2m_\pi^8 w^{-4} + 8m_\pi^6 w^{-2} - q^4 - 2m_\pi^4\right) + \frac{m_\pi^6}{2w^2}\right] \quad (106)$$

$$W_C = -\frac{1}{768\pi^2 M_N^2 f_\pi^4}\Bigg\{L(q)\Bigg[8g_A^2\left(\frac{3}{2}q^4 + 3m_\pi^2 q^2 + 3m_\pi^4 - 6m_\pi^6 w^{-2}\right.$$

$$\left.-k^2(8m_\pi^2 + 5q^2)\right) + 4g_A^4\left(k^2(20m_\pi^2 + 7q^2 - 16m_\pi^4 w^{-2}) + 16m_\pi^8 w^{-4}\right.$$

$$\left.+12m_\pi^6 w^{-2} - 4m_\pi^4 q^2 w^{-2} - 5q^4 - 6m_\pi^2 q^2 - 6m_\pi^4\right) - 4k^2 w^2\Bigg]$$

$$+\frac{16g_A^4 m_\pi^6}{w^2}\Bigg\}\,, \quad (107)$$

$$V_T = -\frac{1}{q^2}V_S = \frac{g_A^4 L(q)}{32\pi^2 M_N^2 f_\pi^4}\left(k^2 + \frac{5}{8}q^2 + m_\pi^4 w^{-2}\right)\,, \quad (108)$$

$$W_T = -\frac{1}{q^2}W_S = \frac{L(q)}{1536\pi^2 M_N^2 f_\pi^4}\Bigg[4g_A^4\left(7m_\pi^2 + \frac{17}{4}q^2 + 4m_\pi^4 w^{-2}\right)$$

$$-32g_A^2\left(m_\pi^2 + \frac{7}{16}q^2\right) + w^2\Bigg]\,, \quad (109)$$

$$V_{LS} = \frac{g_A^4 L(q)}{4\pi^2 M_N^2 f_\pi^4}\left(\frac{11}{32}q^2 + m_\pi^4 w^{-2}\right)\,, \quad (110)$$

$$W_{LS} = \frac{L(q)}{256\pi^2 M_N^2 f_\pi^4}\Bigg[16g_A^2\left(m_\pi^2 + \frac{3}{8}q^2\right)$$

$$+\frac{4}{3}g_A^4\left(4m_\pi^4 w^{-2} - \frac{11}{4}q^2 - 9m_\pi^2\right) - w^2\Bigg]\,, \quad (111)$$

$$V_{\sigma L} = \frac{g_A^4 L(q)}{32\pi^2 M_N^2 f_\pi^4}\,. \quad (112)$$

A.2 Two-loop contributions.

The two-loop contributions are quite intricate. In Fig. 4, we attempt a graphical representation of this class. The gray disk stands for all one-loop πN graphs which are shown in some detail in the lower part of the figure. .Not all of the numerous graphs are displayed. Some of the missing ones are obtained by permutation of the vertices along the nucleon line, others by inverting initial and final states. Vertices denoted by a small dot are from the leading order πN Lagrangian $\widehat{\mathcal{L}}_{\pi N}^{(1)}$, Eq. (33), except for the four-pion vertices which are from $\mathcal{L}_{\pi\pi}^{(2)}$, Eq. (27). The solid square represents vertices proportional to the LECs d_i which are introduced by the third order Lagrangian $\mathcal{L}_{\pi N}^{(3)}$, Eq. (38). The d_i vertices occur actually in one-loop NN diagrams, but we list them among the two-loop NN contributions because they are needed to absorb divergences generated by one-loop πN graphs. Using techniques from dispersion theory, Kaiser [41] calculated the imaginary parts of the NN amplitudes, Im $V_\alpha(i\mu)$ and Im $W_\alpha(i\mu)$, which result from analytic continuation to time-like momentum transfer $q = i\mu - 0^+$ with $\mu \geq 2m_\pi$. From this, the momentum-space amplitudes $V_\alpha(q)$ and $W_\alpha(q)$ are obtained via the subtracted dispersion relations:

$$V_{C,S}(q) = -\frac{2q^6}{\pi} \int_{2m_\pi}^{\infty} d\mu \frac{\text{Im } V_{C,S}(i\mu)}{\mu^5(\mu^2 + q^2)}, \tag{113}$$

$$V_T(q) = \frac{2q^4}{\pi} \int_{2m_\pi}^{\infty} d\mu \frac{\text{Im } V_T(i\mu)}{\mu^3(\mu^2 + q^2)}, \tag{114}$$

and similarly for $W_{C,S,T}$.

In most cases, the dispersion integrals can be solved analytically and the following expressions are obtained [26]:

$$V_C(q) = \frac{3g_A^4 \widetilde{w}^2 A(q)}{1024\pi^2 f_\pi^6} \left[(m_\pi^2 + 2q^2) \left(2m_\pi + \widetilde{w}^2 A(q) \right) + 4g_A^2 m_\pi \widetilde{w}^2 \right];$$

$$\tag{115}$$

$$W_C(q) = W_C^{(a)}(q) + W_C^{(b)}(q), \tag{116}$$

with

$$W_C^{(a)}(q) = \frac{L(q)}{18432\pi^4 f_\pi^6} \left\{ 192\pi^2 f_\pi^2 w^2 \bar{d}_3 \left[2g_A^2 \widetilde{w}^2 - \frac{3}{5}(g_A^2 - 1)w^2 \right] \right.$$

$$+ \left[6g_A^2 \widetilde{w}^2 - (g_A^2 - 1)w^2 \right] \left[384\pi^2 f_\pi^2 \left(\widetilde{w}^2(\bar{d}_1 + \bar{d}_2) + 4m_\pi^2 \bar{d}_5 \right) \right.$$

$$+ L(q) \left(4m_\pi^2(1 + 2g_A^2) + q^2(1 + 5g_A^2) \right)$$

$$\left. \left. - \left(\frac{q^2}{3}(5 + 13g_A^2) + 8m_\pi^2(1 + 2g_A^2) \right) \right] \right\} \tag{117}$$

and

$$W_C^{(b)}(q) = -\frac{2q^6}{\pi} \int_{2m_\pi}^{\infty} d\mu \frac{\text{Im } W_C^{(b)}(i\mu)}{\mu^5(\mu^2 + q^2)}, \tag{118}$$

80

where

$$\text{Im}\, W_C^{(b)}(i\mu) = -\frac{2\kappa}{3\mu(8\pi f_\pi^2)^3} \int_0^1 dx \left[g_A^2(2m_\pi^2 - \mu^2) + 2(g_A^2 - 1)\kappa^2 x^2 \right]$$

$$\times \left\{ -3\kappa^2 x^2 + 6\kappa x \sqrt{m_\pi^2 + \kappa^2 x^2}\, \ln \frac{\kappa x + \sqrt{m_\pi^2 + \kappa^2 x^2}}{m_\pi} \right.$$

$$+ g_A^4 \left(\mu^2 - 2\kappa^2 x^2 - 2m_\pi^2\right) \left[\frac{5}{6} + \frac{m_\pi^2}{\kappa^2 x^2} - \left(1 + \frac{m_\pi^2}{\kappa^2 x^2}\right)^{3/2} \right.$$

$$\left. \left. \times \ln \frac{\kappa x + \sqrt{m_\pi^2 + \kappa^2 x^2}}{m_\pi} \right] \right\} ; \tag{119}$$

$$V_T(q) = V_T^{(a)}(q) + V_T^{(b)}(q)$$

$$= -\frac{1}{q^2} V_S(q) = -\frac{1}{q^2}\left(V_S^{(a)}(q) + V_S^{(b)}(q) \right), \tag{120}$$

with

$$V_T^{(a)}(q) = -\frac{1}{q^2} V_S^{(a)}(q) = -\frac{g_A^2 w^2 L(q)}{32\pi^2 f_\pi^4}(\bar{d}_{14} - \bar{d}_{15}) \tag{121}$$

and

$$V_T^{(b)}(q) = -\frac{1}{q^2} V_S^{(b)}(q) = \frac{2q^4}{\pi} \int_{2m_\pi}^\infty d\mu\, \frac{\text{Im}\, V_T^{(b)}(i\mu)}{\mu^3(\mu^2 + q^2)}, \tag{122}$$

where

$$\text{Im}\, V_T^{(b)}(i\mu) = -\frac{2g_A^6 \kappa^3}{\mu(8\pi f_\pi^2)^3} \int_0^1 dx(1 - x^2) \left[-\frac{1}{6} + \frac{m_\pi^2}{\kappa^2 x^2} \right.$$

$$\left. - \left(1 + \frac{m_\pi^2}{\kappa^2 x^2}\right)^{3/2} \ln \frac{\kappa x + \sqrt{m_\pi^2 + \kappa^2 x^2}}{m_\pi} \right] ; \tag{123}$$

$$W_T(q) = -\frac{1}{q^2} W_S(q) = \frac{g_A^4 w^2 A(q)}{2048\pi^2 f_\pi^6} \left[w^2 A(q) + 2m_\pi(1 + 2g_A^2) \right], \tag{124}$$

where $\kappa \equiv \sqrt{\mu^2/4 - m_\pi^2}$.

Note that the analytic solutions hold modulo polynomials. We have checked the importance of those contributions where we could not find an analytic solution and where, therefore, the integrations have to be performed numerically. It turns out that the combined effect on NN phase shifts from $W_C^{(b)}$, $V_T^{(b)}$, and $V_S^{(b)}$ is smaller than 0.1 deg in F and G waves and smaller than 0.01 deg in H waves, at $T_{\text{lab}} = 300$ MeV (and less at lower energies). This renders these contributions negligible. Therefore, we omit $W_C^{(b)}$, $V_T^{(b)}$, and $V_S^{(b)}$ in the construction of chiral NN potentials at order N^{3}LO.

In Eqs. (117) and (121), we use the scale-independent LECs, $\bar{d}_i$, which are obtained by combining the scale-dependent ones, $d_i^r(\lambda)$, with the chiral logarithm,

$\ln(m_\pi/\lambda)$, or equivalently $\bar{d}_i = d_i^r(m_\pi)$. The scale-dependent LECs, $d_i^r(\lambda)$, are a consequence of renormalization. For more details about this issue, see Ref. [37].

B Partial Wave Decomposition of the Fourth Order Contact Potential

The contact potential contribution of order four, Eq. (80), decomposes into partial-waves as follows.

$$
\begin{aligned}
V^{(4)}(^1S_0) &= \hat{D}_{1S_0}(p'^4 + p^4) + D_{1S_0}p'^2p^2 \\
V^{(4)}(^3P_0) &= D_{3P_0}(p'^3p + p'p^3) \\
V^{(4)}(^1P_1) &= D_{1P_1}(p'^3p + p'p^3) \\
V^{(4)}(^3P_1) &= D_{3P_1}(p'^3p + p'p^3) \\
V^{(4)}(^3S_1) &= \hat{D}_{3S_1}(p'^4 + p^4) + D_{3S_1}p'^2p^2 \\
V^{(4)}(^3D_1) &= D_{3D_1}p'^2p^2 \\
V^{(4)}(^3S_1 - {}^3D_1) &= \hat{D}_{3S_1-{}^3D_1}p^4 + D_{3S_1-{}^3D_1}p'^2p^2 \\
V^{(4)}(^1D_2) &= D_{1D_2}p'^2p^2 \\
V^{(4)}(^3D_2) &= D_{3D_2}p'^2p^2 \\
V^{(4)}(^3P_2) &= D_{3P_2}(p'^3p + p'p^3) \\
V^{(4)}(^3P_2 - {}^3F_2) &= D_{3P_2-{}^3F_2}p'p^3 \\
V^{(4)}(^3D_3) &= D_{3D_3}p'^2p^2
\end{aligned}
\tag{125}
$$

The coefficients in the above expressions are given by:

$$
\begin{aligned}
\hat{D}_{1S_0} &= D_1 + \frac{1}{16}D_2 + \frac{1}{4}D_3 - 3D_5 - \frac{3}{16}D_6 - \frac{3}{4}D_7 - D_{11} - \frac{1}{4}D_{12} - \frac{1}{4}D_{13} \\
&\quad - \frac{1}{16}D_{14} \\
D_{1S_0} &= \frac{10}{3}D_1 + \frac{5}{24}D_2 + \frac{1}{6}D_3 + \frac{2}{3}D_4 - 10D_5 - \frac{5}{8}D_6 - \frac{1}{2}D_7 - 2D_8 - \frac{10}{3}D_{11} \\
&\quad - \frac{1}{6}D_{12} - \frac{1}{6}D_{13} - \frac{5}{24}D_{14} - \frac{2}{3}D_{15} \\
D_{3P_0} &= -\frac{4}{3}D_1 + \frac{1}{12}D_2 - \frac{4}{3}D_5 + \frac{1}{12}D_6 - \frac{2}{3}D_9 - \frac{1}{6}D_{10} + \frac{8}{3}D_{11} + \frac{1}{3}D_{12} - \frac{1}{3}D_{13} \\
&\quad - \frac{1}{6}D_{14} \\
D_{1P_1} &= -\frac{4}{3}D_1 + \frac{1}{12}D_2 + 4D_5 - \frac{1}{4}D_6 + \frac{4}{3}D_{11} - \frac{1}{12}D_{14} \\
D_{3P_1} &= -\frac{4}{3}D_1 + \frac{1}{12}D_2 - \frac{4}{3}D_5 + \frac{1}{12}D_6 - \frac{1}{3}D_9 - \frac{1}{12}D_{10} - 2D_{11} - \frac{1}{6}D_{12} + \frac{1}{6}D_{13}
\end{aligned}
$$

$$+\frac{1}{8}D_{14}$$

$$\hat{D}_{3S_1} = D_1 + \frac{1}{16}D_2 + \frac{1}{4}D_3 + D_5 + \frac{1}{16}D_6 + \frac{1}{4}D_7 + \frac{1}{3}D_{11} + \frac{1}{12}D_{12} + \frac{1}{12}D_{13}$$

$$+\frac{1}{48}D_{14}$$

$$D_{3S_1} = \frac{10}{3}D_1 + \frac{5}{24}D_2 + \frac{1}{6}D_3 + \frac{2}{3}D_4 + \frac{10}{3}D_5 + \frac{5}{24}D_6 + \frac{1}{6}D_7 + \frac{2}{3}D_8 + \frac{10}{9}D_{11}$$

$$+\frac{1}{18}D_{12} + \frac{1}{18}D_{13} + \frac{5}{72}D_{14} + \frac{2}{9}D_{15}$$

$$D_{3D_1} = \frac{8}{15}D_1 + \frac{1}{30}D_2 - \frac{2}{15}D_3 - \frac{2}{15}D_4 + \frac{8}{15}D_5 + \frac{1}{30}D_6 - \frac{2}{15}D_7 - \frac{2}{15}D_8$$

$$+\frac{2}{5}D_9 - \frac{1}{10}D_{10} - \frac{4}{9}D_{11} + \frac{1}{9}D_{12} + \frac{1}{9}D_{13} - \frac{1}{36}D_{14} - \frac{16}{45}D_{15}$$

$$\hat{D}_{3S_1-3D_1} = -\frac{2\sqrt{2}}{3}D_{11} - \frac{\sqrt{2}}{6}D_{12} - \frac{\sqrt{2}}{6}D_{13} - \frac{\sqrt{2}}{24}D_{14}$$

$$D_{3S_1-3D_1} = -\frac{14\sqrt{2}}{9}D_{11} + \frac{\sqrt{2}}{18}D_{12} + \frac{\sqrt{2}}{18}D_{13} - \frac{7\sqrt{2}}{72}D_{14} + \frac{2\sqrt{2}}{9}D_{15}$$

$$D_{1D_2} = \frac{8}{15}D_1 + \frac{1}{30}D_2 - \frac{2}{15}D_3 - \frac{2}{15}D_4 - \frac{8}{5}D_5 - \frac{1}{10}D_6 + \frac{2}{5}D_7 + \frac{2}{5}D_8 - \frac{8}{15}D_{11}$$

$$+\frac{2}{15}D_{12} + \frac{2}{15}D_{13} - \frac{1}{30}D_{14} + \frac{2}{15}D_{15}$$

$$D_{3D_2} = \frac{8}{15}D_1 + \frac{1}{30}D_2 - \frac{2}{15}D_3 - \frac{2}{15}D_4 + \frac{8}{15}D_5 + \frac{1}{30}D_6 - \frac{2}{15}D_7 - \frac{2}{15}D_8$$

$$+\frac{2}{15}D_9 - \frac{1}{30}D_{10} + \frac{4}{5}D_{11} - \frac{1}{5}D_{12} - \frac{1}{5}D_{13} + \frac{1}{20}D_{14} + \frac{4}{15}D_{15}$$

$$D_{3P_2} = -\frac{4}{3}D_1 + \frac{1}{12}D_2 - \frac{4}{3}D_5 + \frac{1}{12}D_6 + \frac{1}{3}D_9 + \frac{1}{12}D_{10} - \frac{2}{15}D_{11} + \frac{1}{30}D_{12}$$

$$-\frac{1}{30}D_{13} + \frac{1}{120}D_{14}$$

$$D_{3P_2-3F_2} = \frac{4\sqrt{6}}{15}D_{11} - \frac{\sqrt{6}}{15}D_{12} + \frac{\sqrt{6}}{15}D_{13} - \frac{\sqrt{6}}{60}D_{14}$$

$$D_{3D_3} = \frac{8}{15}D_1 + \frac{1}{30}D_2 - \frac{2}{15}D_3 - \frac{2}{15}D_4 + \frac{8}{15}D_5 + \frac{1}{30}D_6 - \frac{2}{15}D_7 - \frac{2}{15}D_8$$

$$-\frac{4}{15}D_9 + \frac{1}{15}D_{10} - \frac{2}{15}D_{15} \tag{126}$$

References

[1] H. Yukawa, Proc. Phys. Math. Soc. Japan **17**, 48 (1935).

[2] Prog. Theor. Phys. (Kyoto), Supplement **3** (1956).

[3] M. Taketani, S. Machida, and S. Onuma, Prog. Theor. Phys. (Kyoto) **7**, 45 (1952).

[4] K. A. Brueckner and K. M. Watson, Phys. Rev. **90**, 699; **92**, 1023 (1953).

[5] A. R. Erwin *et al.*, Phys. Rev. Lett. **6**, 628 (1961); B. C. Maglić *et al.*, *ibid.* **7**, 178 (1961).

[6] Prog. Theor. Phys. (Kyoto), Supplement **39** (1967); R. A. Bryan and B. L. Scott, Phys. Rev. **177**, 1435 (1969); M. M. Nagels *et al.*, Phys. Rev. D **17**, 768 (1978).

[7] R. Machleidt, Adv. Nucl. Phys. **19**, 189 (1989).

[8] V. G. J. Stoks *et al.*, Phys. Rev. C **49**, 2950 (1994).

[9] R. Machleidt, Phys. Rev. C **63**, 024001 (2001).

[10] A. D. Jackson, D. O. Riska, and B. Verwest, Nucl. Phys. **A249**, 397 (1975).

[11] R. Vinh Mau, in *Mesons in Nuclei*, edited by M. Rho and D. H. Wilkinson (North-Holland, Amsterdam, 1979), Vol. I, p. 151.

[12] M. Lacombe, B. Loiseau, J. M. Richard, R. Vinh Mau, J. Côté, P. Pires, and R. de Tourreil, Phys. Rev. C **21**, 861 (1980).

[13] R. Machleidt, K. Holinde, and Ch. Elster, Phys. Rep. **149**, 1 (1987).

[14] F. Myhrer and J. Wroldsen, Rev. Mod. Phys. **60**, 629 (1988).

[15] D. R. Entem, F. Fernandez, and A. Valcarce, Phys. Rev. C **62**, 034002 (2000).

[16] G. H. Wu, J. L. Ping, L. J. Teng, F. Wang, and T. Goldman, Nucl. Phys. **A673**, 273 (2000).

[17] S. Weinberg, Physica **96A**, 327 (1979).

[18] S. Weinberg, Phys. Lett. B **251**, 288 (1990); Nucl. Phys. **B363**, 3 (1991); Phys. Lett. B **295**, 114 (1992).

[19] C. Ordóñez, L. Ray, and U. van Kolck, Phys. Rev. Lett. **72**, 1982 (1994); Phys. Rev. C **53**, 2086 (1996).

[20] U. van Kolck, Prog. Part. Nucl. Phys. **43**, 337 (1999).

[21] L. S. Celenza *et al.*, Phys. Rev. C **46**, 2213 (1992); C. A. da Rocha *et al.*, *ibid.* **49**, 1818 (1994); D. B. Kaplan *et al.*, Nucl. Phys. **B478**, 629 (1996).

[22] N. Kaiser, R. Brockmann, and W. Weise, Nucl. Phys. **A625**, 758 (1997).

[23] N. Kaiser, S. Gerstendörfer, and W. Weise, Nucl. Phys. **A637**, 395 (1998).

[24] E. Epelbaum *et al.*, Nucl. Phys. **A637**, 107 (1998); **A671**, 295 (2000).

[25] D. R. Entem and R. Machleidt, Phys. Lett. B **524**, 93 (2002).

[26] D. R. Entem and R. Machleidt, Phys. Rev. C **66**, 014002 (2002).

[27] D. R. Entem and R. Machleidt, Phys. Rev. C **68**, 041001 (2003).

[28] R. Machleidt and D. R. Entem, J. Phys. G: Nucl. Phys. **31**, S1235 (2005).

[29] P. F. Bedaque and U. van Kolck, Ann. Rev. Nucl. Part. Sci. **52**, 339 (2002).

[30] S. Scherer and M. R. Schindler, arXiv:hep-ph/0505265.

[31] Review of Particle Physics, J. Phys. G: Nucl. Part. Phys. **33**, 1 (2006).

[32] S. Coleman, J. Wess, and B. Zumino, Phys. Rev. **177**, 2239 (1969); C. G. Callan, S. Coleman, J. Wess, and B. Zumino, *ibid.* **177**, 2247 (1969).

[33] J. Gasser and H. Leutwyler, Ann. Phys. **158**, 142 (1984).

[34] J. Gasser, M. E. Sainio, and A. Švarc, Nucl. Phys. **B307**, 779 (1988).

[35] V. Bernard, N. Kaiser, and U.-G. Meißner, Int. J. Mod. Phys. E **4**, 193 (1995).

[36] N. Fettes, U.-G. Meißner, M. Mojžiš, and S. Steininger, Ann. Phys. (N.Y.) **283**, 273 (2000); **288**, 249 (2001).

[37] N. Fettes, U.-G. Meißner, and S. Steiniger, Nucl. Phys. **A640**, 199 (1998).

[38] N. Kaiser, Phys. Rev. C **61**, 014003 (1999); **62**, 024001 (2000).

[39] U. van Kolck, Phys. Rev. C **49**, 2932 (1994).

[40] E. Epelbaum et al., Phys. Rev. C **66**, 064001 (2002).

[41] N. Kaiser, Phys. Rev. C **64**, 057001 (2001).

[42] N. Kaiser, Phys. Rev. C **65**, 017001 (2002).

[43] R. Blankenbecler and R. Sugar, *Phys. Rev.* **142**, 1051 (1966).

[44] This section closely follows Ref. [26].

[45] G. Q. Li and R. Machleidt, Phys. Rev. C **58**, 3153 (1998).

[46] V. Stoks, R. Timmermans, and J. J. de Swart, Phys. Rev. C **47**, 512 (1993).

[47] R. A. Arndt, R. L. Workman, and M. M. Pavan, Phys. Rev. C **49**, 2729 (1994).

[48] P. Büttiker and U.-G. Meißner, Nucl. Phys. **A668**, 97 (2000).

[49] V. G. J. Stoks, R. A. M. Klomp, M. C. M. Rentmeester, and J. J. de Swart, Phys. Rev. C **48**, 792 (1993).

[50] R. A. Arndt, I. I. Strakovsky, and R. L. Workman, SAID, Scattering Analysis Interactive Dial-in computer facility, George Washington University (formerly Virginia Polytechnic Institute), solution SM99 (Summer 1999); for more information see, e. g., R. A. Arndt, I. I. Strakovsky, and R. L. Workman, Phys. Rev. C **50**, 2731 (1994).

[51] In fact, preliminary calculations, which take an important class of diagrams of order five into account, indicate that the N^4LO contribution may prevailingly be repulsive (N. Kaiser, private communication).

[52] G. E. Brown and A. D. Jackson, *The Nucleon-Nucleon Interaction*, (North-Holland, Amsterdam, 1976).

[53] N. Kaiser, Phys. Rev. C **63**, 044010 (2001).

[54] K. Erkelenz, R. Alzetta, and K. Holinde, Nucl. Phys. **A176**, 413 (1971); note that there is an error in equation (4.22) of this paper where it should read

$$-W_{LS}^J = 2qq' \frac{J-1}{2J-1} \left[A_{LS}^{J-2,(0)} - A_{LS}^{J(0)} \right]$$

and

$$+W_{LS}^J = 2qq' \frac{J+2}{2J+3} \left[A_{LS}^{J+2,(0)} - A_{LS}^{J(0)} \right].$$

[55] E. E. Salpeter and H. A. Bethe, *Phys. Rev.* **84**, 1232 (1951).

[56] The 1999 NN data base is defined in Ref. [9].

[57] E. Epelbaum, W. Glöckle, and U.-G. Meißner, Eur. Phys. J. **A19**, 401 (2004).

[58] M. Walzl et al., Nucl. Phys. **A693**, 663 (2001).

[59] U. van Kolck *et al.*, Phys. Rev. Lett. **80**, 4386 (1998).

[60] E. Epelbaum, W. Glöckle, and U.-G. Meißner, Nucl. Phys. A747, 362 (2005).

[61] R. B. Wiringa *et al.*, Phys. Rev. C **51**, 38 (1995).

[62] A. Huber *et al.*, Phys. Rev. Lett. **80**, 468 (1998).

[63] A. Nogga, P. Navratil, B. R. Barrett, and J. P. Vary, Phys. Rev. C **73**, 064002 (2006).

[64] A. Nogga *et al.*, Nucl. Phys. **A737**, 236 (2004).

[65] P. Navratil, V. G. Gueorguiev, J. P. Vary, W. E. Ormand, and A. Nogga, arXiv:nucl-th/0701038.

[66] U.-G. Meißner, Proc. 18th International Conference on Few-Body Problems in Physics, Santos, SP, Brazil, August 2006, to be published in Nucl. Phys. **A**.

[67] E. Epelbaum, Phys. Lett. B **639**, 456 (2006).

[68] D. Rozpedzik *et al.*, Acta Phys. Polon. **B37**, 2889 (2006); arXiv:nucl-th/0606017.

[69] L. Coraggio *et al.*, Phys. Rev. C **66**. 021303 (2002).

[70] L. Coraggio *et al.*, Phys. Rev. C **71**. 014307 (2005).

[71] P. Navrátil and E. Caurier (2004) *Phys. Rev. C* **69** 014311.

[72] C. Forssen *et al.*, Phys. Rev. C **71**, 044312 (2005).

[73] J.P. Vary *et al.*, Eur. Phys. J. A **25** s01, 475 (2005).

[74] K. Kowalski *et al.*, Phys. Rev. Lett. **92**, 132501 (2004).

[75] D.J. Dean and M. Hjorth-Jensen (2004) *Phys. Rev.* C **69** 054320.

[76] M. Wloch *et al.*, J. Phys. G **31**, S1291 (2005); Phys. Rev. Lett. **94**, 21250 (2005).

[77] D.J. Dean *et al.*, Nucl. Phys. **752**, 299 (2005).

[78] J.R. Gour *et al.*, Phys. Rev. C **74**, 024310 (2006).

[79] S. Fujii, R. Okamato, and K. Suzuki, Phys. Rev. C **69**, 034328 (2004).

[80] K. Ermisch *et al.*, Phys. Rev. C **71**, 064004 (2005).

[81] H. Witala, J. Golak, R. Skibinski, W. Glöckle, A. Nogga, E. Epelbaum, H. Kamada, A. Kievsky, and M. Viviani, Phys. Rev. C **73**, 044004 (2006).

[82] A. Nogga, R. G. E. Timmermans, and U. van Kolck, Phys. Rev. C **72**, 054006 (2005).

[83] E. Epelbaum, U.-G. Meißner, arXiv:nucl-th/0609037.

Physics and Astrophysics of Hadrons and Hardronic Matter
Editor: A. B. Santra
Copyright © 2008, Narosa Publishing House, New Delhi, India

Pion Reactions on Two–Nucleon Systems

C. Hanhart

Institut für Kernphysik,
Forschungszentrum Jülich GmbH
D–52425 Jülich, Germany
email: c.hanhart@fz-juelich.de

Abstract

We review recent progress in our understanding of elastic and inelastic pion reactions on two nucleon systems from the point of view of effective field theory. The discussion includes πd scattering, $\gamma d \to \pi^+ nn$, and $NN \to NN\pi$. At the end some remarks are made on strangeness production reactions like $\gamma d \to K\Lambda N$ and $NN \to K\Lambda N$.

1 Introduction

Even after several decades of research, the interactions and dynamics of strongly interacting few–nucleon systems are still not fully understood. Although phenomenological approaches are quite often very successful in describing certain sets of data, a coherent overall picture with clear connection to the fundamental theory, QCD, is still lacking.

The only way to a systematic and well controlled understanding of hadron physics is through the use of an effective field theory. The effective field theory (EFT) for the Standard Model at low energies is chiral perturbation theory (ChPT). This EFT already provided deep insights into strong interaction physics at low energies from systematic studies of the $\pi\pi$ [1] and the πN [2, 3] system — reviewed in Ref. [4] — as well as the NN system [5–7]. Here we focus on the field of elastic and inelastic —reactions on few nucleon system.

A first step towards a systematic study of elastic and inelastic reactions on nuclei was taken by Weinberg already in 1992 [8]. He suggested that all that needs to be done is to convolute transition operators, calculated perturbatively in standard ChPT, with proper nuclear wave functions to account for the non–perturbative character of the few–nucleon systems. This procedure looks very similar to the so–called distorted wave born approximation used routinely in phenomenological calculations, but, in contrast to this, opens up the possibility to use a power counting scheme. Within ChPT this idea was already applied to a large number of reactions like $\pi d \to \pi d$ [9], $\gamma d \to \pi^0 d$ [10, 11], $\pi^3\text{He} \to \pi^3\text{He}$ [12], $\pi^- d \to \gamma nn$ [13], and $\gamma d \to \pi^+ nn$ [14], where only the most recent references were given. We start our presentation with a brief description of πd scattering in sec. 2.

Using standard ChPT especially means to use an expansion in inverse powers of M_N — the nucleon mass. However, some pion–few-nucleon diagrams employ few-body singularities that lead to contributions non–analytic in m_π/M_N, with m_π for the pion mass [15]. This is discussed in detail in sec. 3. There we show that the appearance of these contributions is linked to the Pauli principle operative while there is a pion in flight. In this context we discuss also the reaction $\gamma d \to \pi^+ nn$, for in the mentioned sense this reaction is complementary to πd scattering. We show that within ChPT the existing data can be described to high accuracy and therefore the reaction qualifies as a tool to measure the nn scattering length.

A problem was observed when the original scheme by Weinberg was applied to the reactions $NN \to NN\pi$ [16, 17]: the inclusion of potentially higher order corrections were large and lead to even larger disagreement between theory and experiment than found in earlier phenomenological studies [18]. For the reaction $pp \to pp\pi^0$ one loop diagrams that in the Weinberg counting appear only at NNLO where evaluated [19, 20] and they turned out to give even larger corrections putting into question the convergence of the whole series. However, already quite early the authors of Refs. [21, 22] stressed that an additional new scale enters for $NN \to NN\pi$ that needs to be accounted for in the power counting. Since the two nucleons in the initial state need to have sufficiently high kinetic energy to put the pion in the final state on–shell, the initial momentum needs to be larger than

$$p_{thr} = \sqrt{M_N m_\pi} \ . \tag{1}$$

The proper way to include this scale was presented in Ref. [23] — for a recent review see Ref. [24]. As a result, pion p-waves are given by tree level diagrams up to NLO and the corresponding calculations showed satisfying agreement with the data. However, for pion s–waves loops appear already at NLO [23, 25]. In sec. 5 we discuss their effect on the reaction $NN \to d\pi$ near threshold. In some detail we will compare the effective field theory result to that on phenomenological calculations.

The central concept to be used in the construction of the transition operators is that of reducibility, for it allows one to disentangle effects of the wave functions and those from the transition operators. As long as the operators are energy independent, the scheme can be applied straight forwardly [26], however, as we will see below, for energy dependent interactions more care is necessary. This will also be subject of sec. 5.

Once the reaction $NN \to d\pi$ is understood within effective field theory one is in the position to also calculate the so–called dispersive and absorptive corrections to the πd scattering length. This calculation will be presented in section 7.

When switching to systems with strangeness one immediately observes that the initial momentum needs to be quite large in any production reaction where there are two nucleons and no strangeness in the initial state. Strangeness conservation demands that at least two particles with strangeness in the final state. Therefore

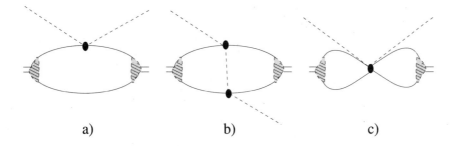

Figure 1: Typical diagrams that contribute to πd scattering. Shown is the one–body term (diagram (a)), the leading two–body correction (diagram (b)) and a possible short–ranged operator (diagram (c)).

the mass produced is at least

$$\Delta = M_\Lambda - M_N + m_K - m_i \, ,$$

where m_i denotes the mass present in the initial state in addition to the two nucleons, e.g., $m_i \simeq m_\pi$ for $\pi d \to K\Lambda N$ and $m_i = 0$ for $NN \to N\Lambda K$. Therefore, in the latter reaction the expansion parameter of the ChPT, namely the initial momentum in units of the nucleon mass, is 0.85, using the analog of Eq. (1). Also in the former reaction we are faced with an initial momentum 0.6 in units of the nucleon mass — again not useful as an expansion parameter. Clearly, in those cases the chiral expansion as proposed above is no longer applicable. In sec. 9 we discuss, how for those large momentum transfer reactions the scattering lengths of the outgoing baryons can still be extracted in a controlled way using dispersion theory.

We close with a brief summary and outlook.

2   Remarks on the πd system

We start our discussion with some remarks on the πd system. Pion-deuteron (πd) scattering near threshold plays an exceptional role in the quest for the isoscalar πN scattering length a_+, since the deuteron is an isoscalar target. Therefore one may write $\mathrm{Re}(a_{\pi d}) = 2a_+ + $ (few-body corrections) . The first term $\sim a_+$ is simply generated from the impulse approximation (scattering off the proton and off the neutron; diagram (a) of Fig. 1) and is independent of the deuteron structure. Thus, if one is able to calculate the few–body corrections in a controlled way, πd scattering is a prime reaction to extract a_+ (most effectively in combination with an analysis of the high accuracy data on pionic hydrogen). In addition, already at threshold the πd scattering length is a complex-valued quantity. It is therefore also important to gain a precise understanding of its imaginary part - this issue will be discussed in sec. 7.

Recently the πd scattering length was measured to be [27]

$$a_{\pi d}^{\exp} = (-26.1 \pm 0.5 + i(6.3 \pm 0.7)) \times 10^{-3} \; m_\pi^{-1} \;,$$

where m_π denotes the mass of the charged pion. In the near future a new measurement with a projected total uncertainty of 0.5% for the real part and 4% for the imaginary part of the scattering length will be performed at PSI [28]. Clearly, performing calculations up to this accuracy poses a challenge to theory that several groups recently took up [29–34]. In addition, an interesting isospin violating effect in pionic deuterium was found, see [35]. For a review on older work we refer to Ref. [36].

A typical few–body correction to the πd scattering length is shown in diagram (b) of Fig. 1. As we will see below, the contribution of this diagram largely exhausts the value of the πd scattering length not leaving much room for a contribution from a_+, or stated differently, pointing at a small value of a_+. Based on calculations within pion less EFT, it was claimed recently that this diagram is sensitive to the short range part of the deuteron wave function [37, 38]. As a consequence field theoretic consistency requires that at the same order there is to be a local operator to absorb this model dependence — the corresponding diagram is shown as diagram (c). Since this diagram comes with an a priori unknown strength not fixed by symmetries, πd scattering would be useless for the extraction of a_+. However, systematic investigations showed that, as soon as the pion exchange is included explicitly in the NN potential, diagram (b) can be evaluated in a controlled way [30, 32–34]. Given this we assume from now on that short–ranged operators contributing to the $\pi NN \to \pi NN$ transition potential scale naturally. In other words, contribute with strength parameters of the order of one. Based on this we may estimate the contribution of diagram (c) of Fig. 1 relative to diagram (b). This kind of analysis gives a relative suppression of the order $\mathcal{O}(\chi^2)$, where $\chi = m_\pi/M_N$ is the standard expansion parameter of ChPT with M_N (m_π) for the nucleon (pion) mass. This, together with the knowledge that diagram (a) largely exhausts the value of the πd scattering length, allows us to estimate the theoretical limit for the extraction of a_+ from a measurement of the πd scattering length. We find

$$\Delta a^{\text{theo}} \sim 5 \times 10^{-4} \; m_\pi^{-1} \;. \tag{2}$$

To meet this theoretical limit we need to include in the calculation all contributions to the πd scattering length lower than $\mathcal{O}(\chi^2)$.

Already in his original work, Weinberg discussed πd scattering at threshold as an illustrative example [8]. As usual, the leading contributions to the transition operators are all those tree–level diagrams that can be constructed from the leading πN and $\pi\pi$ Lagrangians. Those are shown in Fig. 2. Note that it is very important

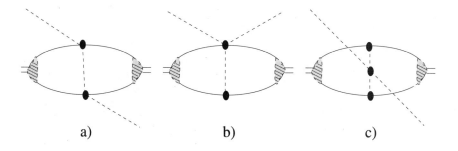

Figure 2: Formally leading few–body corrections to the πd scattering length.

that the complete set of diagrams is considered, since, for example, both diagram (b) as well as diagram (c) are depending on the particular choice made for the pion field. However, the sum of both is independent of this choice, as was first pointed out in Ref. [39].

Since all diagrams of Fig. 2 contribute to the same chiral order, naively one expects them to give similar contributions. However, explicit numerical evaluations showed, that diagram (a) exceeds the sum of the other two by about two orders of magnitude [8]. Can we understand this? One possible explanation could be that the sum of (b) and (c) is small because of significant cancellations between the two, probably due to the mechanism indicated at the end of the previous paragraph. Another possible explanation was given in Ref. [9]: as a consequence of the small binding energy of the deuteron, ϵ, the typical nucleon momentum inside the deuteron, γ, is also small, $\gamma \sim \sqrt{M_N \epsilon}$, which turns out to be numerically about 1/3 of the pion mass. Since diagram (a) is proportional to the expectation value of $1/\vec{q}^2$, where $\vec{q}$ denotes the momentum transfer through the pion, and the sum of diagram (b) and (c) is proportional to the expectation value of $\vec{q}^2/(\vec{q}^2 + m_\pi^2)^2$, we expect the ratio of the two contributions to be of the order of $(\gamma/m_\pi)^4 \sim 10^{-2}$, where $|\vec{q}|$ was identified with the value of γ defined above. Thus, the small binding energy of the deuteron seems to provide a natural explanation of the relative suppression of diagram (b)+(c) to (a). However, more systematic studies are necessary.

Another important issue is the role of nucleon recoil contributions. All calculations mentioned so far use as starting point the limit of infinitely heavy nucleons; corrections due to the finite nucleon mass are then included as a power expansion in $1/M_N$. However, it was already observed in the 70s [40], based on calculations using Gaussian wave–functions, that this way one may miss important terms. This was further investigated in Ref. [15], where the analysis was done model independently and the appearance of these additional contributions was related to the Pauli principle in the intermediate NN state, while the pion is in flight. This will be discussed in detail in the next section.

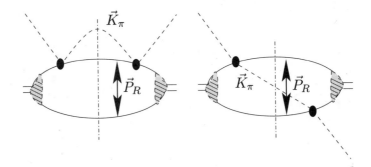

Figure 3: Typical pion loop contributions to πd scattering. Since the exchange of the two nucleons in the intermediate state (at the perpendicular lines) transforms one diagram into the other, the imaginary parts (and their analytic continuation) are linked by the Pauli principle, as described in the text.

3 Role of πNN cuts

Let us investigate the diagrams of Fig. 3 from the point of view of their πNN cut. Then we may write for the corresponding matrix elements

$$I_{\pi NN}(Q) = \int \frac{dK_\pi K_\pi^2 \, dP_R P_R^2}{(2\pi)^6} \frac{f(K_\pi^2, P_R^2)}{Q - K_\pi^2/(2m_\pi) - P_R^2/M_N + i0} \, , \qquad (3)$$

where the function f contains the reaction specific parts like vertex functions, wave functions and transition operators. The only part spelled out explicitly is the πNN propagator, where for simplicity non–relativistic kinematics was chosen. Here Q denotes the excess energy with respect to the πNN threshold. The goal of this study is to compare the full expression given in Eq. 3 with the corresponding one that emerges when static nucleons are used. This corresponds to taking the limit $M_N \to \infty$ prior to integration and we may write

$$I_{\pi NN}^{(static)}(Q) = \int \frac{dK_\pi K_\pi^2 \, dP_R P_R^2}{(2\pi)^6} \frac{f(K_\pi^2, P_R^2)}{Q - K_\pi^2/(2m_\pi) + i0} + O\left(\frac{m_\pi}{M_N}\right) \, . \qquad (4)$$

It was generally assumed that the difference between the integrals of Eqs. (3) and (4) can be accounted for in a polynomial expansion in m_π/M_N. However, this is in general not the case: at pion production threshold $I_{\pi NN}(Q)$ has a branch point singularity. As we will see this leads to a contribution non–analytic in m_π/M_N, even below pion threshold. To see this, let us study in detail the cut contribution by replacing the πNN propagator by its delta function piece

$$(Q - K_\pi^2/(2m_\pi) - P_R^2/M_N + i0)^{-1} \; \to \; -i\pi\delta(Q - K_\pi^2/(2m_\pi) - P_R^2/M_N) \, .$$

Then we may write

$$I_{\pi NN}^{(cut)}(Q) = -i\pi m_\pi \int \frac{dP_R P_R^2}{(2\pi)^6} f\left(K_\pi^{on}(Q, P_R^2)^2, P_R^2\right) K_\pi^{on}(Q, P_R^2) .$$

for the full contribution, where $K_\pi^{on}(Q, P_R^2) = \sqrt{2m_\pi(Q - P_R^2/M_N)}$ and

$$I_{\pi NN}^{(cut,static)}(Q) = -i\pi m_\pi K_\pi^{on}(Q) \int \frac{dP_R P_R^2}{(2\pi)^6} f(K_\pi^{on}(Q)^2, P_R^2)$$

for the static one, with $K_\pi^{on}(Q)|_{static} = \sqrt{2m_\pi Q}$. At this point we see already one important difference between the static and the full treatment: while the imaginary part of the former scales in a completely wrong way, namely according to two–body phase–space, the latter shows the proper scaling as three–body phase space.

But this is not all. Also below the πNN threshold the static contribution gives wrong results. To see this let us focus on the contribution at πNN threshold , $Q = 0$. Although for this kinematics both the above integrals are real, the presence of the πNN cut still plays a significant role. To evaluate the relevant integral the on–shell momentum $K_\pi^{on}(Q, P_R^2)$ needs to be continued analytically to imaginary values using the prescription

$$K_\pi^{on}(0, P_R^2) = \sqrt{-2m_\pi P_R^2/M_N} \to i\sqrt{2m_\pi P_R^2/M_N} .$$

With this we get

$$I_{\pi NN}^{(cut)}(0) = \pi m_\pi \sqrt{\frac{2m_\pi}{M_N}} \int \frac{dP_R P_R^3}{(2\pi)^6} f\left(K_\pi^{on}(0, P_R^2)^2, P_R^2\right) .$$

whereas

$$I_{\pi NN}^{(cut,static)}(0) = 0 .$$

Thus, taking the $M_N \to \infty$ limit prior to integration is in general not allowed in the presence of few–body cuts.

The natural question is: when does this matter? As explained the mentioned effect originates from the opening of the physical πNN threshold. However, this threshold can only matter, if the πNN state is allowed by selection rules. In the isospin limit the two nucleons are identical particles that are to obey the Pauli principle. It therefore depends on the operator that acts on the deuteron wave function to produce the intermediate pion, whether or not the NN state is allowed and, consequently, whether the above contributions matter.

Let us first look at πd scattering. The leading operator that contributes to the $\pi N \to \pi N$ transition is the so–called Weinberg–Tomozawa term $\propto \epsilon^{abc}\tau^c$, which is a vector in isospin space, but spin and momentum independent. Therefore, this operator acting on the deuteron (isospin 0 and spin 1), leads to an NN state that is isospin 1 and spin 1, predominantly in an s–wave due to the momentum independence of the transition operator. This NN state is forbidden by the Pauli principle and therefore all said in the first part of this section does not matter and the static approximation gives a good description of the leading few–body correction. The same holds, e.g., for $\pi d \to \gamma NN$.

However, there are reactions where we expect the above terms to become significant. One example is the reaction $\gamma d \to \pi^+ nn$ that will be discussed in more detail in the next section. Here the operator acting on the deuteron wave function in leading order is the so–called Kroll–Ruderman term, which is a vector in both isospin as well as spin space. Thus, a transition to an NN pair in the $^1S_0$ isovector state, which is allowed by the Pauli principle, is possible. This reaction was studied in detail in Ref. [41], and indeed the pattern sketched above on general grounds was observed. However, for a Pauli allowed intermediate state, the two nucleons will interact. It was found that the inclusion of the two–nucleon intermediate state gives a significant contribution, however, numerically smaller than the static exchange itself. This is different to the case of πd scattering with an isoscalar πN interaction that also leads to a Pauli allowed intermediate state. In this case the inclusion of the NN interaction at threshold numerically restored the contribution of the static exchange [42].

4   The reaction $\gamma d \to \pi^+ nn$

The reaction $\gamma d \to \pi^+ nn$ was studied intensively already in the 70s — for a review see Ref. [43]. At this time only diagrams $(a1)$ and $(a2)$ of Fig. 4 were included. In Ref. [43] there is only one comment to an unpublished work, where pion rescattering (diagrams $(c1)$ and $(c2)$ of Fig. 4) was calculated, however, in the static approximation. It is stated that the inclusion of this contribution destroys the nice agreement of the calculation based on the one–body terms only and therefore it will no longer be considered. Based on the discussion of the previous section we now understand, why the static pion exchange diagram gave a contribution way too large: in a complete calculation it would have been largely canceled by the recoil corrections, since the two–nucleon state in diagrams (b), (c) and (d) can go on–shell while the pion is in flight.

The results of our calculation are shown in Fig. 5. At leading order and next–to–leading order only one–body terms contribute (c.f. Fig. 4) with their strength fixed by the chiral Lagrangian. The relevant $\gamma p \to \pi^+ n$ vertices are momentum independent in both cases and therefore their energy dependence is identical. At NNLO there is a counter term for the transition $\gamma p \to \pi^+ n$ and the strength of the one–body operator can be adjusted to data [44]. This gives a large fraction of the shift in strength when going from NLO to NNLO. In addition the amplitude

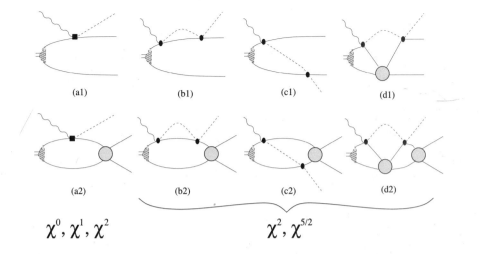

$$\chi^0, \chi^1, \chi^2 \qquad\qquad \chi^2, \chi^{5/2}$$

Figure 4: Diagrams contributing to $\gamma d \to \pi^+ nn$ up to order $\chi^{5/2}$. As before, solid lines denote nucleons, and dashed lines pions. The wavy lines denote photons. The hatched areas denote the deuteron wave function and the filled circles the NN interaction.

gets energy dependent [45]. Another source of energy dependence comes from the few–body corrections as well as higher partial waves that start to contribute at this order. As can be seen from the figure, the data is described very nicely in the whole low energy region considered.

It seems as if the few body corrections, when treated properly, only have a minor effect on the event rates for $\gamma d \to \pi^+ nn$, however, this is correct only for the total cross section. Especially the neutron momentum distributions are sensitive to higher order corrections and those are to be understood to very high accuracy, in order to make use of this reaction to extract the nn scattering length.

On the level of neutron momentum distributions diagram $(a1)$ leads to very specific signals due to the so–called quasi–free production. When all particles in the final state go forward, the intermediate proton is very near on–shell and the diagram gives a large contribution, which decreases quickly, however, as we go away from forward kinematics. In addition, the quasi–free production favors large relative momenta of the two neutrons. Near threshold this is clear, as — in the center of mass system — the spectator neutron keeps on going with half the deuteron momentum, whereas the reaction neutron gets decelerated to almost at rest through the production process.

All diagrams with a final–state interaction, on the other hand, give contributions peaked at small relative nn momenta and almost insensitive to their orientation. This is illustrated in Fig. 6, where the differential rate is shown for two different angles as a function of the relative nn momentum $\vec{P}_R$. The dashed line denotes the

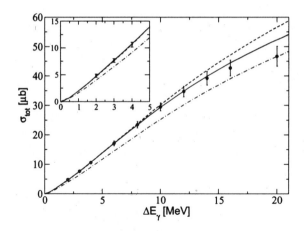

Figure 5: Total cross section of the reaction $\gamma d \to \pi^+ nn$ at LO (dashed line), NLO (dash–dotted line) and $\chi^{5/2}$ order (solid line) together with experimental data from Ref. [46].

distribution where $\vec{P}_R$ is directed along the beam axis, whereas for the solid line it is perpendicular to the beam. In the former case the distribution shows a clear two–peak structure — the quasi–free production shows up at large P_R and the final state peak shows up at small P_R. In the latter case, on the other hand, the quasi free contribution has disappeared and only the final state interaction piece remains with basically identical strength.

The hight and shape of the FSI peak is sensitive to the value of a_{nn}—the neutron–neutron scattering length. A systematic study revealed that high accuracy data on $\gamma d \to \pi^+ nn$ will allow one to extract the nn scattering length with an uncertainty of the order of 0.1 fm which is compatible with that estimated for the competing reactions, $pd \to nnp$ [47] and $\pi^- d \to \gamma nn$ [13].

5 $NN \to d\pi$

As sketched in the introduction, for reactions of the type $NN \to NN\pi$ a simultaneous expansion in the large initial momentum $p_{\text{thr}} \sim \sqrt{m_\pi M_N}$, that also sets the scale for the typical momenta in the loops, and the pion mass is compulsory. Before we go into details in discussing a particular reaction channel we would like to briefly illustrate the impact of this. In practice this means that momenta and pion masses are to be treated independently in the power counting.

To see how this works let us for example estimate the contributions of the loops shown in Fig. 7. For diagram (a) we then estimate

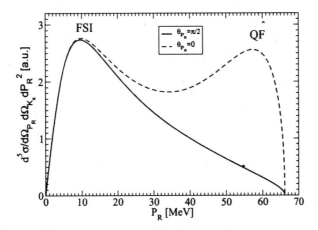

Figure 6: Predicted event rate for the reaction $\gamma d \to \pi^+ nn$ at an excitation energy of 5 MeV as a function of the nn relative momentum P_R for two different orientations of it. The region of small values of P_R are dominated by the nn final state interaction (FSI) and that of large values of it by the quasi-free production (QF).

$$\frac{p}{f_\pi^2} \left(\frac{p}{f_\pi} \right)^3 \left(\frac{1}{p^2} \right)^2 \left(\frac{1}{p} \right)^2 \frac{p^4}{(4\pi)^2} \sim \frac{p^2}{M_N^2},$$

where the different terms refer to the $\pi N \to \pi N$ vertex, the three πNN vertices, the two pion propagators, the two nucleon propagators, and the integral measure, in order. Each individual piece was expressed by the dimensionful parts, where momenta were identified with their typical values. For more details we refer to Appendix E of Ref. [24]. To come to the order estimate we used $4\pi f_\pi \sim M_N$ and dropped an overall factor of $1/f_\pi^3$ common to all production amplitudes. On the other hand we find for diagram (b)

$$\left\{ \left(\frac{m_\pi}{f_\pi^2} \right)^2 \frac{1}{m_\pi} \frac{1}{m_\pi^2} \frac{m_\pi^4}{(4\pi)^2} \right\} \frac{1}{p^2} \left(\frac{p}{f_\pi} \right) \sim \frac{m_\pi^3}{p M_N^2}.$$

The expression in the curly bracket refers to the pion loop — it contains two $\pi N \to \pi N$ vertices, one nucleon and one pion propagator as well as the integral measure — however, in this case the typical momentum is of the order of m_π instead of p_{thr} as in the previous example. The reason is simply that one may choose in the loop the momenta such that the large momentum does not run through the pion. Then the large scale does not appear in the loop at all since the leading heavy baryon

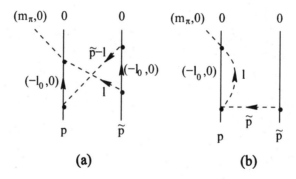

Figure 7: Two typical pion loops that contribute to $NN \to NN\pi^+$. Diagram (a) starts to contribute at NLO whereas diagram (b) starts to contribute at N^4LO.

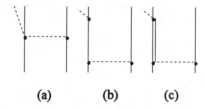

Figure 8: Tree level diagrams that contribute to $pp \to d\pi^+$ up to NLO. Solid lines denote nucleons, dashed ones pions and the double line the propagation of a Delta-isobar.

propagator feels energies only. Outside the curly bracket the momentum transfer is large and therefore the pion propagator as well as the pion vertex appear with $p \sim p_{\text{thr}}$.

If we compare the two order asignments we observe that under the assumption that momenta are of order m_π, both expressions appear at the same order and are therefore expected to be of similar order of magnitude. However, if we assume p to be of order p_{thr}, then diagram (a) is of order m_π/M_N and diagram (b) is of order $(m_\pi/M_N)^{5/2}$, which corresponds to a relative suppression of $(m_\pi/M_N)^{3/2} \sim 1/20$. An explicit calculation [19] revealed an even larger suppression of diagram (b), which turned out to be suppressed by a factor of 50 compared to (a). For more examples we refer to Ref. [24].

As already discussed in the context of πd scattering, in some cases only the sum of various diagrams can give meaningful results, since the individual diagrams depend on the choice made for the pion field. Due to the reordering of loop diagrams, the members of those invariant subgroups are not necessarily of the same chiral order anymore, in contrast to the original Weinberg counting. That, regardless this, also

Figure 9: Irreducible pion loops with nucleons only that start to contribute to $NN \to NN\pi$ at NLO that were considered in Ref. [25].

$$V_{\pi\pi NN} = \qquad \begin{array}{l} (E + l_0 - m_\pi, \vec{p} + \vec{l}) \\ (m_\pi, \vec{0}) \\ \\ (l_0, \vec{l}) \\ (E, \vec{p}) \end{array}$$

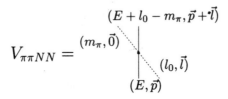

Figure 10: The $\pi N \to \pi N$ transition vertex: definition of kinematic variables as used in the text.

the new scheme gives meaningful results is demontrated in Ref. [48].

The tree level amplitudes that contribute to $pp \to d\pi^+$ are shown in Fig. 8. In Ref. [25] all NLO contributions of loops that start to contribute to $NN \to NN\pi$ at NLO were[1] calculated in threshold kinematics — that is neglecting the distortions from the NN final– and initial state interaction and putting all final states at rest. At threshold only two amplitudes contribute, namely the one with the nucleon pair in the final and initial state in isospin 1 (measured, e.g., in $pp \to pp\pi^0$) and the one where the total NN isospin is changed from 1 to 0 (measured, e.g., in $pp \to d\pi^+$). It was found that the sum all loops that contain Δ–excitations vanish in both channels. This was understood, since the loops were divergent and at NLO no counter term is allowed by chiral symmetry. On the other hand the nucleonic loops were individually finite. It was found that the sum of all nucleonic loops that contribute to $pp \to pp\pi^0$ vanish, whereas the sum of those that contribute to $pp \to d\pi^+$ gave a finite answer. The resulting amplitude grows linear with the initial momentum. In Ref. [49] it was pointed out that this growth of the amplitude is problematic: when evaluated for finite outgoing NN momenta, the transition amplitudes turned out to scale as the momentum transfer. Especially, the amplitudes then grew linearly with the external NN momenta. As a consequence, once convoluted with the NN wave functions, a large sensitivity to those was found, in conflict with general requirements from field theory. The solution to this puzzle was presented in Ref. [50] and will be reported now.

---

[1] In a scheme with two expansion parameters — here m_π and p_{thr} — loops no longer contribute at a single order but at all orders higher than where they start to contribute.

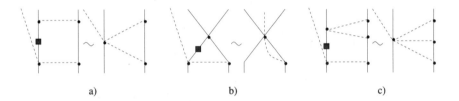

a) b) c)

Figure 11: Induced irreducible topologies, when the off–shell terms of Eq. (5) hit the NN potential in the final state. The filled box on the nucleon line denotes the propagator canceled by the off–shell part of the vertex.

The observation central to the analysis is that the leading $\pi N \to \pi N$ transition vertex, as it appears in Fig. 8a, is energy dependent. Using the notation of Fig. 10 its momentum and energy dependent part may be written as [2]

$$
\begin{aligned}
V_{\pi\pi NN} &= l_0 + m_\pi - \frac{\vec{l} \cdot (2\vec{p} + \vec{l})}{2M_N} \\[2mm]
&= \underbrace{2m_\pi}_{\text{on-shell}} + \underbrace{\left(l_0 - m_\pi + E - \frac{(\vec{l} + \vec{p})^2}{2M_N} \right)}_{(E'-H_0)=(S')^{-1}} - \underbrace{\left(E - \frac{\vec{p}^2}{2M_N} \right)}_{(E-H_0)=S^{-1}} .
\end{aligned} \tag{5}
$$

For simplicity we skipped the isospin part of the amplitude. The first term in the last line denotes the transition in on–shell kinematics, the second the inverse of the outgoing nucleon propagator and third the inverse of the incoming nucleon propagator. First of all we observe that for on–shell incoming and outgoing nucleons, the the $\pi N \to \pi N$ transition vertex takes its on–shell value $2m_\pi$ — even if the incoming pion is off–shell, as it is for diagram (a) of Fig. 8. This is in contrast to standard phenomenological treatments [51], where l_0 was identified with $m_\pi/2$ — the energy transfer in on–shell kinematics — and the recoil terms were not considered. Note, since $p_{thr}^2/M_N = m_\pi$ the recoil terms are to be kept.

A second consequence of Eq. (5) is even more interesting: when the $\pi N \to \pi N$ vertex gets convoluted with NN wave functions, only the first term leads to a reducible diagram. The second and third term, however, lead to irreducible contributions, since one of the nucleon propagators gets canceled. This is illustrated in Fig. 11, where those induced topologies are shown that appear, when one of the nucleon propagators is canceled (marked by the filled box) in the convolution of typical diagrams of the NN potential with the $NN \to NN\pi$ transition operator. Power counting gives that diagram (b) and (c) appear only at order N^{4}LO and N^{3}LO, respectively. However, diagram (a) starts to contribute at NLO and it was found in Ref. [50] that those induced irreducible contributions cancel the finite remainder of the NLO loops in the $pp \to d\pi^+$ channel. Thus, up to NLO only the diagrams

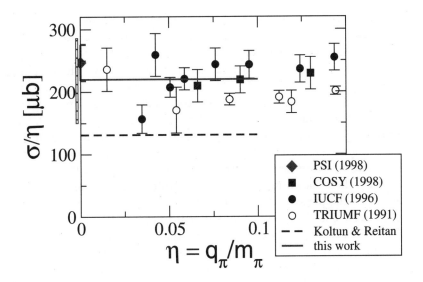

Figure 12: Comparison of the results of Ref. [50] to experimental data for $NN \to d\pi$. The dashed line corresponds to the model of Koltun and Reitan [51], whereas the solid line is the result of the ChPT calculation of Ref. [50]. The estimated theoretical uncertainty (see text) is illustrated by the narrow box. The data is from Refs. [52] (open circles), [53] (filled circles) and [54] (filled squares). The first data set shows twice the cross section for $pn \to d\pi^0$ and the other two the cross section for $pp \to d\pi^+$.

of Fig. 8 contribute to $pp \to d\pi^+$, with the rule that the $\pi N \to \pi N$ vertex is put on–shell.

The result found in Ref. [50] is shown in Fig. 12 as the solid line, where the total cross section (normalized by the energy dependence of phase space) is plotted against the normalized pion momentum. The dashed line is the result of the model by Koltun and Reitan [51], as described above. The data sets are from TRIUMF [52], IUCF [53], and COSY [54].

6   Comparison to phenomenological works

All recent phenomenological calculations for $NN \to NN\pi$ add additional diagrams to the model of Ref. [51]. Here we will focus only on $pp \to d\pi^+$. Phenomenological calculations for this reaction in near threshold kinematics are given, e.g., in Ref. [55] and Ref. [56]. In both works in addition to the diagrams of Ref. [51] some Δ–loops as well as additional short range contributions are included — heavy meson exchanges for the former and off–shell πN scattering[2] for the latter. Based on this the cross section for $pp \to d\pi^+$ is overestimated near threshold. How can we interpret this discrepancy in light of the discussion above?

First of all, the NLO parts of the Δ–loops cancel, as was shown already in Ref. [25]. However, in both Refs. [55, 56] only one of these diagrams was included and, especially for Ref. [55], gave a significant contribution. In addition, in the effective field theory short ranged operators start to contribute only at N^2LO. The only diagram of those NLO loops shown in Fig. 9 that is effectively included in Ref. [56] is the fourth, since the pion loop there can be regarded as part of the $\pi N \to \pi N$ transition T–matrix. However, as described, the contribution of this diagram gets canceled by the others shown in Fig. 9 and the induced irreducible pieces described above. Therefore, the physics that enhances the cross section compared to the work of Ref. [51] in Refs. [55, 56] is completely different to that of Ref. [50]. However, only the last one is field theoretically consistent as explained in the previous section.

The natural question that arises is that for observable consequences. As explained, in the effective field theory calculation the near threshold cross section for $pp \to d\pi^+$ is basically given by a long–ranged pion exchange diagram, whereas the phenomenological calculations rely on short ranged operators with respect to the NN system. Obviously those observables are sensitive to this difference that get prominent contributions from higher partial waves in the final NN system. We therefore need to look at the reaction $pp \to pn\pi^+$. Unfortunately, the total cross section for this reaction is largely saturated by NN S–waves in the final state (see, e.g., Fig. 17 in Ref. [24]). On the other hand, linear combinations of double polarization observables allow one to remove the prominent components and the sub-leading amplitudes should be visible. We therefore expect from the above considerations that the phenomenological calculations give good results for polarization observables for $pp \to d\pi^+$, whereas there should be deviations for some of those for $pp \to pn\pi^+$. Predictions for these observables were presented in Ref. [57] and indeed the π^+ observables with the deuteron in the final state, reported in Ref. [58], are described well whereas there are discrepancies for the pn final state (see Fig. 24 of Ref. [24]). For the corresponding data see Ref. [59].

It remains to be seen how well the same data can be described in the effective field theory framework. Up to NNLO the number of counter terms is quite low: there are two counter terms for pion s–waves, that can be arranged to contribute to $pp \to pp\pi^0$ and $pp \to d\pi^+$ individually, and then there is one counter term for pion p–waves, that contributes only to a small amplitude in charged pion production [23]. On the other hand there is a huge amount of even double polarized data available [58–60] — and there is more to come especially for $pn \to pp\pi^-$ [61].

---

[2]That those are also short range contributions is discussed in Ref. [24].

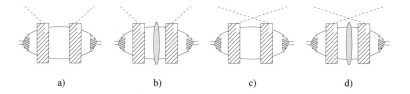

a)   b)   c)   d)

Figure 13: Dispersive corrections to the πd scattering length.

7   Dispersive corrections to $a_{\pi d}$

Let us come back to πd scattering at threshold. The corresponding scattering length was presented in Eq. (2). In the introductory sections we exclusively focused on the real part. However, now we are in the position to also discuss the imaginary part, which is closely linked to the reaction $NN \to d\pi$ through unitarity and detailed balance. One may write

$$4\pi\mathrm{Im}(a_{\pi d}) = \lim_{q \to 0} q \left\{ \sigma(\pi d \to NN) + \sigma(\pi d \to \gamma NN) \right\} , \tag{6}$$

where q denotes the relative momentum of the initial πd pair. The ratio $R = \lim_{q \to 0} (\sigma(\pi d \to NN)/\sigma(\pi d \to \gamma NN))$ was measured to be 2.83 ± 0.04 [62]. At low energies diagrams that lead to a sizable imaginary part of some amplitude are expected to also contribute significantly to its real part. Those contributions are called dispersive corrections. As a first estimate Brückner speculated that the real and imaginary part of these contributions should be of the same order of magnitude [63]. This expectation was confirmed within Faddeev calculations in Refs. [64]. Given the high accuracy of the measurement and the size of the imaginary part of the scattering length, another critical look at this result is called for as already stressed in Refs. [65, 66]. A consistent calculation is only possible within a well defined effective field theory — the first calculation of this kind was presented in Ref. [67] and is briefly sketched here.

To identify the diagrams that are to contribute we first need to specify what we mean by a dispersive correction. We define dispersive corrections as contributions from diagrams with an intermediate state that contains only nucleons, photons and at most real pions. Therefore, all the diagrams shown in Fig. 13 are included in our work. On the other hand, all diagrams that, e.g., have Delta excitations in the intermediate state do not qualify as dispersive corrections, although they might give significant contributions [31].

The hatched blocks in the diagrams of Fig. 13 refer to the relevant transition operators for the reaction $NN \to NN\pi$ depicted in Fig. 8. Also in the kinematics of relevance here the $\pi N \to \pi N$ transitions are to be taken with their on–shell value

$2m_\pi$. Using the CD–Bonn potential [68] for the NN distortions we found for the dispersive correction from the purely hadronic transition

$$a_{\pi d}^{disp} = (-6.3 + 2 + 3.1 - 0.4) \times 10^{-3}\, m_\pi^{-1} = -1.6 \times 10^{-3}\, m_\pi^{-1}\,, \qquad (7)$$

where the numbers in the first bracket are the individual results for the diagrams shown in Fig. 13, in order. Note that the diagrams with intermediate NN interactions and the crossed ones (diagram (c) and (d)), neither of them included in most of the previous calculations, give significant contributions. The latter finding might come as a surprise on the first glance, however, please recall that in the chiral limit all four diagrams of Fig. 13 are kinematically identical and chiral perturbation theory is a systematic expansion around exactly this point. Thus, as a result we find that the dispersive corrections to the πd scattering length are of the order of 6 % of the real part of the scattering length. This number is fully in line with the expectations from power counting, which predicted a relative suppression of the dispersive corrections compared to the leading double scattering term — diagram (b) of Fig. 1 — of the order of $(m_\pi/M_N)^{3/2} \sim 5$ %. Note that the same calculation gave very nice agreement for the corresponding imaginary part [67].

In Ref. [67] also the electro–magnetic contribution to the dispersive correction was calculated. It turned out that the contribution to the real part was tiny — $-0.1 \times 10^{-3}\, m_\pi^{-1}$ — while the sizable experimental value for the imaginary part (c.f. Eqs. (2) and 6) was described well.

To get a reliable estimate of the uncertainty of the calculation just presented a NNLO calculation is necessary. At that order a counter term appears for pions at rest that can be fixed from $NN \to NN\pi$, as indicated above. For now we can only present a conservative estimate for the uncertainty by using the uncertainty of order $2\, m_\pi/M_N$ one has for, e.g., the sum of all direct diagrams to derive a $\Delta a_{\pi d}^{disp}$ of around $1.4 \times 10^{-3}\, m_\pi^{-1}$, which corresponds to about 6% of $\mathrm{Re}\left(a_{\pi d}^{\mathrm{exp}}\right)$. However, given that the operators that contribute to both direct and crossed diagrams are almost the same and that part of the mentioned cancellations is a direct consequence of kinematics, this number for $\Delta a_{\pi d}^{disp}$ is probably too large.

In Ref. [67] a detailed comparison to previous works is given. Differences in the values found for the dispersive corrections were traced to the incomplete sets of diagrams included in those phenomenological studies.

8   Summary and Outlook for pion reactions

In the lectures recent progress in our understanding of elastic and inelastic pion reactions on the two–nucleon system was presented. The central reaction discussed was πd scattering at threshold. Arguments were given that in the years to come one should be able to calculate the πd scattering length with sufficient accuracy to use the reaction as one of the prime sources for the isoscalar scattering length a_+. To reach this not only significant progress was necessary for the coherent πd scattering but also the reactions $NN \to NN\pi$ need to understood. In the future also the role of isospin violation on πd scattering needs to be investigated further as stressed in Ref. [35].

The process $NN \to NN\pi$ is a puzzle already since more than a decade. Given the progress presented above we have now reason to believe that this puzzle will be solved soon. This mentioned results could only be found, because a consistent effective field theory was used. For example, the potential problem with the transition operators of Ref. [25], pointed at in Ref. [49], would always be hidden in phenomenological calculations, since the form factors routinely used there always lead to finite, well behaved amplitudes. The very large number of observables available for the reactions $NN \to NN\pi$ will provide a non–trivial test to the approach described.

Once the scheme is established, the same field theory can be used to analyze the isospin violating observables measured in $pn \to d\pi^0$ [69] and $dd \to \alpha\pi^0$ [70]. First steps in this direction were already done in Ref. [71] for the former and in Refs. [72, 73] for the latter.

In the lectures also the reaction $\gamma d \to \pi^+ nn$ was discussed. Not only gave those studies a further confirmation that we understand the few body dynamics well within ChPT, but it also promises to become an ideal reaction for the extraction of the nn scattering length with high accuracy. The corresponding measurements could be performed at HIGS [74].

9 Some remarks on strangeness production

As described in the previous sections a lot is already known about the properties and dynamics of systems composed of light quarks only. However, much less is known about the scattering of systems with strangeness, especially for low energies. The reason is of experimental nature: the lifetime of particles with strangeness is typically too short to allow for secondary scatterings at low energies, necessary to get information on low energy scattering.

Here we will focus on the hyperon–nucleon (YN) system. The poor status of our information on the YN interaction is most obviously reflected in the present knowledge of the ΛN scattering lengths. Attempts in the 1960's to pin down the low energy parameters for the S-waves led to results that were afflicted by rather large uncertainties [75, 76]. In Ref. [76] the following values are given for the singlet scattering length a_s and the triplet scattering length a_t

$$a_s = -1.8 \begin{Bmatrix} +2.3 \\ -4.2 \end{Bmatrix} \text{ fm and } a_t = -1.6 \begin{Bmatrix} +1.1 \\ -0.8 \end{Bmatrix} \text{ fm,} \tag{8}$$

where the errors are strongly correlated. The situation of the corresponding effective ranges is even worse: for both spin states values between 0 and 16 fm are allowed

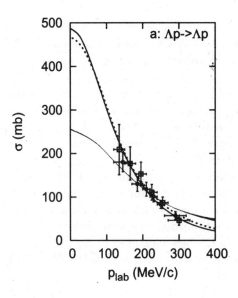

Figure 14: Comparison of different variants for the ΛN interaction to the available data at low energies. The solid line, dashed line, and the light area are the results of Refs. [77–79], in order. The data are from Refs. [75, 76].

by the data. Later, the application of microscopic models for the extrapolation of the data to the threshold, was hardly more successful to pin down the low energy parameters. For example, in Ref. [77] one can find six different models that equally well describe the available data but whose (S-wave) scattering lengths range from -0.7 to -2.6 fm in the singlet channel and from -1.7 to -2.15 fm in the triplet channel. To illustrate this point in Fig. 14 we show a comparison of model f of Ref. [77] (dark solid line), the Jülich '04 model [78] (dashed curve), and the result from the recent effective field theory approach of Ref. [79] to the world data.

The natural alternative to scattering experiments are production reactions. However, the central insights of the previous sections were that only within a consistent field theory reliable calculations can be performed for the reactions under considerations. On the other hand, as stressed in the introduction, any strangeness production reaction involves momenta that do not allow for an expansion along the lines just discussed, since the corresponding expansion parameter in this case would be larger than 1/2. Does this mean that one can learn nothing from a study of strangeness production off two nucleon systems?

Not at all. For one thing, an investigation of baryon and meson resonances does not need any detailed knowledge on the production mechanism — most of the relevant information is contained in the Dalitz plots — and the comment of the previous paragraph applies only to this part. But one can learn even more from the production reactions by using that the momentum transfer in those reactions are large. Since this leads to an effectively point–like production operator, one may employ dispersion theory to relate invariant mass spectra of production reactions to elastic scattering data. Obviously, for this no knowledge on the production operator is necessary whatsoever. Then, in contrast to above, one works with the typical outgoing relative momentum in units of the momentum transfer as expansion parameter.

In the remainder of this text I will only focus on how to extract scattering parameters from production reactions. One aspect of spectroscopy, also discussed in the lectures, namely that of scalar mesons, was described in the recent conference proceeding [80] in very much detail and it will not be repeated here.

The use of dispersion theory was very common in the 50s and the basis for the study to be described now was worked out already then [81]. A controlled method of extraction of scattering lengths from production reactions opens up the opportunity to measure scattering parameters also for unstable states. As an example we will discuss the option to measure the ΛN scattering lengths from $NN \to K\Lambda N$ and $\gamma d \to K\Lambda N$. In Refs. [82–84] it was shown, how to derive an integral representation of the scattering length of a pair of outgoing particles (here we show the formula relevant for neutral final states, as is relevant for ΛN; in the presence of Coulomb interactions the equation has to be modified — see Ref. [83]):

$$
a_S = \lim_{m^2 \to m_0^2} \frac{1}{2\pi} \left(\frac{m_\Lambda + m_N}{\sqrt{m_\Lambda m_N}} \right) \mathbf{P} \int_{m_0^2}^{m_{max}^2} dm'^2 \sqrt{\frac{m_{max}^2 - m^2}{m_{max}^2 - m'^2}}
$$
$$
\times \frac{1}{\sqrt{m'^2 - m_0^2}\,(m'^2 - m^2)} \log \left\{ \frac{1}{p'} \left(\frac{d^2\sigma_S}{dm'^2 dt} \right) \right\}, \quad (9)
$$

where σ_S denotes the spin cross section for the production of a Λ–nucleon pair with invariant mass m'^2—corresponding to a relative momentum p'—and total spin S. In addition $t = (p_1 - p_{K^+})^2$, with p_1 being the beam momentum, $m_0^2 = (m_\Lambda + m_N)^2$, where m_Λ (m_N) denotes the mass of the Lambda hyperon (nucleon), and m_{max} is some suitably chosen cutoff in the mass integration. In Ref. [82] it was shown that it is sufficient to include relative energies of the final ΛN system of at most 40 MeV in the range of integration to get accurate results. $\mathbf{P}$ denotes that the principal value of the integral is to be used and the limit has to be taken from above.

The formula as given is applicable if there are no significant effects from crossed channels — this can be monitored by a Dalitz plot analysis — and only a single partial wave contributes. With respect to the angular momentum this can be achieved by a proper choice of m_{max}. In order to select a single spin state for the outgoing two–particle system polarization observables are necessary. In Refs. [82, 84] the relevant observables are identified for the reactions $NN \to NKY$ and $\gamma d \to K\Lambda N$, respectively. The former class of reactions can be measured at COSY [61, 87] and the latter either at J-Lab [88], MAMI [89], or ELSA [90].

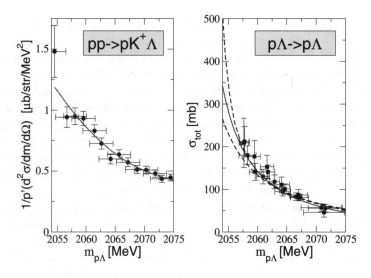

Figure 15: Illustration of the uncertainties of extraction of scattering parameters from scattering data (left panel), where an extrapolation is necessary, and from production data (right panel), where the data needs to be interpolated. The data was taken from Refs. [76, 85] for the right panel and from Ref. [86] for the left one.

The essential advantages of the use of Eq. (9) to extract the scattering lengths of unstable particles compared to a determination from scattering experiments are, e.g., for ΛN scattering:

- Instead of an extrapolation of data, the scattering length is found from an interpolation of an invariant mass spectrum, which is theoretically much better controlled. This is illustrated in Fig. 15.

- The integral representation gives a result for the scattering length without any assumption on the energy dependence of the hyperon–nucleon interaction. This opens the possibility to fix the scattering lengths from production reactions and then use scattering data to fix, e.g., the effective range.

- Since Eq. (9) is derived from a dispersion integral, a controlled error estimate is possible. A systematic study revealed that for the kinematics relevant for the production of the ΛN system at low energies, the uncertainty of Eq. (9) was found to be 0.5 fm [82, 84]. Sources of this uncertainty are the possible energy dependence of the production operator, the influence of the upper limit of integration, and the possible influence of crossed channel effects.

10 Summary

In these proceedings various reactions on few nucleon systems were discussed. It was demonstrated that due to significant advances in the technologies of effective field theories, high precision calculations became possible for hadronic reactions even on few–nucleon systems. At the same time a clear connection to QCD is provided.

For the production of more heavy systems, like those that contain the strangeness degree of freedom, off two nucleons no effective field theory is developed yet and therefore many reactions are still being analyzed using models. However, also for this class of reactions some aspects can be analyzed in a model independent way. The ΛN scattering lengths were discussed as an example.

We are now in a phase, were both theory and experiment are advanced sufficiently that we should understand much better how QCD influences low energy nuclear dynamics in the upcoming years. Especially the symmetry breaking sector — violation of isospin as well as violation of the flavor $SU(3)$ — promises deep insights into the mechanisms of strong interactions.

Acknowledgments

I thank the organizers for a superb job and V. Lensky, V. Baru, A. Gasparyan, J. Haidenbauer, A. Kudryavtsev, and U.-G. Meißner for a very fruitful collaboration that lead to the results presented.

References

[1] G. Colangelo, J. Gasser and H. Leutwyler, Nucl. Phys. B **603**, 125 (2001) [arXiv:hep-ph/0103088].

[2] V. Bernard, N. Kaiser and U.-G. Meißner, Int. J. Mod. Phys. E **4** (1995) 193.

[3] N. Fettes and U.-G. Meißner, Nucl. Phys. A **693**, 693 (2001) [arXiv:hep-ph/0101030].

[4] B. Kubis, these proceedings.

[5] P. F. Bedaque and U. van Kolck, Ann. Rev. Nucl. Part. Sci. **52** (2002) 339 [arXiv:nucl-th/0203055].

[6] E. Epelbaum, Prog. Part. Nucl. Phys. **57** (2006) 654 [arXiv:nucl-th/0509032].

[7] R. Machleidt, these proceedings.

[8] S. Weinberg, Phys. Lett. B **295** (1992) 114.

[9] S. R. Beane, V. Bernard, E. Epelbaum, U.-G. Meißner and D. R. Phillips, Nucl. Phys. A **720** (2003) 399. [arXiv:hep-ph/0206219].

[10] S. R. Beane, V. Bernard, T. S. H. Lee, U.-G. Meißner and U. van Kolck, Nucl. Phys. A **618**, 381 (1997) [arXiv:hep-ph/9702226].

[11] H. Krebs, V. Bernard and U.-G. Meißner, Eur. Phys. J. A **22** (2004) 503 [arXiv:nucl-th/0405006].

[12] V. Baru, J. Haidenbauer, C. Hanhart and J. A. Niskanen, Eur. Phys. J. A **16**, 437 (2003) [arXiv:nucl-th/0207040].

[13] A. Gardestig and D. R. Phillips, Phys. Rev. C **73** (2006) 014002 [arXiv:nucl-th/0501049]; A. Gardestig and D. R. Phillips, arXiv:nucl-th/0603045.

[14] V. Lensky, V. Baru, J. Haidenbauer, C. Hanhart, A. E. Kudryavtsev and U.-G. Meißner, Eur. Phys. J. A **26**, 107 (2005) [arXiv:nucl-th/0505039].

[15] V. Baru, C. Hanhart, A. E. Kudryavtsev and U.-G. Meißner, Phys. Lett. B **589**, 118 (2004) [arXiv:nucl-th/0402027].

[16] B.Y. Park et al., Phys. Rev. C **53** (1996) 1519 [arXiv:nucl-th/9512023].

[17] C. Hanhart, J. Haidenbauer, M. Hoffmann, U.-G. Meißner and J. Speth, Phys. Lett. B **424** (1998) 8 [arXiv:nucl-th/9707029].

[18] H. Machner and J. Haidenbauer, J. Phys. G **25**, R231 (1999).

[19] V. Dmitrašinović, K. Kubodera, F. Myhrer and T. Sato, Phys. Lett. B **465** (1999) 43 [arXiv:nucl-th/9902048].

[20] S. I. Ando, T. S. Park and D. P. Min, Phys. Lett. B **509** (2001) 253 [arXiv:nucl-th/0003004].

[21] T.D. Cohen, J.L. Friar, G.A. Miller and U. van Kolck, Phys. Rev. C **53** (1996) 2661 [arXiv:nucl-th/9512036].

[22] C. da Rocha, G. Miller and U. van Kolck, Phys. Rev. C **61** (2000) 034613 [arXiv:nucl-th/9904031].

[23] C. Hanhart, U. van Kolck, and G.A. Miller, Phys. Rev. Lett. **85** (2000) 2905 [arXiv:nucl-th/0004033].

[24] C. Hanhart, Phys. Rep. **397** (2004) 155 [arXiv:hep-ph/0311341].

[25] C. Hanhart and N. Kaiser, Phys. Rev. C **66** (2002) 054005 [arXiv:nucl-th/0208050].

[26] D. R. Phillips, S. J. Wallace and N. K. Devine, Phys. Rev. C **72** (2005) 0140061. [arXiv:nucl-th/0411092].

[27] P. Hauser et al., Phys. Rev. C **58** (1998) 1869; D. Chatellard et al., Nucl. Phys. A **625** (1997) 855.

[28] D. Gotta et al., PSI experiment R-06.03; D. Gotta, private communication.

[29] S. R. Beane, V. Bernard, E. Epelbaum, U.-G. Meißner and D. R. Phillips, Nucl. Phys. A **720** (2003) 399 [arXiv:hep-ph/0206219].

[30] U.-G. Meißner, U. Raha and A. Rusetsky, Eur. Phys. J. C **41** (2005) 213 [arXiv:nucl-th/0501073].

[31] M. Döring, E. Oset and M. J. Vicente Vacas, Phys. Rev. C **70** (2004) 045203 [arXiv:nucl-th/0402086].

[32] M. Pavon Valderrama and E. Ruiz Arriola, Phys. Rev. C **72** (2005) 054002 [arXiv:nucl-th/0504067]; M. P. Valderrama and E. R. Arriola, arXiv:nucl-th/0605078.

[33] A. Nogga and C. Hanhart, Phys. Lett. B **634** (2006) 210 [arXiv:nucl-th/0511011].

[34] L. Platter and D. R. Phillips, arXiv:nucl-th/0605024.

[35] U.-G. Meißner, U. Raha and A. Rusetsky, Phys. Lett.B **639** (2006) 478 [arXiv:nucl-th/0512035].

[36] A. W. Thomas and R. H. Landau, Phys. Rept. **58** (1980) 121.

[37] B. Borasoy and H. W. Grießhammer, Int. J. of Mod. Phys. **E** 12, 65 (2003) [arXiv:nucl-th/0105048].

[38] S. R. Beane and M. J. Savage, Nucl. Phys. A **717** (2003) 104 [arXiv:nucl-th/0204046].

[39] M. R. Robilotta and C. Wilkin, J. Phys. G **4**, L115 (1978).

[40] V. M. Kolybasov and A. E. Kudryavtsev, Nucl. Phys. B **41** (1972) 510.

[41] V. Lensky, V. Baru, J. Haidenbauer, C. Hanhart, A. E. Kudryavtsev and U G. Meißner, Eur. Phys. J. A **26**, 107 (2005) [arXiv:nucl-th/0505039].

[42] G. Fäldt, Phys. Scripta **16**, 81 (1977); A. Rusetsky, private communication.

[43] J. M. Laget, Phys. Rep. **69** (1981) 1.

[44] V. Bernard, N. Kaiser and U.-G. Meißner, Nucl. Phys. B **383** (1992) 442.

[45] H. W. Fearing, T. R. Hemmert, R. Lewis, and C. Unkmeir, Phys. Rev. C **62** (2000) 054006 [arXiv:hep-ph/0005213].

[46] E. C. Booth et al., Phys. Rev. **C20** (1979) 1217.

[47] D.E. Gonzalez–Trotter et al., Phys. Rev. Lett. **83** (1999) 3788; V. Huhn et al., Phys. Rev. Lett. **85** (2000) 1190.

[48] C. Hanhart and A. Wirzba, arXiv:nucl-th/0703012.

[49] A. Gårdestig, talk presented at ECT* workshop 'Charge Symmetry Breaking and Other Isospin Violations', Trento, June 2005; A. Gårdestig, D. R. Phillips and C. Elster, arXiv:nucl-th/0511042.

[50] V. Lensky, V. Baru, J. Haidenbauer, C. Hanhart, A. E. Kudryavtsev and U.-G. Meißner, Eur. Phys. J. A **27**, 37 (2006) [arXiv:nucl-th/0511054].

[51] D. Koltun and A. Reitan, Phys. Rev. **141** (1966) 1413.

[52] D. A. Hutcheon et al., Nucl. Phys. A **535** (1991) 618.

[53] P. Heimberg et al., Phys. Rev. Lett. **77** (1996) 1012.

[54] M. Drochner et al., Nucl. Phys. A **643** (1998) 55.

[55] J. A. Niskanen, Phys. Rev. C **53**, 526 (1996) [arXiv:nucl-th/9502015].

[56] C. Hanhart, J. Haidenbauer, O. Krehl and J. Speth, Phys. Lett. B **444** (1998) 25 [arXiv:nucl-th/9808020].

[57] C. Hanhart, J. Haidenbauer, O. Krehl and J. Speth, Phys. Rev. C **61** (2000) 064008 [arXiv:nucl-th/0002025].

[58] B. von Przewoski et al., Phys. Rev. C **61** (2000) 064604.

[59] W. W. Daehnick et al., Phys. Rev. C **65** (2002) 024003.

[60] H. O. Meyer et al., Phys. Rev. C **63** (2001) 064002.

[61] A. Kacharava et al., arXiv:nucl-ex/0511028.

[62] V. C. Highland et al., Nucl. Phys. A **365** (1981) 333.

[63] K. Brückner, Phys. Rev. **98** (1955) 769.

[64] I.R. Afnan and A.W. Thomas, Phys. Rev. C**10** (1974) 109; D.S. Koltun and T. Mizutani, Ann. Phys. (N.Y.) **109** (1978) 1.

[65] T. E. O. Ericson, B. Loiseau, A. W. Thomas, Phys. Rev. C **66** (2002) 014005 [arXiv:hep-ph/0009312].

[66] V.Baru, A. Kudryavtsev, Phys. Atom. Nucl., **60** (1997) 1476.

[67] V. Lensky, V. Baru, J. Haidenbauer, C. Hanhart, A. E. Kudryavtsev and U.-G. Meißner, arXiv:nucl-th/0608042; Phys. Lett. B, in print.

[68] R. Machleidt, Phys. Rev. C **63** (2001) 024001 [arXiv:nucl-th/0006014].

[69] A. K. Opper et al., Phys. Rev. Lett. **91** (2003) 212302 [arXiv:nucl-ex/0306027].

[70] E. J. Stephenson et al., Phys. Rev. Lett. **91** (2003) 142302 [arXiv:nucl-ex/0305032].

[71] U. van Kolck, J. A. Niskanen, and G. A. Miller, Phys. Lett. B **493** (2000) 65 [arXiv:nucl-th/0006042].

[72] A. Gårdestig et al., Phys. Rev. C **69** (2004) 044606[arXiv:nucl-th/0402021].

[73] A. Nogga et al., Phys. Lett. B **639**, 465 (2006) [arXiv:nucl-th/0602003].

[74] A. Bernstein, private communication.

[75] G. Alexander et al., Phys. Lett. **19**, 715 (1966); B. Sechi-Zorn et al., Phys. Rev. **175**, 1735 (1968).

[76] G. Alexander et al., Phys. Rev. **173**, 1452 (1968);

[77] T. A. Rijken, V. G. J. Stoks, Y. Yamamoto, Phys. Rev. C 59 (1999) 21.

[78] J. Haidenbauer, U.-G. Meißner, Phys. Rev. C 72 (2005) 044005.

[79] H. Polinder, J. Haidenbauer and U. G. Meissner, Nucl. Phys. A **779** (2006) 244 [arXiv:nucl-th/0605050]; J. Haidenbauer, U. G. Meissner, A. Nogga and H. Polinder, arXiv:nucl-th/0702015.

[80] C. Hanhart, arXiv:hep-ph/0609136.

[81] N. I. Muskhelishvili, *Singular Integral Equations*, (P. Noordhof N. V., Groningen, 1953); R. Omnes, Nuovo Cim. **8**, 316 (1958); W. R. Frazer and J. R. Fulco, Phys. Rev. Lett. **2**, 365 (1959).

[82] A. Gasparyan, J. Haidenbauer, C. Hanhart and J. Speth, Phys. Rev. C **69**, 034006 (2004) [arXiv:hep-ph/0311116].

[83] A. Gasparyan, J. Haidenbauer and C. Hanhart, Phys. Rev. C **72**, 034006 (2005) [arXiv:nucl-th/0506067].

[84] A. Gasparyan, J. Haidenbauer, C. Hanhart and K. Miyagawa, arXiv:nucl-th/0701090.

[85] F. Eisele et al. Phys. Lett. B **37**, 1971 (204).

[86] R. Siebert et al., Nucl. Phys. **A567**, 819 (1994).

[87] A. Gillizer et al., in preparation.

[88] B.L. Berman et al., CEBAF proposal PR-89-045 (1989).

[89] R. Beck and A. Starostin, Eur. Phys. J. A **S19**, 279 (2004); R. Beck, Prog. Part. Nucl. Phys. **55**, 91 (2005).

[90] V. Kleber and H. Schmieden, private communication.

Physics and Astrophysics of Hadrons and Hardronic Matter
Editor: A. B. Santra

Applications of Effective Field Theory in Nuclear Physics

Jiunn-Wei Chen

Department of Physics and Center for Theoretical Sciences
National Taiwan University, Taipei 10617, Taiwan
email: jwc@phys.ntu.edu.tw

Abstract

Three lectures are given on nuclear effective field theory (EFT). In the first lecture, I will talk about nuclear EFT in the continuum (with lattice spacing $a_{latt} = 0$). In the second lecture, I will talk about nuclear physics on the lattice ($a_{latt} \neq 0$) using (i) lattice QCD when $0 < a_{latt} < 1/\Lambda_{QCD}$ and (ii) using lattice NEFT when $1/\Lambda_\chi < a_{latt} < 1/m_\pi$, where $\Lambda_\chi \sim 1$ GeV is the chiral symmetry breaking scale and m_π is the pion mass. In the third lecture, I will address the issue of the fermion sign problem in many body systems. Then I will talk about two interesting problems related to QCD viscosity and the BEC-BCS cross over in cold atoms.

1   Nuclear Effective Field Theory in the Continuum

The history of physics seems to be a history of discovering the old theories to be the effective theories of the new in certain limits. In the beginning of the 20th century, Newtonian dynamics was found to be an effective theory of Special Relativity in the limit that the speed of light is infinite. By the 1920's, it was established that Newtonian dynamics is an effective theory of Quantum Mechanics in the limit of infinite Planck mass. In the 1940's, Special Relativity and Quantum Mechanics were both incorporated in Quantum Field Theory which provided an unified description of particles (such as electrons) and fields (such as photons). Today, physicists are trying to find a theory which includes General Relativity and Quantum Field Theory as effective theories in the limits of zero Planck constant and infinite plank mass respectively.

It is important to realize that every known theory is an effective theory of some other more complete theory. However, the unknown short distance physics can be systematically parameterized and one can actually go beyond the original theories without model dependent assumptions. The usefulness of EFT depends on if there is a hierarchical structure in the physics being parameterized. If the hierarchical structure does not exist, one would need to know every detail of the unknown physics which is equivalent to complete knowledge of the underlying theory. In general, if there are highly separable energy scales in the physical systems, then the hierarchy in the parameterization in terms of momentum expansions can exist. Thus for calculations to a certain precision level, there involve only finite numbers of parameters associated with the leading terms in the expansions. This implies the low energy behavior of such systems will not be sensitive to the details of physics associated with the high energy scales.

In the case of strong interaction, QCD (Quantum Chromo-dynamics) is known to be the underlying theory. However, QCD has not yet been solved in the low energy regime ($\lesssim 1$ GeV). Chiral perturbation based on the chiral symmetry of QCD parameterizes the low lying meson (pion and kaon) and baryon (nucleon, Δ and hyperon) dynamics in terms of momentum and chiral (quark mass) expansions. In the meson and single baryon systems, the parameterization is hierarchical because the typical energy scales in the systems are separable. Low energy scales in the systems are the pion mass (~ 140 MeV), kaon mass (~ 500 MeV) and characteristic momentum transfer. High energy scales are baryon masses (≥ 940 MeV) and a chiral dynamics scale $\Lambda_\chi (\sim 1$ GeV) set by strong interaction dynamics. In the multi-nucleon systems, nucleons can form bound states. For ordinary nuclei, the typical binding momentum per nucleon is ~ 100 MeV which is much smaller than Λ_χ. This makes it very attractive to attempt to construct an EFT in nuclear system.

On the other hand, nuclear physicists have accumulated a large amount of experience in the description of nuclear force via nuclear potential models. A natural question to ask is what EFT can offer to improve the understanding? The answer is, given the goal of both potential models and EFT is to build in the known physics (such as chiral dynamics) and parameterize the unknown, EFT provides a systematic way to implement this. To be more explicit, the benefits of EFT are, first, field theory approaches do not suffer from off-shell ambiguity which is generally not true for potential model calculations. Second, EFT has (special) relativity, chiral symmetry and gauge symmetry built in, thus it naturally describe relativistic corrections, pion retardation effects, pion exchange and insertions of external currents. Third, power counting allows one to reliably estimate errors from the omission of higher order effects. And fourth, the calculations are more easily performed than in the underlying theories. Usually analytic results are obtained in EFT calculations including the results presented in this thesis. The advantage of getting analytic results is that the expression is suitable for the whole momentum complex plane.

The development of nuclear EFT dates back to Weinberg's pioneering work in the early 90's [1]. He proposed an EFT with long distance (or low energy) physics described by chiral perturbation and short distance (or high energy) physics parameterized by nucleon-nucleon (NN) contact interactions. However, the unnaturally large scattering lengths ("unnatural" means the scattering length is not set by typical strong interaction scale, m_π or Λ_χ) in the S-wave channels complicate the power counting (counting the powers of the small expansion parameter associated with each Feynman diagram). The success in the single nucleon sector does not extend to the multi-nucleon sector straight forwardly.

In an elastic two body scattering process, the quantity that is natural to calculate in a field theory is the sum of Feynman graphs, which gives the amplitude $i\mathcal{A}$, related to the S-matrix by

$$S = 1 + i\frac{Mp}{2\pi}\mathcal{A} .$$ (1)

For S-wave scattering, $\mathcal{A}$ is related to the phase shift δ through the relation

$$\mathcal{A} = \frac{4\pi}{M}\frac{1}{p\cot\delta - ip} .$$ (2)

From quantum mechanics it is well known that it is not $\mathcal{A}$, but rather the quantity $p\cot\delta$, which has a nice momentum expansion for $p \ll \Lambda$ (the effective range expansion):

$$p\cot\delta = -\frac{1}{a} + \frac{1}{2}\Lambda^2 \sum_{n=0}^{\infty} r_n \left(\frac{p^2}{\Lambda^2}\right)^{n+1} .$$ (3)

In the spin triplet channel, the scattering length $a = -23.714\pm0.013$ fm, the effective range $r_0 = 2.73$ fm, and the shape parameter $r_1 = -0.48$ fm^3. The coefficients r_n are generally $O(1/\Lambda)$ for all n, but a can take on any value. In NN scattering $|a| \gg 1/\Lambda$, $|r_n| \sim 1/\Lambda$. If one performs a p expansion to the $\mathcal{A}$, as in chiral perturbation, one has

$$\begin{aligned}\mathcal{A} &\simeq -\frac{4\pi}{M}\frac{1}{\frac{1}{a}+ip}\left[1 + \mathcal{O}(\frac{1}{a\Lambda})\right]\\ &= -\frac{4\pi a}{M}\left[1 - iap + (-iap)^2 + \cdots\right].\end{aligned}$$ (4)

However, the radius of convergence is $1/|a| \ll \Lambda$, which is not very useful. Alternatively, if the terms in Eq.(4) with powers of $|ap|$ can be resummed, then the amplitude becomes

$$\mathcal{A} = -\frac{4\pi}{M}\frac{1}{(1/a+ip)}\left[1 + \frac{r_0/2}{(1/a+ip)}p^2 + \frac{(r_0/2)^2}{(1/a+ip)^2}p^4 + \frac{(r_1/2\Lambda^2)}{(1/a+ip)}p^4 + \cdots\right]$$ (5)

This is the desired expansion because the radius of convergence is extended from $1/a$ to Λ.

The above expression for the amplitude is clearly a non-perturbative one because there is a pole in the momentum plan. It is known that the sum of the bubble diagrams of Fig.1 gives rise to the Schrodinger equation with the bubble identified as the potential. If one uses delta functions or the derivatives of delta functions as the potential, the result is just the effective range expansion developed by Bethe in the 60's. In the early 90's, Weinberg developed a power counting scheme (W counting) to power count the potentials using the chiral perturbation theory power counting rules. The leading order (LO) potential contains the one pion exchange and the nonderivative contact interaction diagrams shown in Fig 2. However, potentials

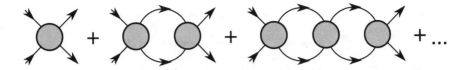

Figure 1: The bubble chain giving rise to the Schrodinger equation. The gray blob can be interpreted as the potential.

Figure 2: Diagrams of the LO potential in Weinberg's power counting.

are not physical observables. It makes more sense to power count the amplitudes than the potentials. For the LO diagram in the W counting shown in Fig. 3, the diagram is proportional to m_π^2/ϵ. However, the counterterm needed to regularize exists at higher order. This inconsistency drove Kaplan, Savage and Wise (KSW) to devise their power counting scheme in the late 90's. In the KSW power counting, the LO diagrams for NN scattering contain bubble diagrams with insertions of non-derivative contact interactions (see Fig. 4). The one pion exchange is treated as a perturbation. It contributes at the next-to-leading order (NLO). Fig. 5 gives all the NLO diagrams in the KSW power counting.

When the typical momentum scale p in the problem is much smaller than the pion mass $m_\pi (\simeq 140$ MeV$)$, pions do not need to be treated as dynamical particles. This is because they only propagate over distances $\sim 1/m_\pi$, much shorter than the scale set by the typical momentum of the problem. Thus the pionless nuclear effective field theory, EFT($\not\pi$) [2–7], is applicable when $p \ll m_\pi$. In this theory, the dynamical degrees of freedom are nucleons and non-hadronic external currents. Massive hadronic excitations such as pions and the delta resonance are "integrated out," resulting in contact interactions between nucleons. Nucleon-nucleon interactions are

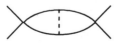

Figure 3: A LO amplitude in Weinberg's power counting.

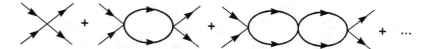

Figure 4: Graphs for the LO amplitude in the KSW power counting.

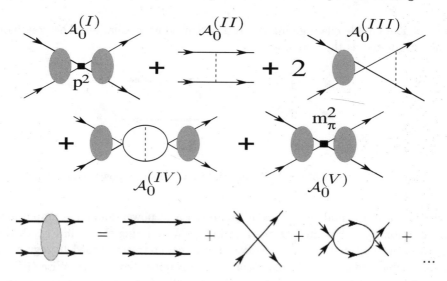

Figure 5: Graphs for the NLO amplitude in the KSW power counting.

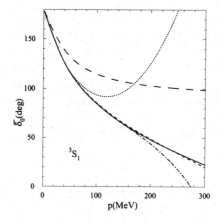

Figure 6: The NNLO $^3S_1$ phase shift for NN scattering calculated in Ref. [19]. The solid line is the Nijmegen multi-energy fit, the long dashed line is the LO effective field theory result, the short dashed line is the NLO result, and the dotted line is the NNLO result.

118

calculated perturbatively with the small expansion parameter

$$Q \equiv \frac{(1/a, \gamma, p)}{\Lambda} \qquad (6)$$

which is the ratio of the light to heavy scales. The light scales include the inverse S-wave nucleon-nucleon scattering length $1/a(\lesssim 12$ MeV$)$ in the $^1S_0$ channel, the deuteron binding momentum $\gamma(= 45.7$ MeV$)$ in the $^3S_1$ channel, and the typical nucleon momentum p in the center-of-mass frame. The heavy scale Λ is set by the pion mass m_π. This formalism has been applied successfully to many processes involving the deuteron [7, 8], including Compton scattering [9, 10], $np \rightarrow d\gamma$ for big-bang nucleosynthesis [11, 12], νd scattering for SNO physics [13], the solar pp fusion process [14, 15], and parity violating observables [16]. Also studies on three-nucleon systems [17] have revealed highly non-trivial renormalizations associated with three body forces in the $s_{1/2}$ channel (e.g., ^{3}He and the triton). For other channels, precision calculations were carried out to higher orders [6, 18]. Although the small expansion parameter in EFT($\not\pi$) is only $\sim 1/3$, one can carry out the calculations to high orders and get $\sim 1\%$ accuracy. This theory is by now well established. There is no significant difference between the KSW and W power counting schemes. The difference is of higher order.

At higher energies, pions should be included as a dynamical degree of freedom. However, when carried out to NNLO, the theoretically sound KSW power counting does not give convergent result for the $^3S_1$ NN scattering phase shift [19] (see Fig. 6) while the W power counting is stil going strong even though there ware doubts about its consistency with chiral symmetry. The puzzle was solved in Ref. [20], where it was found that for $^1S_0$ channel, KSW counting is consistent with chiral symmetry, but the W counting is not. However, in the $^3S_1$ channel, one can resum the tensor part of the one pion exchange potential, which is quite singular, to get better numerical convergence and still being consistent with chiral symmetry. The technique they used is to see whether a mass dependent square well potential is required at LO to absorbed the cut-off (the range of the square well potential) dependence. It was found that the non-perturbative regularization gives qualitatively different result with the KSW power counting in the $^3S_1$ channel and saved the W counting. Also, even the W counting is formally inconsistent with chiral symmetry in the $^1S_0$ channel, the violation is numerically small—the pion mass dependent potential required at LO to absorb the cut-off dependence is numerically much smaller than the pion mass independent one. This also explains why the W counting is numerically successful on both channels.

In conclusion, when $p \ll m_\pi$, EFT($\not\pi$) works. When $p \gtrsim m_\pi$, the we need the KSW power counting for the $^1S_0$ channel and Weinberg's power counting for the $^3S_1$ channel. This is called the BBSvK counting.

2 Nuclear Physics on the Lattice

2.1 Lattice QCD ($0 < a_{latt} < 1/\Lambda_{QCD}$)

Due to the limitation of computer resources, lattice QCD calculations are usually carried out with finite lattice spacing $a_{latt} \neq 0$, finite space-time volume (V_4 finite) and with light quark masses heavier than their physical values ($m_\pi^{latt} > m_\pi^{phys}$). It is crucial to perform the $a_{latt} \to 0$, $V_4 \to \infty$, and $m_\pi^{latt} \to m_\pi^{phys}$ extrapolations. It is by now widely appreciated that EFT can provide good theoretical guidelines for these extrapolations.

For example, for Wilson fermions, the LO finite a_{latt} correction comes from the dimension five operator

$$\delta\mathcal{L} = c a_{latt}\overline{\psi}\sigma^{\mu\nu}G_{\mu\nu}\psi. \tag{7}$$

This operator breaks chiral symmetry in the same way as the quark mass matrix. Thus, it is very easy to construct the corresponding finite a_{latt} effects in the corresponding chiral lagrangian [21, 22]:

$$\delta\mathcal{L}_{ChPT} = Btr[m_q\Sigma^\dagger + h.c.] + W a_{latt}tr[\Sigma^\dagger + h.c.], \tag{8}$$

where we have only shown the leading finite a_{latt} effect.

For finite volume effects, the effects do not change the UV physics. Thus, the corresponding low energy EFT will have the same set of counterterms. The only effect in computing the diagrams is to replace the loop integrals by discrete momentum sums:

$$\int dp \to \sum_i p_i, \tag{9}$$

with, e.g., $p_i = 2\pi n_i/L$ for periodic boundary conditions. The volume dependence for physical quantities should vanish in the infinite volume limit. Typically, the volume dependence goes like $e^{-m_\pi L}/L^n$ or $1/L^n$, where L is the length of the four dimensional box.

For the quark mass dependence, chiral perturbation theory is a natural tool to use. In lattice QCD, it is also useful to use different masses for the valence quarks (quarks connected to external legs) and sea quarks (quarks not connected to external legs). These effects can also be implemented in chiral perturbation theory to give the so called quenched [23] or partially quenched [24, 25] chiral perturbation theory.

To study nuclear physics, one would like to put more than one nucleon on the lattice and measure the particle scattering phase shifts and effective range parameters. However, one immediately realizes it might not be a good idea to study scattering processes on a Euclidean lattice. This is because if a state created at one space-time point, and when the state is propagating in Euclidean time, the contributions from higher energy states will be eventually suppressed exponentially. Thus, what one could measure is just the properties at the kinematic threshold. However, in this case, the finite volume effect saves the day. With a finite space-time box, the momentum eigenstates are quantized which will allow extractions of scattering phase shifts away from the kinematic threshold.

Shown in Fig. 7 is the $I = 2$ $\pi\pi$ scattering length a_2 calculated by the NPLQCD group [26]. The extrapolation using the NLO ChPT formula has a very small error at the physical pion mass. This is because $m_\pi a_2$ is exactly zero in the chiral limit in ChPT. Thus, the error band becomes smaller as it moves toward the chiral limit. However, this is incorrect. One should not expect $m_\pi a_2 = 0$ in the chiral limit in this calculation. This is because different versions of fermion discretization methods are used for the sea and valence quarks in this calculation. Thus, while the valence quarks still have nice chiral symmetry on the lattice, the sea quarks (called staggered quarks) break chiral symmetry at $\mathcal{O}(a^2)$. The corresponding ChPT can be constructed to study the $m_\pi^{latt} \to m_\pi^{phys}$ and $a_{latt} \to 0$ extrapolations. The result of Ref. [27, 28] shows that if one uses the m_π and f_π measured on the lattice to express $m_\pi a_2$, then at one loop level, the a_{latt} dependent counterterm will not show up. In fact, the number of counterterms is the same as that in ChPT. The finite a_{latt} effect will shift the $m_\pi a_2$ curve such that $m_\pi a_2 \neq 0$ in the chiral limit. However, the shift is very small. For example, for the lightest lattice point the shift is only a few percent. Thus the previous extrapolation using ChPT is still very robust.

To study NN scattering, one can use EFT($\not{\pi}$) to extract the effective range parameters from lattice QCD data. However, this requires $|\mathbf{p}|^{max} < m_\pi/2$ for EFT($\not{\pi}$)

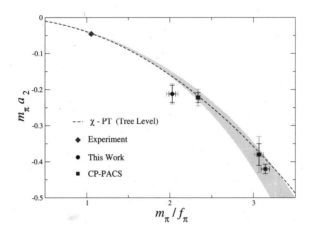

Figure 7: The $\pi\pi$ scattering length in the $I = 2$ channel as a function of m_π^2/f_π^2. The dashed curve corresponds to the unique prediction of tree-level χPT, while the shaded region is the fit to the results of the NPLQCD calculations [26]. The NPLQCD calculations are performed at a single lattice spacing of $a_{latt} \sim 0.125$ fm.

to be valid. This implies the box size L should be $\gtrsim 10$ fm with physical m_π [29], which is quite demanding. However, currently available field configurations typically have $m_\pi^{latt} \geq 300$ MeV. With such quark masses, the scattering length is of natural size. Thus, $L = 2.6$ fm is enough to measure the scattering length . Fig. 8 shows the scattering length measurements for the $^1S_0$ and $^3S_1$ channels respectively. The extrapolations are done using the lattice points with $m_\pi^{latt} = 350$ MeV and the physical data. With future measurements at smaller m_π^{latt}, one would eventually be able to answer whether the deuteron is bounded or not in the chiral limit.

2.2 Nuclear EFT on the Lattice ($1/\Lambda_\chi < a_{latt} < 1/m_\pi$)

When it comes to larger nuclei, direct computations of nuclei properties with lattice QCD will be difficult. In that case, putting nuclear EFT on the lattice becomes an interesting alternative. The major obstacle to this approach would be the so called fermion sign problem, which means the fermion determinant is not positive definite. This will make the Monte Carlo method which was used for important sampling no longre useful. In EFT(π), the LO nuclear EFT has an approximate SU(4) spin isospin symmetry. In the SU(4) limit, the action does not have the fermion sign problem. Using this property, some general inequalities can be proved based on the positive definite property of the fermion determinant. Including the pion will reintroduce the sign problem. However, the study of Ref. [31] suggests that the problem is relatively mild.

Here we use the example of inequalities proved for the LO EFT(π) action to illustrate how the nuclear EFT can be put on the lattice.

In the non-relativistic limit and below the threshold for pion production, we can write the lowest order terms in the effective Lagrangian as

$$\mathcal{L} = \bar{N}[i\partial_0 + \frac{\vec{\nabla}^2}{2m_N} - (m_N^0 - \mu)]N - \frac{1}{2}C_S\bar{N}N\bar{N}N$$
$$- \frac{1}{2}C_{odd}\left[\bar{N}\vec{\sigma}N \cdot \bar{N}\vec{\sigma}N - \bar{N}\vec{\tau}N \cdot \bar{N}\vec{\tau}N\right], \tag{10}$$

where N represents the nucleon fields. We use $\vec{\tau}$ to represent Pauli matrices acting in isospin space and $\vec{\sigma}$ to represent Pauli matrices acting in spin space. We assume exact isospin symmetry. We have neglected three-nucleon terms for now but will consider them later. We have written the Lagrangian so that the operator multiplying C_{odd} flips sign under the exchange of isospin and spin degrees of freedom.

We see that C_{odd} is much smaller in magnitude than C_S. In the limit $C_{odd} \to 0$ the $SU(2) \times SU(2)$ spin-isospin symmetry is elevated to an $SU(4)$ symmetry. This symmetry was first studied by Wigner [32–34], and arises naturally in the limit of large number of colors, N_c [35, 36]. Although the $^1S_0$ and $^3S_1$ scattering lengths are

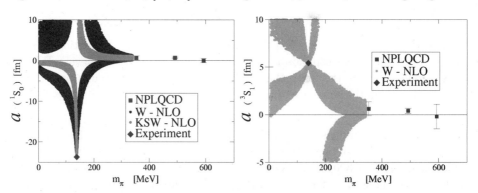

Figure 8: Scattering lengths in the $^1S_0$ and $^3S_1 - {}^3D_1$ NN channels as a function of pion mass. The experimental value of the scattering length and NDA have been used to constrain the extrapolation in both BBSvK and W power-countings at NLO.

quite different, the fact that both scattering lengths are large suggests we are close to the Wigner limit [37].

When $C_{odd} = 0$ the grand canonical partition function is given by

$$Z_G = \int DN D\bar{N} \exp\left(-S_E\right) = \int DN D\bar{N} \exp\left(\int d^4x\, \mathcal{L}_E\right), \qquad (11)$$

where

$$\mathcal{L}_E = -\bar{N}[\partial_4 - \tfrac{\vec{\nabla}^2}{2m_N} + (m_N^0 - \mu)]N - \tfrac{1}{2}C_S \bar{N}N\bar{N}N. \qquad (12)$$

Using Hubbard-Stratonovich transformations, we can rewrite Z_G as

$$Z_G \propto \int DN D\bar{N} Df \exp\left(\int d^4x\, \mathcal{L}_E^f\right), \qquad (13)$$

where

$$\mathcal{L}_E^f = -\bar{N}[\partial_4 - \tfrac{\vec{\nabla}^2}{2m_N} + (m_N^0 - \mu)]N + C_S f \bar{N}N + \tfrac{1}{2}C_S f^2. \qquad (14)$$

Since $C_S < 0$, the f integration is convergent.

Let M be the nucleon matrix. M has the block diagonal form;

$$M = M_{\text{block}} \oplus M_{\text{block}} \oplus M_{\text{block}} \oplus M_{\text{block}}, \qquad (15)$$

where we have one block for each of the four nucleon states and

$$M_{\text{block}} = -\left[\partial_4 - \tfrac{\vec{\nabla}^2}{2m_N} + (m_N^0 - \mu)\right] + C_S f. \qquad (16)$$

We note that M is real valued and therefore $\det M \geq 0$.

Consider the two-nucleon operator $A_2(x) = [N]_i[N]_j(x)$, where $i \neq j$ are indices for two different nucleon states. The two-point correlation function for A_2 is

$$\left\langle A_2(x) A_2^\dagger(0) \right\rangle_{\mu,T} = \langle [N]_i[N]_j(x)\, [N^*]_j[N^*]_i(0) \rangle_{\mu,T}. \qquad (17)$$

Using our Euclidean functional integral representation, we have

$$\left\langle A_2(x) A_2^\dagger(0) \right\rangle_{\mu,T} = \int D\Theta\; \left[M_{\text{block}}^{-1}(x,0)\right]^2 \qquad (18)$$

where $D\Theta$ is the positive normalized measure defined by

$$D\Theta = \frac{Df\, \det M \exp\left(\tfrac{1}{2}C_S \int d^4x\, f^2\right)}{\int Df\, \det M \exp\left(\tfrac{1}{2}C_S \int d^4x\, f^2\right)}. \qquad (19)$$

We note that since M_{block} is real valued, M_{block}^{-1} is also real valued.

Next we consider the three-nucleon and four-nucleon operators $A_3(x) = [N]_i[N]_j[N]_k(x)$ and $A_4(x) = [N]_i[N]_j[N]_k[N]_l(x)$, where i, j, k, l are all distinct. We have

$$\left\langle A_3(x) A_3^\dagger(0) \right\rangle_{\mu,T} = \int D\Theta\; \left[M_{\text{block}}^{-1}(x,0)\right]^3, \qquad (20)$$

$$\left\langle A_4(x) A_4^\dagger(0) \right\rangle_{\mu,T} = \int D\Theta\; \left[M_{\text{block}}^{-1}(x,0)\right]^4. \qquad (21)$$

123

We note that

$$\int D\Theta \; \left| M_{\text{block}}^{-1}(x,0) \right|^3 = \int D\Theta \; \left| M_{\text{block}}^{-1}(x,0) \right| \left[M_{\text{block}}^{-1}(x,0) \right]^2$$

$$\leq \sqrt{\int D\Theta \; \left[M_{\text{block}}^{-1}(x,0) \right]^2} \sqrt{\int D\Theta \; \left[M_{\text{block}}^{-1}(x,0) \right]^4}, \quad (22)$$

where the second line is from the Cauchy-Schwarz inequality. Therefore

$$\left| \left\langle A_3(x) A_3^\dagger(0) \right\rangle_{\mu,T} \right| \leq \sqrt{\left\langle A_2(x) A_2^\dagger(0) \right\rangle_{\mu,T} \left\langle A_4(x) A_4^\dagger(0) \right\rangle_{\mu,T}}. \quad (23)$$

Let E_{A_2} be the energy of the lowest state that couples to A_2, and E_{A_4} be the energy of the lowest state that couples to A_4. Taking the limit $x \to \infty$ in the temporal direction we conclude that any state with the quantum numbers of A_3 must have energy less than the average of E_{A_2} and E_{A_4},

$$E_{A_3} \geq \tfrac{1}{2} \left[E_{A_2} + E_{A_4} \right]. \quad (24)$$

In the real world C_{odd} is small but nonzero. We can measure the shift in the energy of a given state $|A\rangle$ using first-order perturbation theory, $\Delta E_A = \langle A| H' |A\rangle$, where

$$H' = \tfrac{1}{2} C_{odd} \int d^3\vec{x} \; \left[\bar{N} \vec{\sigma} N \cdot \bar{N} \vec{\sigma} N - \bar{N} \vec{\tau} N \cdot \bar{N} \vec{\tau} N \right]. \quad (25)$$

We can now adjust for the first-order energy corrections due to H',

$$E_{^3\text{He}}, E_T \geq \tfrac{1}{2} \left[\tfrac{1}{2} \left(E_D + E_{^1S_0} \right) + E_{^4\text{He}} \right]. \quad (26)$$

The physical binding energies are shown in Table 2 [38].

Table 2: Binding energies for light nuclides

$^1S_0$	~ 0 MeV (nearly bound)
D	-2.224 MeV
^{3}He	-7.718 MeV
T	-8.481 MeV
^{4}He	-28.296 MeV

Plugging these values into (26), we find that the inequality is satisfied, -7.7 MeV, -8.5 MeV ≥ -14.7 MeV. An analogous relation can be derived:

$$\tfrac{1}{2} \left(E_D + E_{^1S_0} \right) \geq \tfrac{1}{2} E_{^4\text{He}}. \quad (27)$$

This is also satisfied, -1.1 MeV ≥ -14.1 MeV.

Up to this point we have ignored three and four-nucleon forces. It has been shown that the dominant three-nucleon force is Wigner-symmetric [39, 40]. We now show that introducing Wigner-symmetric three and four-nucleon forces do not spoil positivity of the Euclidean functional integral so long as the three-nucleon force is not too strong and the four-nucleon force is not too repulsive. We want to find a Hubbard-Stratonovich transformation that reproduces a contribution to the action of the form,

$$\prod_x \exp\left[c_2[\bar{N}N(x)]^2 + c_3[\bar{N}N(x)]^3 + c_4[\bar{N}N(x)]^4\right]. \tag{28}$$

Let us just concentrate on what happens at a single point x, and in our notation we suppress writing the x explicitly. We note that $\bar{N}N$ raised to any power greater than 4 mush vanish. So we have

$$\exp\left[c_2\left(\bar{N}N\right)^2 + c_3\left(\bar{N}N\right)^3 + c_4\left(\bar{N}N\right)^4\right]$$

$$= a_0 + a_1\bar{N}N + \tfrac{a_2}{2!}\left(\bar{N}N\right)^2 + \tfrac{a_3}{3!}\left(\bar{N}N\right)^3 + \tfrac{a_4}{4!}\left(\bar{N}N\right)^4, \tag{29}$$

$$a_0 = 1, \qquad a_1 = 0, \qquad a_2 = 2c_2, \qquad a_3 = 6c_3, \qquad a_4 = 12c_2^2 + 24c_4. \tag{30}$$

We now to try find a real function $g(f)$ such that

$$\int_{-\infty}^{\infty} df\, \exp\left[f\bar{N}N + g(f)\right] = a_0 + a_1\bar{N}N + \tfrac{a_2}{2!}\left(\bar{N}N\right)^2 + \tfrac{a_3}{3!}\left(\bar{N}N\right)^3 + \tfrac{a_4}{4!}\left(\bar{N}N\right)^4. \tag{31}$$

We observe that a Hubbard-Stratonovich transformation of this form maintains the positive functional integral measure. Expanding the left-hand side, we have

$$a_n = \int_{-\infty}^{\infty} df\, f^n \exp\left[g(f)\right], \quad n = 0, 1, 2, 3, 4. \tag{32}$$

Finding sufficient and necessary conditions for the existence of $g(f)$ is known in the mathematics literature as the truncated Hamburger moment problem. This problem has been solved [41][42], and in our case $g(f)$ exists if and only if the so-called block-Hankel matrix,

$$\begin{bmatrix} a_0 & a_1 & a_2 \\ a_1 & a_2 & a_3 \\ a_2 & a_3 & a_4 \end{bmatrix} = \begin{bmatrix} 1 & 0 & 2c_2 \\ 0 & 2c_2 & 6c_3 \\ 2c_2 & 6c_3 & 12c_2^2 + 24c_4 \end{bmatrix}, \tag{33}$$

is positive semi-definite, with the added condition that if $c_2 = 0$ then $c_4 = 0$. The determinant of this matrix is $16c_2^3 - 36c_3^2 + 48c_2c_4$. With an attractive two-nucleon force and small three and four-nucleon forces, the conditions are clearly satisfied. Whether or not these conditions are satisfied in the real world and at which lattice spacings is beyond the scope of this letter. But hopefully this will be numerically determined in the near future.

125

3 Aspects of the Nuclear Many Problems from EFT

It is well known that lattice QCD with finite fermion number chemical potential has a fermion sign problem, meaning the fermion determinant is not positive definite. This will make the Monte Carlo method which was used for important sampling no longre useful. There are no practical ways to solve this problem yet. So it is still impossible to use lattice QCD to study phases with color super conductivity. In the hadronic phase, however, one could hope that the nuclear matter problems can be studied by lattice nuclear EFT as discussed in the previous section. Work in this direction is still limited. In the section I will discuss two problems in many body systems using analytic approaches.

3.1 QCD Viscosity to Entropy Density Ratio in the Hadronic Phase

3.1.1 Zero baryon number density

Shear viscosity η characterizes how strongly particles interact and move collectively in a many body system. In general, strongly interacting systems have smaller η than the weakly interacting ones. This is because η is proportional to $\epsilon \tau_{mft}$, where ϵ is the energy density and τ_{mft} is the mean free time, which is inversely proportional to particle scattering cross section. Recently a universal minimum bound for the ratio of η to entropy density s was proposed by Kovtun, Son, and Starinets [43]. The bound,

$$\frac{\eta}{s} \geq \frac{1}{4\pi} \,, \tag{34}$$

is found to be saturated for a large class of strongly interacting quantum field theories whose dual descriptions in string theory involve black holes in anti-de Sitter space [44–47].

Recently, η/s close to the minimum bound were found in relativistic heavy ion collisions (RHIC) [48–50]. This discovery came as a surprise. Traditionally, quark gluon plasma (QGP)—the phase of QCD above the deconfinement temperature $T_c (\sim 170$ MeV at zero baryon density [51])—was thought to be weakly interacting. Partly because lattice QCD simulations of the QGP equation of state above $2T_c$ were not inconsistent with that of an ideal gas of massless particles, $e = 3p$, where e is the the energy density and p is the pressure of the system [51]. However, recent analyses of the elliptic flow generated by non-central collisions in RHIC [49, 50] and lattice simulations of a gluon plasma [52] yielded η/s close to the the minimum bound at just above T_c. This suggests QGP is strongly interacting at this temperature. (However, see Ref. [53] for a different interpretation.)

Given this situation, one naturally wonders if η/s of QCD was already close to the minimum bound at just above T_c, what would happen if we keep reducing the temperature such that the coupling constant of QCD gets even stronger? Will the η/s minimum bound hold up below T_c? If the bound does hold up, what is the mechanism? Is the change of degrees of freedom through a phase transition or cross over sufficient to save the bound? If the bound does not hold up, what is the implication to string theory?

To explore these issues, we use chiral perturbation theory (χPT) and the linearized Boltzmann equation to perform a model independent calculation to the η/s of QCD in the confinement phase [54]. Earlier attempts to compute meson matter viscosity using the Boltzmann equation and phenomenological phase shifts in the context of RHIC hydrodynamical evolution after freeze out can be found in Refs. [55–57]. In the deconfinement phase, state of the art perturbative QCD calculations of η can be found in Refs. [58, 59].

In the hadronic phase of QCD with zero baryon-number density, the dominant degrees of freedom are the lightest hadrons—the pions. The pion mass $m_\pi = 139$ MeV is much lighter than the mass of the next lightest hadron—the kaon whose mass is 495 MeV. Given that T_c is only ~ 170 MeV, it is sufficient to just consider the pions in the calculation of thermodynamical quantities and transport coefficients for $T \ll T_c$.

The interaction between pions can be described by chiral perturbation theory (χPT) in a systematic expansion in energy and quark (u and d quark) masses. χPT is a low energy effective field theory of QCD. It describes pions as Nambu-Goldstone bosons of the spontaneously broken chiral symmetry. At $T \ll T_c$, the temperature dependence in $\pi\pi$ scattering can be calculated systematically. At $T = T_c$, however, the theory breaks down due to the restoration of chiral symmetry.[1]

The shear viscosity η of the pion gas can be calculated either using the Boltzmann equation or the Kubo formula. Since the Boltzmann equation requires semi-classical descriptions of particles with definite position, energy and momentum except during brief collisions, the mean free path is required to be much greater than the range of interaction. Thus the Boltzmann equation is usually limited to low temperature systems. The Kubo formula does not have this restriction. In this approach η can be calculated through the linearized response function

$$\eta = -\frac{1}{5} \int_{-\infty}^{0} dt' \int_{-\infty}^{t'} dt \int dx^3 \langle [T^{ij}(0), T^{ij}(\mathbf{x}, t)] \rangle \qquad (35)$$

with T^{ij} the spacial part of the off-diagonal energy momentum tensor. One might think a perturbative calculation of the above two point function will give the answer for η. But this can not be true if $\eta \propto \tau_{mft}$, as mentioned above, for $\tau_{mft} \to \infty$ in the

---

[1]The QCD chiral restoration temperature and the deconfinement temperature happen to be close to each other at zero baryon density. We do not distinguish the two in this paper.

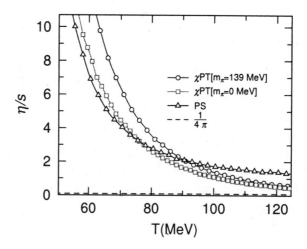

Figure 9: Shear viscosity to entropy density ratios as functions of temperature. Line with circles (rectangles) is the LO χPT result with $m_\pi = 139\,(0)$ MeV and $f_\pi = 93(87)$ MeV. Line with triangles is the result using $\pi\pi$ phase shifts (PS). Dashed line is the conjectured minimum bound $1/4\pi \simeq 0.08$.

free case. Indeed, the Kubo formula involves an infinite number of diagrams at the leading order (LO) [60]. However, in a weak coupling ϕ^4 theory, it is proven that the summation of LO diagrams is equivalent to solving the linearized Boltzmann equation with temperature dependent particle masses and scattering amplitudes [60]. This proof extended the applicable range of the Boltzmann equation to higher temperature but is restricted to weak coupling theories. In the case we are interested (QCD with $T < 140$ MeV), the pion mean free path is always greater than the range of interaction (~ 1 fm) by a factor of 10^3. Thus, even though the coupling in χPT is too strong to use the result of Ref. [60], the temperature is still low enough that the use of the Boltzmann equation is justified.

The calculation of the entropy density s is more straightforward since s, unlike η, does not diverge in a free theory. In χPT, the interaction contributions are all higher order in our LO calculation. Thus we just compute the s for a free pion gas:

$$s = -g_\pi \beta^2 \frac{\partial}{\partial \beta} \frac{\log Z}{\beta} \ , \tag{36}$$

where the partition function Z for free pions is

$$\frac{\log Z}{\beta} = -\frac{1}{\beta} \int \frac{d^3\mathbf{p}}{(2\pi)^3} \log\left\{1 - e^{-\beta E(p)}\right\} \ , \tag{37}$$

up to temperature independent terms.

The LO χPT result for η/s is shown in Fig. 9 (line with rectangles). The error is estimated to be $\sim 50\%$ up to 120 MeV. η/s is monotonically decreasing and reaches 0.6 at $T = 120$ MeV. This is similar to the behavior in the $m_\pi = 0$ case (shown as the line with rectangles) where $\eta/s \propto f_\pi^4/T^4$ with $s \propto T^3$ from dimensional analysis and $f_\pi = 87$ MeV in the chiral limit.

For comparison, we also show the result using phenomenological $\pi\pi$ phase shifts [61] for η but free pions for s. (Our result for η is in good agreement with that of [57] for T between 60 and 120 MeV. For an earlier calculation using the Chapman-Enskog approximation, see Ref. [62].) This amounts to take into account part of the NLO $\pi\pi$ scattering effects but ignore its temperature dependence and the interaction in s. Since not all the NLO effects are accounted for, this η/s is not necessarily more accurate than the one using LO χPT. The comparison, however, gives us some feeling of the size of error for the LO result we present here. Thus, an error of $\sim 100\%$ at $T = 120$ MeV for the LO result might be more realistic.

Naive extrapolations of the three η/s curves show that the $1/4\pi = 0.08$ minimum bound conjectured from string theory might never be reached as in phase shift result (the first scenario), or more interestingly, be reached at $T \sim 200$ MeV, as in the LO χPT result (the second scenario). In both scenarios, we see no sign of violation of the universal minimum bound for η/s below T_c. But to really make sure the bound is valid from 120 MeV to T_c, a lattice computation as was performed to gluon plasma above T_c [52] is needed. In the second scenario, assuming the bound is valid for QCD, then either a phase transition or cross over should occur before the minimum bound is reached at $T \sim 200$ MeV. Also, in this scenario, it seems natural for η/s to stay close to the minimum bound around T_c as was recently found in heavy ion collisions.

In the second scenario, one might argue that the existence of phase transition is already known, otherwise we will not have spontaneous symmetry breaking and the corresponding Nambu-Golstone boson theory at low temperature in the first place. Indeed, it is true in the case of QCD. However, if the η/s bound is really set by Nature, then a phase transition is inevitable in the vicinity of the temperature where the bound is reached. For a spontaneous symmetry breaking theory, the general feature of η/s we see here seems generic. At very high T, collective motion is weak, thus η/s gets smaller at lower T. At very low T in the symmetry breaking phase, the Nambu-Goldstone bosons are weakly interaction at low temperature, thus η/s gets smaller at higher T. A phase transition should occur before the extrapolated η/s curve coming from high T reaches the bound at T_1. Similarly, a phase transition should occur before the extrapolated η/s curve coming from low T reaches the bound at T_2. Thus the range of phase transition is $T_1 \leq T_c \leq T_2$. However, it is also possible that the first scenario takes place and η/s bounces back to higher values without a phase transition. In this case, it is less clear what makes η/s non-monotonic and it certainly deserves further study.

It is interesting to note that the degeneracy factor g_π drops out of η while the entropy s is proportional to g_π. This suggests the η/s bound might be violated if a system has a large particle degeneracy factor. For QCD, large g_π can be obtained by having a large number of quark flavors N_f with $g_\pi \sim N_f^2$. However, the existence of confinement demands that the number of colors N_c should be of order N_f to have a negative QCD beta function. After using $f_\pi \propto \sqrt{N_c}$, the combined N_c and N_f scaling of η/s is

$$\frac{\eta}{s} \propto \frac{f_\pi^4}{g_\pi T^4} \propto \frac{N_c^2}{N_f^2} , \qquad (38)$$

which is of order one. Thus QCD with large N_c and N_f can still be consistent with the η/s bound below T_c.

3.1.2 Finite baryon number density: η/s and the QCD Phase Diagram

In this subsection, we extend the discussion of the η/s of QCD in the confinement phase at zero baryon chemical potential μ to finite μ [63] and study its relation to the QCD phase diagram. String theory methods give $\eta/s = 1/4\pi$ for $\mathcal{N} = 4$ supersymmetric theories with finite R-charge density, suggesting the minimum bound is independent of μ. For QCD, the fermion sign problem (the fermion determinant is not positive definite) makes the current lattice QCD methods inapplicable in the low T and finite μ regime. An alternative is to use effective field theory (EFT). Reliable results using EFT in hadronic degrees of freedom can be obtained when both T and μ are small. At higher μ (with $|k_F a| \gg 1$, where k_F denotes the fermi momentum and a is the nucleon-nucleon scattering length) the problem becomes non-perturbative in coupling and mean-field treatments are not sufficient. (It is essentially the same type of problem as in cold fermionic atoms near the infinite scattering length limit.) Non-perturbative computations of the EFT on the lattice is free from the fermion sign problem at the leading order (LO) with only non-derivative contact interactions [64]. But this theory is suitable only in low T and low density systems. For the nuclear matter problem, the inclusion of one pion exchange will re-introduce the sign problem. However, the sign problem is claimed to be mild and lattice simulations are still possible [31]. The computation of η using lattice nuclear EFT has not been carried out before. Although η is associated with real time response to perturbations, it can be reconstructed through the spectral function computed on the Euclidean lattice [52, 65].

As an exploratory work, we compute η using coupled Boltzmann equations for a system of pion π and nucleon N while the entropy s is computed only for free particles. This approach will not give accurate η/s in the regime dominated by near threshold NN interaction. However, for most of the regime we are exploring, our result should be robust.

In Fig. 11, η/s as a function of T and μ is shown as a 3-D plot and a contour plot. Note that at the corner of large μ and large T, the system is no longer in

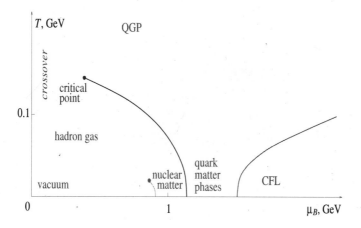

Figure 10: A semi-quantitative sketch of the QCD phase diagram [66].

the hadronic phase. Thus, the result should be discarded there. In general η/s is decreasing in T except when $\mu \simeq m_N$ and $T < 30$ MeV. (This regime is blown up in Fig. 12 and will be studied later.) There are some interesting structures at larger μ, but η/s is μ independent when $\mu < 500$ MeV . This is because when $\mu \ll m_N$ the nucleons only exist through particle-antiparticle pair creations, thus they are highly suppressed. The η/s is determined by the pion gas which is μ independent. Our result just reproduces the $\mu = 0$ result of Ref. [54] (see [55–57, 62] for earlier results) in this regime.

When $\mu > m_N - m_\pi = 800$ MeV, the nucleon population is no longer suppressed compared with the pion population and when $\mu \gtrsim m_N$ the nucleons become the dominant degrees of freedom. Numerically η/s is dominated by the nucleon contributions when $\mu > 800$ MeV. It is decreasing in both T and μ until $\mu \simeq m_N$. This is because s is increasing in both T and μ while η is getting smaller at higher μ (larger nucleon population) and lower T (stronger interaction, closer to the interaction threshold). 500-800 MeV in μ is the transition between the π and N dominant regimes.

Now let us focus on the $\mu \simeq m_N$ and $T < 30$ MeV region in the η/s plot. Two 3-D plots viewed from different angles are shown in Fig. 12(a) and 12(b) and a contour plot is shown in Fig. 12(c). One clearly sees that η/s maps out the nuclear gas-liquid phase transition shown in Fig.1 by forming a valley tracing the nuclear gas-liquid phase transition line in the T-μ plane. When the phase transition turns into a crossover at larger T, the valley also gradually disappears at around 30 MeV. This result is encouraging.

However, even though the gross features of the phase transition are mapped out by the η/s valley nicely, some details are not correct. First, since the density is discontinuous across the first-order phase transition, η, s, and η/s are likely to be

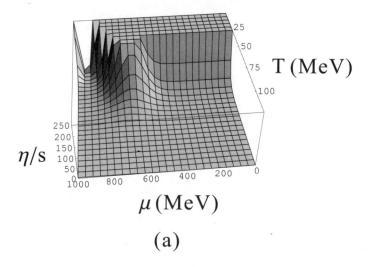

(a)

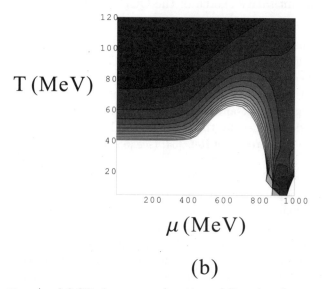

(b)

Figure 11: η/s of QCD shown as a function of T and μ shown as a 3-D plot (a) and a contour plot (b). Note that the corner of large μ and large T should be discarded since it is not in the hadronic phase.

discontinues across the phase transition as observed in H_2O, He and N systems in Ref. [67]. This discontinuity, which defines the critical chemical potential μ_c, should lie at the bottom of the η/s valley. Second, the position of μ_c suggested by our result is not correct. Near $T = 0$, one expects $\mu_c \simeq m_N - \langle B \rangle$, where $\langle B \rangle$ is the binding energy per nucleon, but we have $\mu_c > m_N$.

It is quite obviously that our free particle treatment of s is very poor near the phase transition. However, we have not pursued other treatments like the mean field approximation in this work because it is known that the approximation is insufficient when $|k_F a| \gg 1$. For the same reason, the computation of η using the Boltzmann equation might not be justified near the bottom the valley even though the mean free path is still bigger than the range of potential (~ 1 fm). However, the valley of η is located at $\mu < m_N$ near $T = 0$; thus, it is possible that after reliable s is used, μ_c for η/s will be in the correct position. Furthermore, the regime in Fig. 3 is completely dominated by the nucleon degree of freedom (η/s hardly changes with the thermal pions completely ignored). This simplifies the problem significantly and makes the system exhibit universal properties shared by dilute fermionic systems with large scattering lengths such as cold atoms tuned to be near a Feshbach resonance.

As mentioned above, it was observed that below the critical pressure, η/s has a discontinuity at the critical temperature for H_2O, He and N, and above the critical pressure, η/s has a smooth minimum near the crossover temperature (defined as the temperature where the density changes rapidly) [67]. For QCD at $\mu = 0$, η/s also

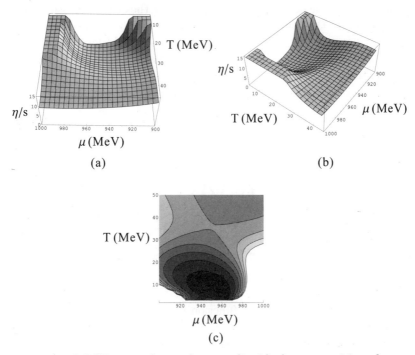

(a)

(b)

(c)

Figure 12: η/s of QCD near the nuclear gas-liquid phase transition shown as 3-D plots (a) and (b), viewed from different angles, and a contour plot (c). The η/s maps out the nuclear gas-liquid phase transition shown in Fig.1 by forming a valley tracing the nuclear gas-liquid phase transition line in the T-μ plane. When the phase transition turns into a crossover at larger T, the valley also gradually disappears at around 30 MeV. There should be a discontinuity that looks like a fault in the bottom of the η/s valley that is not seen in our approximation. The fault would lie on top of the phase transition line and end at the critical point ($T \sim 10$-15 MeV in our result). Beyond the critical point, η/s turns into a smooth valley. The valley could disappear far away from the critical point. Similar behavior is also seen in water shown in Fig. 4. We suspect these are general features for first-order phase transitions.

133

has a valley near the crossover temperature [54, 67]. But there is no evidence yet to show the valley is smooth or has a discontinuity.

We suspect η/s should be smooth in a crossover and should have a discontinuity across a first-order phase transition. If this is correct, then in nuclear gas-liquid transition, there should be a discontinuity looking like a fault in the bottom of the η/s valley. The fault would lie on top of the phase transition line and end at the critical point where the first-order phase transition turns into a crossover. If we look at Fig. 3(c), we would conclude that the critical point is at $T \sim 10$-15 MeV, agreeing with 7-16 MeV from experimental extractions [68–70]. Near the critical point, a smooth η/s valley is seen in the crossover (like the confinement-deconfinement crossover of QCD at $\mu = 0$); however, the valley could disappear far away from the critical point. We suspect these are general features of first-order phase transitions. Indeed, similar behavior is also seen in H_2O. The η/s of H_2O as a function of T and P is shown in Fig. 4. Below the critical pressure 22.06 MPa, η/s has a discontinuity at the bottom of the valley. Above the critical pressure, the valley becomes smooth and the bottom (minimum) of the valley moves toward the large T and large P direction. In a limited range of T, η/s could look like a monotonic function without a valley far above the critical pressure. Thus, one might use η/s measurements to identify the first-order phase transition and the critical point (see Refs. [67, 71] for a similar point of view).

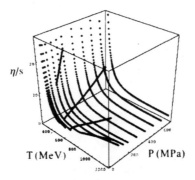

Figure 13: η/s of water shown as a function of temperature and pressure. Below the critical pressure 22.06 MPa, η/s has a discontinuity at the bottom of the valley. Above the critical pressure, the valley becomes smooth and the bottom (minimum) of the valley moves toward the large T and large P direction. In a limited range of T, η/s could look like a monotonic function without a valley far away from the critical pressure.

3.2 BEC-BCS Crossover in the ϵ Expansion

BEC-BCS crossover is a field attracting lots of attention recently [72–76]. The simplest systems to study the crossover are dilute Fermion systems with attractive interactions, for which the effective range for two-body scattering is much less than the inter-particle spacing. At zero temperature, such systems are characterized by a dimensionless number $\eta = 1/(ak_F)$, where a is the two-body scattering length, and k_F is the corresponding Fermi momentum in non-interacting systems. For η large and negative (weak attraction) one finds the BCS solution with pairing and superfluidity. With η large and positive, corresponding to strong attraction with a two-body bound state well below threshold, the bound pairs will Bose-Einstein condense (BEC). Experimentally, η can be changed between the BEC and BCS limits at will using the technique of Feshbach resonance [77–83] Theoretically, physical quantities are expected to be a smooth function of η, since there is no phase transition between the BEC and BCS limits. In each of these limits the behavior is nonperturbative, however, the effective interaction is weak; the system can be successfully described in a mean field approximation. On the other hand, dilute fermion systems near the unitarity limit (the $\eta = 0$ limit) require treatments beyond the mean field approximation such as numerical simulations [84–90].

Recently, Nishida and Son have proposed an analytical approach called "ϵ expansion" to deal with problems near the unitarity limit [91, 92]. In this approach one computes physical quantities in d spacial dimensions, then treats $\epsilon = 4 - d$ as a small expansion parameter, and then sets $\epsilon = 1$ at the end to obtain the three dimensional physics. It is similar to the dimensional expansion method used by Wilson to study critical exponents of second order phase transitions [93]. Nishida and Son's idea was inspired by the observation of Nussinov and Nussinov [94] that the ground state of a two-component fermion system in the unitarity limit is a free Bose gas for $d \geq 4$ (also see Refs. [95, 96] for simplification of certain diagrams with $d \rightarrow \infty$). It is because the two Fermion bound state (with zero binding energy) wave function behaves as $\Phi(r) \propto 1/r^{d-2}$, where r is the separation between the two Fermions. The probability integral $\int d^d r \, |\Phi(r)|^2$ has a singularity at $r \rightarrow 0$ when $d \geq 4$. Thus the bound state has zero size and will not interact with each other. This feature was explicitly implemented in the theory set up near the unitarity limit in Refs. [91, 92]. As we will see, away from the unitarity limit, one also expects the localization of the bound state wave function all the way to the BEC limit but not to the BCS limit.

Now the ϵ expansion has been applied to few-body scattering [97], polarized fermions (i.e., with uneven chemical potential between two fermion species) in the unitarity limit [98], and near the unitarity limit [92] to identify the "splitting point" in the phase diagram where three different phases meet [99]. The critical temperature in the unitarity limit has also been computed [100]. Recently, the first next-to-next-to-leading order (NNLO) calculation in the ϵ expansion was carried out for μ/ε_F in the unitarity limit, where μ is the Fermion chemical potential and ε_F is the Fermi energy in non-interacting systems [101]. Contrary to the nice convergence seen previously at next-to-leading order (NLO), the NNLO of μ/ε_F is as large as the leading order (LO) with an opposite sign. Even so, it is still premature to claim that the large NNLO correction cannot be tamed through reorganizing the series. Thus, we consider it is still worth while to explore the ϵ expansion at lower orders.

In our work, we apply the ϵ expansion to NLO to study the BEC-BCS transition with an arbitrary η. In the BEC side, an interesting relation between μ and the two fermion binding energy B is implemented:

$$2\mu + B = \mathcal{O}(\epsilon). \tag{39}$$

This comes from the chemical equilibrium between two fermions (with chemical potential 2μ) and one boson (with chemical potential $-B$) in a 4-d system where the bosons do not interact with each other. We also perform the resummation of the $(\epsilon \log \eta)^n /n!$ terms arising from $\eta^\epsilon - 1$. These logarithms are formally higher order in ϵ but numerically important near the unitarity limit ($\epsilon = 1$, $|\eta| \ll 1$), and near the BEC and BCS limits ($\epsilon = 1$, $|\eta| \gg 1$). For example, this resummation gives the linear term in μ/ε_F when $|\eta| \ll 1$:

$$\frac{\mu}{\varepsilon_F} = 0.475 - 0.707\eta - 0.5\eta^2. \tag{40}$$

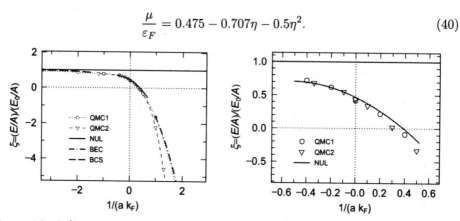

Figure 14: ξ (energy per particle relative to the value of a non-interacting fermi gas) shown as functions of $(ak_F)^{-1}$. QMC1 [84] and QMC2 [85] are Quantum Monte Carlo calculations with zero and finite range, respectively. NUL is our near unitarity limit result in the ϵ expansion. The right panel is the blow-up of the near unitarity regime.

Without this resummation, physical quantities are always even in η which would prevent us from reaching the BCS side.

In order to compare with the Quantum Monte Carlo (QMC) results [84, 85], μ/ε_F can be converted to energy per particle relative to the value of a non-interacting fermi gas

$$\xi = \frac{E/A}{E_0/A} \tag{41}$$

with $E_0/A = 3/5\varepsilon_F$ in 3-d. Near the unitarity limit $|\eta| \to 0$, we found $\mu/\varepsilon_F = 0.475 - 0.707\eta - 0.5\eta^2$. Near the BEC limit $\eta \to \infty$, $\mu/\varepsilon_F = 0.062/\eta - \eta^2$, while near the BCS limit $\eta \to -\infty$, $\mu/\varepsilon_F = 1 + 0.707/\eta$. This is compared with two sets of QMC results denoted as QMC1[84] and QMC2 [85] in Fig. 2. Overall good agreement with Quantum Monte Carlo results is found.

Acknowledgements

I thank the organizers for inviting me to such a wonderful workshop. It was also a remarkable and enjoyable experience for me to visit India, a truly remarkable country, for the first time. This work was supported by the NSC and NCTS of Taiwan, ROC.

References

[1] S. Weinberg, *Phys. Lett.* B **251**, 288 (1990); *Nucl. Phys.* B **363**, 3 (1991); *Phys. Lett.* B **295**, 114 (1992).

[2] D.B. Kaplan, M.J. Savage and M.B. Wise, *Nucl. Phys.* B**478**, 629 (1996).

[3] D.B. Kaplan, *Nucl. Phys.* B**494**, 471 (1997).

[4] U. van Kolck, hep-ph/9711222; *Nucl. Phys.* A**645** 273 (1999).

[5] T.D. Cohen, *Phys. Rev.* C**55**, 67 (1997); D.R. Phillips and T.D. Cohen, *Phys. Lett.* B**390**, 7 (1997); S.R. Beane, T.D. Cohen, and D.R. Phillips, *Nucl. Phys.* A**632**, 445 (1998).

[6] P.F. Bedaque and U. van Kolck, *Phys. Lett.* B**428**, 221 (1998).

[7] J.W. Chen, G. Rupak and M.J. Savage, *Nucl. Phys.* **A653**, 386 (1999).

[8] J.W. Chen, G. Rupak and M.J. Savage, *Phys. Lett.* **B464**, 1 (1999).

[9] S.R. Beane and M.J. Savage, *Nucl. Phys.* **A694**, 511 (2001).

[10] H.W. Griesshammer and G. Rupak, *Phys. Lett.* **B529**, 57 (2002) .

[11] J.W. Chen and M.J. Savage, *Phys. Rev.* C **60**, 065205 (1999).

[12] G. Rupak, *Nucl. Phys.* **A678**, 405 (2000).

[13] M.N. Butler and J.W. Chen, *Nucl. Phys.* **A675**, 575 (2000); M.N. Butler, J.W. Chen and X. Kong, *Phys. Rev.* C **63**, 035501 (2001).

[14] X. Kong and F. Ravndal, *Nucl. Phys.* **A656**, 421 (1999); *Nucl. Phys.* **A665**, 137 (2000); *Phys. Lett.* **B470**, 1 (1999); *Phys. Rev.* C **64**, 044002 (2001).

[15] M. Butler and J.W. Chen, *Phys. Lett.* **B520**, 87 (2001).

[16] M.J. Savage, *Nucl. Phys.* **A695**, 365 (2001).

[17] P.F. Bedaque, H.W. Hammer and U. van Kolck, *Phys. Rev. Lett.* **82**, 463 (1999); *Nucl. Phys.* A **676**, 357 (2000); H.-W. Hammer and T. Mehen, *Phys. Lett.* B **516**, 353 (2001).

[18] P.F. Bedaque, H.-W. Hammer and U. van Kolck, *Phys. Rev.* C**58**, R641 (1998); F. Gabbiani, P.F. Bedaque and H.W. Grießhammer, *Nucl. Phys.* A **675**, 601 (2000);

[19] S. Fleming, T. Mehen and I. W. Stewart, Nucl. Phys. A **677**, 313 (2000) [arXiv:nucl-th/9911001].

[20] S. R. Beane, P. F. Bedaque, M. J. Savage and U. van Kolck, Nucl. Phys. A **700**, 377 (2002) [arXiv:nucl-th/0104030].

[21] S. R. Sharpe and R. L. . Singleton, Phys. Rev. D **58**, 074501 (1998) [arXiv:hep-lat/9804028].

[22] G. Rupak and N. Shoresh, Phys. Rev. D **66**, 054503 (2002) [arXiv:hep-lat/0201019].

[23] S. R. Sharpe, *Nucl. Phys.* **B17** (Proc. Suppl.), 146 (1990); *Phys. Rev.* **D46**, 3146 (1992); C. Bernard and M. F. L. Golterman, *Phys. Rev.* **D46**, 853 (1992); J. N. Labrenz and S. R. Sharpe, *Phys. Rev.* **D54**, 4595

[24] C. W. Bernard and M. F. L. Golterman, Phys. Rev. D **46**, 853 (1992) [arXiv:hep-lat/9204007].

[25] S. R. Sharpe, Phys. Rev. D **56**, 7052 (1997) [Erratum-ibid. D **62**, 099901 (2000)] [arXiv:hep-lat/9707018].

[26] S. R. Beane, P. F. Bedaque, K. Orginos and M. J. Savage [NPLQCD Collaboration], Phys. Rev. D **73**, 054503 (2006) [arXiv:hep-lat/0506013].

[27] J. W. Chen, D. O'Connell, R. S. Van de Water and A. Walker-Loud, Phys. Rev. D **73**, 074510 (2006) [arXiv:hep-lat/0510024].

[28] J. W. Chen, D. O'Connell and A. Walker-Loud, Phys. Rev. D **75**, 054501 (2007) [arXiv:hep-lat/0611003].

[29] S. R. Beane, P. F. Bedaque, A. Parreno and M. J. Savage, Phys. Lett. B **585**, 106 (2004) [arXiv:hep-lat/0312004].

[30] S. R. Beane, P. F. Bedaque, K. Orginos and M. J. Savage, Phys. Rev. Lett. **97**, 012001 (2006) [arXiv:hep-lat/0602010].

[31] B. Borasoy, E. Epelbaum, H. Krebs, D. Lee and U. G. Meissner, Eur. Phys. J. A **31**, 105 (2007) [arXiv:nucl-th/0611087].

[32] E. Wigner, Phys. Rev. **51**, 106 (1939).

[33] E. Wigner, Phys. Rev. **51**, 947 (1939).

[34] E. Wigner, Phys. Rev. **56**, 519 (1939).

[35] D. B. Kaplan and M. J. Savage, Phys. Lett. **B365**, 244 (1996), hep-ph/9509371.

[36] D. B. Kaplan and A. V. Manohar, Phys. Rev. **C56**, 76 (1997), nucl-th/9612021.

[37] T. Mehen, I. W. Stewart, and M. B. Wise, Phys. Rev. Lett. **83**, 931 (1999), hep-ph/9902370.

[38] G. Audi and A. H. Wapstra, Nucl. Phys. **A595**, 409 (1995).

[39] P. F. Bedaque, H. W. Hammer, and U. van Kolck, Phys. Rev. Lett. **82**, 463 (1999), nucl-th/9809025.

[40] P. F. Bedaque, H. W. Hammer, and U. van Kolck, Nucl. Phys. **A676**, 357 (2000), nucl-th/9906032.

[41] V. Adamyan, J. Alcober, and I. Tkachenko, American Mathematics Research eXpress **2**, 33 (2003).

[42] R. Curto and L. Fialkow, Houston J. Math. **17**, 603 (1991).

[43] P. Kovtun, D.T. Son, and A.O. Starinets, Phys.Rev.Lett. **94**,111601 (2005).

[44] G. Policastro, D.T. Son, and A.O. Starinets, Phys. Rev. Lett. **87**, 081601 (2001).

[45] G. Policastro, D. T. Son and A. O. Starinets, JHEP **0209**, 043 (2002).

[46] C.P. Herzog, J. High Energy Phys. **0212**, 026 (2002).

[47] A. Buchel and J.T. Liu, Phys. Rev. Lett. **93**, 090602 (2004).

[48] I. Arsene *et al.*, Nucl. Phys. A **757**, 1 (2005); B. B. Back *et al.*, *ibid.* **757**, 28 (2005); J. Adams *et al.*, *ibid.* **757**, 102 (2005); K. Adcox *et al.*, *ibid.* **757**, 184 (2005).

[49] D. Molnar and M. Gyulassy, Nucl. Phys. A **697**, 495 (2002) [Erratum *ibid.* **703**, 893 (2002)].

[50] D. Teaney, Phys. Rev. C **68**, 034913 (2003).

[51] F. Karsch and E. Laermann, in *Quark-Gluon Plasma 3*, edited by R. C. Hwa and X.-N. Wang (World Scientific, Singapore, 2004), p. 1 [arXiv:hep-lat/0305025].

[52] A. Nakamura and S. Sakai, Phys. Rev. Lett. **94**, 072305 (2005).

[53] M. Asakawa, S. A. Bass and B. Muller, arXiv:hep-ph/0603092.

[54] J. W. Chen and E. Nakano, Phys. Lett. B **647**, 371 (2007) [arXiv:hep-ph/0604138].

[55] D. Davesne, Phys. Rev. C53, 3069 (1996).

[56] A. Dobado and S.N. Santalla, Phys. Rev. D **65**, 096011 (2002).

[57] A. Dobado and F.J. Llanes-Estrada, Phys. Rev. D **69**, 116004 (2004).

[58] P. Arnold, G. D. Moore and L. G. Yaffe, JHEP **0305**, 051 (2003).

[59] P. Arnold, G. D. Moore and L. G. Yaffe, JHEP **0011**, 001 (2000).

[60] S. Jeon, Phys. Rev. D **52**, 3591 (1995); S. Jeon and L. Yaffe, Phys. Rev. D **53**, 5799 (1996).

[61] A. Schenk, Nucl. Phys. B **363**, 97 (1991). G. M. Welke, R. Venugopalan and M. Prakash, Phys. Lett. B **245**, 137 (1990).

[62] M. Prakash, M. Prakash, R. Venugopalan and G. M. Welke, Phys. Rev. Lett. **70** (1993) 1228; Phys. Rept. **227** (1993) 321.

[63] J. W. Chen, Y. H. Li, Y. F. Liu and E. Nakano, arXiv:hep-ph/0703230.

[64] J. W. Chen and D. B. Kaplan, Phys. Rev. Lett. **92**, 257002 (2004).

[65] F. Karsch and H. W. Wyld, Phys. Rev. D **35**, 2518 (1987).

[66] M. A. Stephanov, arXiv:hep-lat/0701002.

[67] L. P. Csernai, J. I. Kapusta and L. D. McLerran, Phys. Rev. Lett. **97**, 152303 (2006) [arXiv:nucl-th/0604032].

[68] J. B. Natowitz *et al.*, Phys. Rev. C **65**, 034618 (2002) [arXiv:nucl-ex/0106016].

[69] J. B. Elliott *et al.* [EOS Collaboration], Phys. Rev. C **67**, 024609 (2003) [arXiv:nucl-ex/0205004].

[70] L. G. Moretto, J. B. Elliott and L. Phair, Phys. Rev. C **72**, 064605 (2005) [arXiv:nucl-ex/0507015].

[71] R. A. Lacey et al., Phys. Rev. Lett. **98**, 092301 (2007) [arXiv:nucl-ex/0609025]. [arXiv:hep-lat/0308016].

[72] A. J. Leggett, *Modern trends in the Theory of Condensed Matter*, Springer-verlag, Berlin (1980), p. 13.

[73] P. Nozières and S. Schmitt-Rink, J. Low Temp. Phys. **59**, 195 (1985).

[74] M. Randeria, *Bose-Einstein Condensation*, ed. A. Griffin, D. W. Snoke, and S. Stringari, Campride, N. Y., (1995), p. 355.

[75] Y. Ohashi and A. Griffin, Phys. Rev. Lett. **89**, 130402 (2002) [cond-mat/0201262].

[76] For a review, Q. Chen, J. Stajic, S. Tan, and K. Levin, Phys. Rep. **412**, 1 (2005).

[77] K. M. O'Hara et al., Science **298**, 2179 (2002).

[78] C. A. Regal, M. Greiner, and D. S. Jin, Phys. Rev. Lett. **92**, 040403 (2004).

[79] M. Bartenstein et al., Phys. Rev. Lett. **92**, 120401 (2004).

[80] M. W. Zwierlein et al., Phys. Rev. Lett. **92**, 120403 (2004).

[81] J. Kinast et al., Phys. Rev. Lett. **92**, 150402 (2004).

[82] T. Bourdel et al., Phys. Rev. Lett. **93**, 050401 (2004).

[83] J. Kinast et al., Science **307**, 1296 (2005).

[84] G. E. Astrakharchik, J. Boronat, J. Casulleras, and S. Giorgini, Phys. Rev. Lett. **93**, 200404 (2004) [cond-mat/0406113].

[85] J. Carlson, S. Y. Chang, V. R. Pandharipande, and K. E. Schmidt, Phys. Rev. Lett. **91**, 050401 (2003);
S. Y. Chang, V. R. Pandharipande, J. Carlson, and K. E. Schmidt, Phys. Rev. A **70**, 043602 (2004).

[86] J. W. Chen and D. B. Kaplan, Phys. Rev. Lett. **92**, 257002 (2004).

[87] M. Wingate, Nucl. Phys. Proc. Suppl. 140:592-594, 2005; cond-mat/0502372; hep-lat/0609054.

[88] J. Carlson and S. Reddy, Phys. Rev. Lett. **95**, 060401 (2005) [cond-mat/0503256].

[89] A. Bulgac, J. E. Drut, and P. Magierski, Phys. Rev. Lett. **96**, 090404 (2006).

[90] D. Lee, Phys. Rev. B **73**: 115112 (2006); cond-mat/0606706; Phys. Rev. B**73**:115112 (2006); Phys. Rev. C**73**:015202 (2006).

[91] Y. Nishida and D. T. Son, Phys. Rev. Lett. **97** [cond-mat/0604500].

[92] Y. Nishida and D. T. Son, cond-mat/0607835.

[93] K. G. Wilson and J. Kogut, Phys. Rep. **12**, 75 (1974).

[94] Z. Nussinov and S. Nussinov, cond-mat/0410597.

[95] J. V. Steele, nucl-th/0010066.

[96] T. Schafer, C. W. Kao, and S. R. Cotanch, Nucl. Phys. **A 762**, 82 (2005) [nucl-th/0504088].

[97] G. Rupak, nucl-th/0605074.

[98] G. Rupak, T. Schafer, and A. Kryjevski, e-Print Archive: cond-mat/0607834.

[99] D. T. Son and M. A. Stephanov, e-Print Archive: cond-mat/0507586.

[100] Y. Nishida, e-Print Archive: cond-mat/0608321.

[101] P. Arnold, J. E. Drut, and D. T. Son, cond-mat/0608477.

Physics and Astrophysics of Hadrons and Hardronic Matter
Editor: A. B. Santra

Neutron Star Structure with Hyperons and Quarks

M. Baldo, F. Burgio, H.-J. Schulze*

*INFN Sezione di Catania, Via Santa Sofia 64
I-95123 Catania, Italy*
* email: schulze@ct.infn.it

Abstract

We discuss the high-density nuclear equation of state within the Brueckner-Hartree-Fock approach. Particular attention is paid to the effects of nucleonic three-body forces, the presence of hyperons, and the joining with an eventual quark matter phase. The resulting properties of neutron stars, in particular the mass-radius relation, are determined. It turns out that stars heavier than 1.3 solar masses contain necessarily quark matter and that the maximum mass remains below 1.7 solar masses in our approach.

1 Brueckner theory

Over the last two decades the increasing interest for the equation of state (EOS) of nuclear matter has stimulated a great deal of theoretical activity. Phenomenological and microscopic models of the EOS have been developed along parallel lines with complementary roles. The former models include nonrelativistic mean field theory based on Skyrme interactions [1] and relativistic mean field theory based on meson-exchange interactions (Walecka model) [2]. Both of them fit the parameters of the interaction in order to reproduce the empirical saturation properties of nuclear matter extracted from the nuclear mass table. The latter ones include nonrelativistic Brueckner-Hartree-Fock (BHF) theory [3] and its relativistic counterpart, the Dirac-Brueckner (DB) theory [4], the nonrelativistic variational approach also corrected by relativistic effects [5], and more recently the chiral perturbation theory [6]. In these approaches the parameters of the interaction are fixed by the experimental nucleon-nucleon (NN) and/or nucleon-meson scattering data.

For states of nuclear matter with high density and high isospin asymmetry the experimental constraints on the EOS are rather scarse and indirect. Different approaches lead to different or even contradictory theoretical predictions for the nuclear matter properties. The interest for these properties lies, to a large extent, in the study of dense astrophysical objects, i.e., supernovae and neutron stars. In particular, the structure of a neutron star is very sensitive to the compressibility and the symmetry energy. The neutron star mass, measured in binary systems, has been proposed as a constraint for the EOS of nuclear matter [7].

One of the most advanced microscopic approaches to the EOS of nuclear matter is the Brueckner theory. In the recent years, it has made a rapid progress in several aspects: (i) The convergence of the Brueckner-Bethe-Goldstone (BBG) expansion has been firmly established [8]. (ii) Important relativistic effects have been incorporated by including into the interaction the virtual nucleon-antinucleon excitations, and the relationship with the DB approach has been numerically clarified [9]. (iii) The addition of microscopic three-body forces (TBF) based on nucleon excitations via pion and heavy meson exchanges, permitted to improve to a large extent the agreement with the empirical saturation properties [9–11]. (iv) Finally, the BHF approach has been extended in a fully microscopic and self-consistent way to describe nuclear matter containing also hyperons [12], opening new fields of applications such as hypernuclei [13] and a more realistic modeling of neutron stars [14, 15].

In the present report we review these issues and present our results for neutron star structure based on the resulting EOS of dense hadronic matter, also supplemented by an eventual transition to quark matter at high density.

1.1 Convergence of the hole-line expansion

The nonrelativistic BBG expansion of the nuclear matter correlation energy E/A can be cast as a power series in terms of the number of hole lines contained in the corresponding diagrams, which amounts to a density power expansion [3]. The two hole-line truncation is named the Brueckner-Hartree-Fock (BHF) approximation. At this order the energy D_2 is very much affected by the choice of the auxiliary single-particle (s.p.) potential, as shown in Fig. 1 (solid curves), where the numerical results obtained with the gap and the continuous choice are compared for symmetric nuclear matter as well as neutron matter. But, as also shown in the same figure [8], adding the three-hole line contributions D_3, the resulting EOS is almost insensitive to the choice of the auxiliary potential, and very close to the result D_2 with the continuous choice.

In spite of the satisfactory convergence, the saturation density misses the empirical value $\rho_0 = 0.17$ fm^{-3} extracted from the nuclear mass tables. This confirms the belief that the concept of a many-nucleon system interacting with only a two-body force is not adequate to describe nuclear matter, especially at high density.

1.2 Relativistic corrections

Before the possible effects of TBF are examined, one should introduce relativistic corrections in the preceding nonrelativistic BHF predictions. This is done in the Dirac-Brueckner approach [4], where the nucleons, instead of propagating as plane waves, propagate as spinors in a mean field with a scalar component U_S and a vector component U_V, self-consistently determined together with the G-matrix. The nucleon self-energy can be expanded in terms of the scalar field,

$$\Sigma(\boldsymbol{k}) = \sqrt{(m+U_S)^2 + \boldsymbol{k}^2} + U_V \approx e_k + U_V + \frac{M}{e_k}U_S + \frac{\boldsymbol{k}^2}{2e_k^3}U_S^2 + \ldots \qquad (1)$$

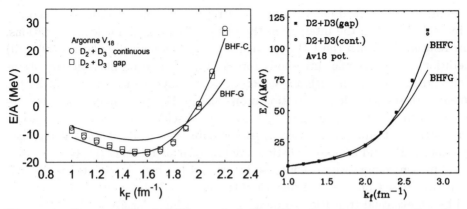

Figure 1: Comparison of BHF two hole-line (lines) and three hole-line (markers) results for symmetric nuclear matter (left plot) and pure neutron matter (right plot), using continuous or gap choice for the s.p. potentials and the Argonne V_{18} nucleon-nucleon potential.

with $e_k = \sqrt{m^2 + k^2}$. The second-order term can be interpreted as due to the interaction between two nucleons with the virtual excitation of a nucleon-antinucleon pair [16]. This interaction is a TBF with the exchange of a scalar (σ) meson, as illustrated by the diagram (b) of Fig. 2. Actually this diagram represents a class of TBF with the exchange of light (π, ρ) and heavy (σ, ω) mesons. There are, however, other diagrams representing TBF, Fig. 2, which should be evaluated as well in a consistent treatment of TBF.

2   Three-body forces

Since long it is well known that two-body forces are not enough to explain some nuclear properties, and TBF have to be introduced. Typical examples are: the binding energy of light nuclei, the spin dynamics of nucleon-deuteron scattering, and the saturation point of nuclear matter. Phenomenological and microscopic TBF have been widely used to describe the above mentioned properties.

In the framework of the Brueckner theory a rigorous treatment of TBF would require the solution of the Bethe-Faddeev equation, describing the dynamics of three bodies embedded in the nuclear matter. In practice a much simpler approach is employed, namely the TBF is reduced to an effective, density-dependent, two-body force by averaging over the third nucleon in the medium, taking account of the nucleon-nucleon correlations by means of the BHF defect function g_{ij},

$$\langle 12|V(\rho)|1'2'\rangle = \sum_{33'} \Psi^*_{123} \langle 123|V|1'2'3'\rangle \Psi_{1'2'3'} . \qquad (2)$$

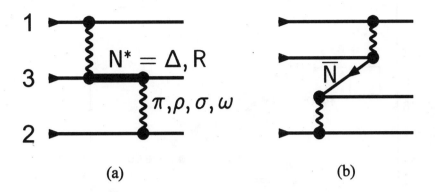

Figure 2: Diagrams contributing to the microscopic TBF.

Here $\Psi_{123} = \phi_3(1-g_{13})(1-g_{23})$ and ϕ_3 is the free wave function of the third particle. This effective two-body force is added to the bare two-body force and recalculated at each step of the iterative procedure.

2.1 Microscopic TBF

The microscopic TBF of Refs. [9–11] is based on meson-exchange mechanisms accompanied by the excitation of nucleonic resonances, as represented by the diagrams plotted in Fig. 2. Besides the TBF arising from the excitation of a $N\bar{N}$ pair [diagram (b)], already discussed in the preceding section, another important class of TBF [diagram (a)] is due to the excitation of the isobar $\Delta(1232)$ resonance via the exchange of light (π, ρ) mesons, or the lowest non-isobar nucleon excitation $N^*(1440)$ excited by heavy meson (σ and ω) exchanges.

The combined effect of these TBF is a remarkable improvement of the saturation properties of nuclear matter [11]. This is illustrated in Fig. 3, showing the saturation points of nuclear matter obtained with different NN potentials and theoretical approaches [17]. Without TBF (black dots) one observes the typical Coester band, i.e., a linear correlation between saturation density and energy, yielding either a too large saturation density or too small binding energy with the different potentials [18]. The inclusion of TBF (stars) shifts the saturation points towards the empirical value, as is the case for the DBHF results (triangles). The spin and isospin properties with TBF exhibit also quite satisfactory behaviour [19].

2.2 Phenomenological TBF

A second class of TBF that are widely used in the literature, in particular for variational calculations of finite nuclei and nuclear matter [5], are the phenomenological

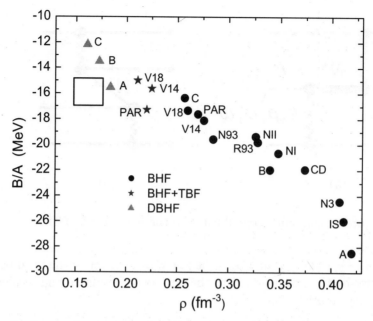

Figure 3: Saturation points of symmetric nuclear matter obtained with different potentials and theoretical approaches. The square indicates the empirical region.

Urbana TBF [20]. We remind that the Urbana IX TBF model contains a two-pion exchange potential $V_{ijk}^{2\pi}$ supplemented by a phenomenological repulsive term V_{ijk}^{R},

$$V_{ijk} = V_{ijk}^{2\pi} + V_{ijk}^{R} \, , \tag{3}$$

where

$$V_{ijk}^{2\pi} = A \sum_{\text{cyc}} \left[\{X_{ij}, X_{jk}\} \{\boldsymbol{\tau}_i \cdot \boldsymbol{\tau}_j, \boldsymbol{\tau}_j \cdot \boldsymbol{\tau}_k\} + \frac{1}{4} [X_{ij}, X_{jk}] [\boldsymbol{\tau}_i \cdot \boldsymbol{\tau}_j, \boldsymbol{\tau}_j \cdot \boldsymbol{\tau}_k] \right] \, , \tag{4}$$

$$V_{ijk}^{R} = U \sum_{\text{cyc}} T^2(m_\pi r_{ij}) T^2(m_\pi r_{jk}) \, . \tag{5}$$

The two-pion exchange operator X_{ij} is given by

$$X_{ij} = Y(m_\pi r_{ij})\boldsymbol{\sigma}_i \cdot \boldsymbol{\sigma}_j + T(m_\pi r_{ij})S_{ij} \, , \tag{6}$$

where $S_{ij} = 3(\boldsymbol{\sigma}_i \cdot \hat{\boldsymbol{r}}_{ij})(\boldsymbol{\sigma}_j \cdot \hat{\boldsymbol{r}}_{ij}) - \boldsymbol{\sigma}_i \cdot \boldsymbol{\sigma}_j$ is the tensor operator and $\boldsymbol{\sigma}$ and $\boldsymbol{\tau}$ are the Pauli spin and isospin operators. Y and T are the Yukawa and tensor functions associated to the one-pion exchange, respectively [20]. The potential $V_{ijk}^{2\pi}$ can be identified with the simplest 2π TBF involving Delta excitation shown in Fig. 2(a), and the parameter A can be related to the relevant meson parameters [21],

$$A = -\frac{2}{81} \left(\frac{g_{\pi NN} g_{\pi N\Delta}}{4\pi} \right)^2 \left(\frac{m_\pi}{2m_N} \right)^4 \frac{m_\pi^2}{m_\Delta - m_N}, \qquad (7)$$

whereas the component V_{ijk}^R and its parameter U are completely phenomenological.

After reducing this TBF to an effective, density-dependent, two-body force by the averaging procedure described earlier, the resulting effective two-nucleon potential assumes a simple structure,

$$\overline{V}_{ij}^{\text{pheno}}(\boldsymbol{r}) = (\boldsymbol{\tau}_i \cdot \boldsymbol{\tau}_j) \Big[(\boldsymbol{\sigma}_i \cdot \boldsymbol{\sigma}_j) V_C^{2\pi}(r) + S_{ij}(\hat{\boldsymbol{r}}) V_T^{2\pi}(r) \Big] + V^R(r), \qquad (8)$$

containing central and tensor two-pion exchange components as well as a central repulsive contribution. For comparison, the averaged microscopic TBF [9] involves five different components:

$$\begin{aligned} \overline{V}_{ij}^{\text{micro}}(\boldsymbol{r}) &= (\boldsymbol{\tau}_i \cdot \boldsymbol{\tau}_j)(\boldsymbol{\sigma}_i \cdot \boldsymbol{\sigma}_j) V_C^{\tau\sigma}(r) + (\boldsymbol{\sigma}_i \cdot \boldsymbol{\sigma}_j) V_C^{\sigma}(r) + V_C(r) \\ &\quad + S_{ij}(\hat{\boldsymbol{r}}) \Big[(\boldsymbol{\tau}_i \cdot \boldsymbol{\tau}_j) V_T^{\tau}(r) + V_T(r) \Big]. \end{aligned} \qquad (9)$$

In the variational approach the two parameters A and U are determined by fitting the triton binding energy together with the saturation density of nuclear matter (yielding however too little attraction, $E/A \approx -12$ MeV, in the latter case [5]). In the BHF calculations they are instead chosen to reproduce the empirical saturation density together with the binding energy of nuclear matter. The resulting parameter values are $A = -0.0293$ MeV and $U = 0.0048$ MeV in the variational Urbana IX model, whereas for the optimal BHF+TBF calculations we require $A = -0.0500$ MeV and $U = 0.00038$ MeV, yielding a saturation point at $k_F \approx 1.38$ fm^{-1}, $E/A \approx -15.3$ MeV, and an incompressibility $K \approx 210$ MeV.

These values of A and U have been obtained by using the Argonne v_{18} two-body force [22] both in the BHF and in the variational many-body theories. However, the required repulsive component ($\sim U$) is much weaker in the BHF approach, consistent with the observation that in the variational calculations usually heavier nuclei as well as nuclear matter are underbound. Indeed, less repulsive TBF became available recently [23] in order to address this problem.

3 EOS of nuclear matter from different TBF

In Fig. 4 we compare the different components $V_C^{2\pi}, V_T^{2\pi}, V^R$, Eq. (8), and $V_C^{\tau\sigma}$, $V_C^{\sigma}, V_C, V_T^{\tau}, V_T$, Eq. (9), of the averaged phenomenological and microscopic TBF potentials in symmetric matter at normal density. One notes that the attractive components $V_C^{2\pi}, V_T^{2\pi}$ and $V_C^{\tau\sigma}, V_T^{\tau}$ roughly correspond to each other, whereas the

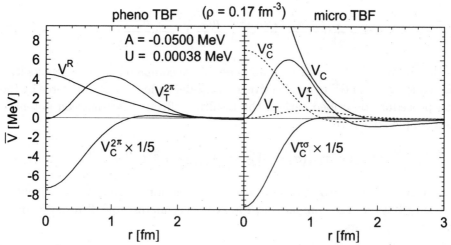

Figure 4: Comparison of the different components of averaged phenomenological and microscopic TBF, Eqs. (8) and (9), at saturation density.

repulsive part (V^R vs. V_C) is much larger for the microscopic TBF. With the choice of parameters A and U given above, one would therefore expect a more repulsive behaviour of the microscopic TBF, which is indeed confirmed in the following.

Let us now confront the EOS predicted by the phenomenological TBF and the microscopic one. In both cases the BHF approximation has been adopted with same two-body force (Argonne v_{18}). In the left panel of Fig. 5 we display the EOS both for symmetric matter (lower curves) and pure neutron matter (upper curves). We show results obtained for several cases, i.e., i) only two-body forces are included (dotted lines), ii) TBF implemented within the phenomenological Urbana IX model (dashed lines), and iii) TBF treated within the microscopic meson-exchange approach (solid lines). We notice that the EOS for symmetric matter with TBF reproduces the correct nuclear matter saturation point. Moreover, the incompressibility turns out to be compatible with the values extracted from phenomenology, i.e., $K \approx 210$ MeV. Up to a density of $\rho \approx 0.4$ fm^{-3} the microscopic and phenomenological TBF are in fair agreement, whereas at higher density the microscopic TBF turn out to be more repulsive.

Within the BHF approach, it has been verified [14, 15, 24] that a parabolic approximation for the binding energy of nuclear matter with arbitrary proton fraction $x = \rho_p / (\rho_n + \rho_p)$ is well fulfilled,

$$\frac{E}{A}(\rho, x) \approx \frac{E}{A}(\rho, x = 0.5) + (1 - 2x)^2 E_{\text{sym}}(\rho), \qquad (10)$$

where the symmetry energy E_{sym} can be expressed in terms of the difference of the energy per particle between pure neutron ($x = 0$) and symmetric ($x = 0.5$) matter:

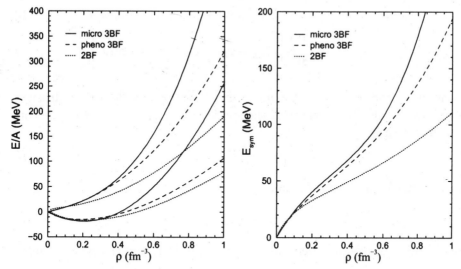

Figure 5: Left plot: Binding energy per nucleon of symmetric nuclear matter (lower curves using a given linestyle) and pure neutron matter (upper curves), employing different TBF. Right plot: Corresponding symmetry energy of nuclear matter.

$$E_{sym}(\rho) = -\frac{1}{4}\frac{\partial(E/A)}{\partial x}(\rho, 0) \approx \frac{E}{A}(\rho, 0) - \frac{E}{A}(\rho, 0.5) \, . \tag{11}$$

In the right panel of Fig. 5 we display the symmetry energy as a function of the nucleon density ρ for different choices of the TBF. We observe results in agreement with the characteristics of the EOS shown in the left panel. Namely, the stiffest EOS, i.e., the one calculated with the microscopic TBF, yields larger symmetry energies compared to the ones obtained with the Urbana phenomenological TBF. Moreover, the symmetry energy calculated (with or without TBF) at the saturation point yields a value $E_{sym} \approx 30$ MeV, compatible with nuclear phenomenology.

4   Neutron star structure

In order to study the effects of different TBF on neutron star structure, we have to calculate the composition and the EOS of cold, catalyzed matter. We require that the neutron star contains charge-neutral matter consisting of neutrons, protons, and leptons (e^-, μ^-) in beta equilibrium. Using the various TBF discussed above, we compute the proton fraction and the EOS for charge-neutral and beta-stable matter in the following standard way [25, 26]: The Brueckner calculation yields the energy density of hadron/lepton matter as a function of the different partial densities,

$$\epsilon(\rho_n, \rho_p, \rho_e, \rho_\mu) = \epsilon_H(\rho_n, \rho_p) + \epsilon_e(\rho_e) + \epsilon_\mu(\rho_\mu) \,, \tag{12}$$

$$\epsilon_H(\rho_n, \rho_p) = (\rho_n m_n + \rho_p m_p) + (\rho_n + \rho_p)\frac{E}{A}(\rho_n, \rho_p) \,, \tag{13}$$

$$\epsilon_l(\rho_l) = \frac{m_l^4}{8\pi^2}\left[(2x_l^3 + x_l)\sqrt{1 + x_l^2} - \text{arsinh}\, x_l\right] \,, \tag{14}$$

where m_l is the lepton mass, $x_l = k_F^{(l)}/m_l$, and $\rho_l = k_F^{(l)^3}/3\pi^2$ is the lepton density.

The various chemical potentials (of the species $i = n, p, e, \mu$) can then be computed straightforwardly,

$$\mu_i = \frac{\partial \epsilon}{\partial \rho_i} \,, \tag{15}$$

and the equations for beta-equilibrium,

$$\mu_i = b_i \mu_n - q_i \mu_e \,, \tag{16}$$

(b_i and q_i denoting baryon number and charge of species i) and charge neutrality,

$$\sum_i \rho_i q_i = 0 \,, \tag{17}$$

allow to determine the equilibrium composition $\rho_i(\rho)$ at given baryon density $\rho = \rho_n + \rho_p$ and finally the EOS,

$$p(\rho) = \rho^2 \frac{d}{d\rho}\frac{\epsilon(\rho_i(\rho))}{\rho} = \rho\frac{d\epsilon}{d\rho} - \epsilon = \rho\mu_n - \epsilon \,. \tag{18}$$

In order to calculate the mass-radius relation, one has then to solve the well-known Tolman-Oppenheimer-Volkov equations [25],

$$\frac{dp}{dr} = -\frac{Gm}{r^2}\frac{(\epsilon + p)(1 + 4\pi r^3 p/m)}{1 - 2Gm/r} \,, \tag{19}$$

$$\frac{dm}{dr} = 4\pi r^2 \epsilon \,, \tag{20}$$

with the newly constructed EOS for the charge-neutral and beta-stable case as input (supplemented by the EOS of Feynman-Metropolis-Teller [27], Baym-Pethick-Sutherland [28], and Negele-Vautherin [29] for the outer part of the neutron star, $\rho \lesssim 0.08$ fm^{-3}). The solutions provide information on the interior structure of a star, $\rho_i(r)$, as well as the mass-radius relation, $M(R)$.

The results are shown in Fig. 6. We notice that the EOS calculated with the microscopic TBF produces the largest gravitational masses, with the maximum mass

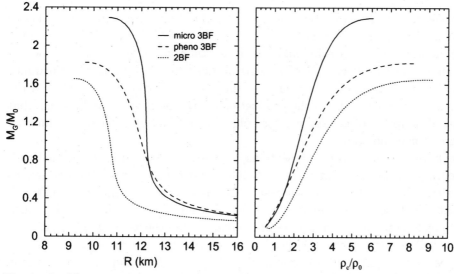

Figure 6: The neutron star gravitational mass (in units of the solar mass $M_\odot$) is displayed versus the radius (left panel) and the normalized central baryon density ρ_c ($\rho_0 = 0.17$ fm^{-3}) (right panel).

of the order of 2.3 $M_\odot$, whereas the phenomenological TBF yields a maximum mass of about 1.8 $M_\odot$. In the latter case, neutron stars are characterized by smaller radii and larger central densities, i.e., the Urbana TBF produce more compact stellar objects. For completeness, we also show a sequence of stellar configurations obtained using only two-body forces. In this unrealistic case the maximum mass is slightly above 1.6 $M_\odot$, with a radius of 9 km and a central density equal to 9 times the saturation value.

However, these results should be considered as only provisory, since it is well known that the inclusion of hyperons [14, 15] or quark matter [30–32] may strongly affect the structure of the star, in particular reducing substantially the maximum mass. We discuss this point now in detail.

5 Hyperons in nuclear matter

While at moderate densities $\rho \approx \rho_0$ the matter inside a neutron star consists only of nucleons and leptons, at higher densities several other species of particles may appear due to the fast rise of the baryon chemical potentials with density. Among these new particles are strange baryons, namely, the Λ, Σ, and Ξ hyperons. Due to its negative charge, the Σ^- hyperon is the first strange baryon expected to appear with increasing density in the reaction $n + n \rightarrow p + \Sigma^-$, in spite of its substantially larger mass compared to the neutral Λ hyperon ($M_{\Sigma^-} = 1197$ MeV, $M_\Lambda = 1116$ MeV).

Other species might appear in stellar matter, like Δ isobars along with pion and kaon condensates. It is therefore mandatory to generalize the study of the nuclear EOS with the inclusion of the possible hadrons, other than nucleons, which can spontaneously appear in the inner part of a neutron star, just because their appearance is able to lower the ground state energy of the nuclear matter dense phase. In the following we will concentrate on the production of strange baryons and assume that a hadronic description of nuclear matter holds up to densities as those encountered in the core of neutron stars.

As we have seen in the previous sections, the nuclear EOS can be calculated with good accuracy in the Brueckner two hole-line approximation with the continuous choice for the s.p. potential, since the results in this scheme are quite close to the full convergent calculations which include also the three hole-line contribution. It is then natural to include the hyperon degrees of freedom within the same approximation in order to calculate the nuclear EOS needed to describe the neutron star interior. To this purpose, one requires in principle nucleon-hyperon (NY) and hyperon-hyperon (YY) potentials. In our work we use the Nijmegen soft-core NY and YY potentials (either the NSC89 [33] or the NSC97e model of Ref. [34]) that are well adapted to the existing experimental NY scattering data and also compatible with Λ hypernuclear levels [35, 36]. Unfortunately, up to date no YY scattering data exist and therefore no reliable YY potentials are available. Thus the NSC89 potentials contain no YY components, whereas the NSC97 potentials comprise extensions to the YY sector based on SU(3) symmetry. Nevertheless the importance of YY potentials should be minor as long as the hyperonic partial densities remain limited. Also, for the nucleonic sector, the v_{18} NN potential is used together with the phenomenological TBF introduced previously.

In the following we give a short review of the BHF approach including hyperons. Detailed accounts can be found in Refs. [12, 37, 38]. The basic input quantities in the Bethe-Goldstone equation are the NN, NY, and YY potentials. With these potentials, the various G-matrices are evaluated by solving numerically the Bethe-Goldstone equation, which can be written in operatorial form as

$$G_{ab}[W] = V_{ab} + \sum_c \sum_{p,p'} V_{ac} |pp'\rangle \frac{Q_c}{W - E_c + i\epsilon} \langle pp' | G_{cb}[W] \,, \tag{21}$$

where the indices a, b, c indicate pairs of baryons and the Pauli operator Q and energy E determine the propagation of intermediate baryon pairs. The pair energy in a given channel $c = (B_1 B_2)$ is

$$E_{(B_1 B_2)} = T_{B_1}(k_{B_1}) + T_{B_2}(k_{B_2}) + U_{B_1}(k_{B_1}) + U_{B_2}(k_{B_2}) \tag{22}$$

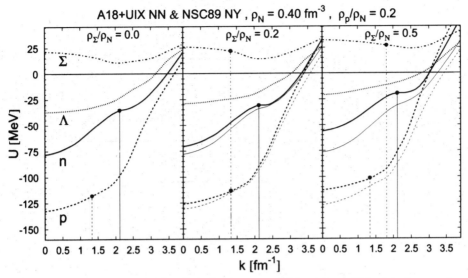

Figure 7: The s.p. potentials of nucleons n, p and hyperons Σ^-, Λ in baryonic matter of fixed nucleonic density $\rho_N = 0.4\,\mathrm{fm}^{-3}$, proton density $\rho_p/\rho_N = 0.2$, and varying Σ^- density $\rho_\Sigma/\rho_N = 0.0, 0.2, 0.5$. The vertical lines indicate the corresponding Fermi momenta of n, p, and Σ^-. For the nucleonic curves, the thick lines represent the total s.p. potentials U_N, whereas the thin lines show the values excluding the Σ^- contribution, i.e., $U_N^{(n)} + U_N^{(p)}$.

with $T_B(k) = m_B + k^2/2m_B$, where the various s.p. potentials are given by

$$U_B(k) = \sum_{B'=n,p,\Lambda,\Sigma^-} U_B^{(B')}(k) \tag{23}$$

and are determined self-consistently from the G-matrices,

$$U_B^{(B')}(k) = \sum_{k'<k_F^{(B')}} \mathrm{Re}\langle kk'|G_{(BB')(BB')}[E_{(BB')}(k,k')]|kk'\rangle . \tag{24}$$

The coupled equations (21) to (24) define the BHF scheme with the continuous choice of the s.p. energies. In contrast to the standard purely nucleonic calculation, the additional coupled channel structure renders the calculations quite time-consuming.

Once the different s.p. potentials are known, the total nonrelativistic hadronic energy density, ϵ_H, can be evaluated:

$$\epsilon_H = \sum_{B=n,p,\Lambda,\Sigma^-} \sum_{k<k_F^{(B)}} \left[T_B(k) + \frac{1}{2}U_B(k)\right] . \tag{25}$$

153

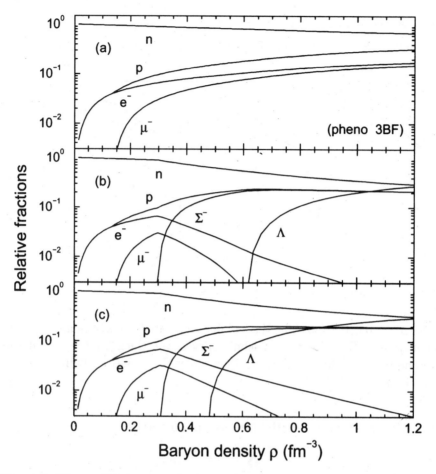

Figure 8: The equilibrium composition of asymmetric and beta-stable nuclear matter comprising (a) only nucleons, (b) noninteracting hyperons, (c) interacting hyperons.

The different s.p. potentials involved in the previous equations are illustrated in Fig. 7, where neutron and proton densities are fixed, given by $\rho_N = 0.4\,\mathrm{fm}^{-3}$ and $\rho_p/\rho_N = 0.2$, and the Σ^- density is varied. Under these conditions the Σ^- s.p. potential is sizeably repulsive, while U_Λ is still attractive (see also Ref. [14]) and the nucleons are both strongly bound. The Σ^- s.p. potential has a particular shape with an effective mass m^*/m slightly larger than 1, whereas the Λ effective mass is typically about 0.8 and the nucleon effective masses are much smaller.

The knowledge of the energy density allows then to compute EOS and neutron star structure as described before, now making allowance for the species $i = n, p, \Sigma^-, \Lambda, e^-, \mu^-$. The main physical features of the nuclear EOS which determine the resulting composition are essentially the symmetry energy of the nucleon part of

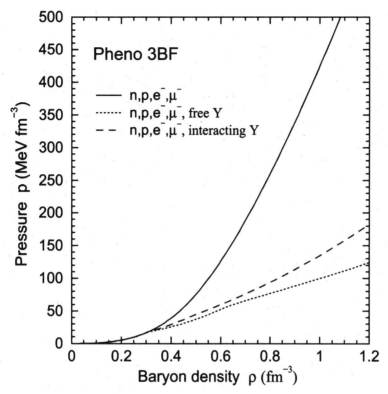

Figure 9: The EOS for hyperon-free (solid curve) and hyperon-rich (broken curves) matter.

the EOS and the hyperon s.p. potentials inside nuclear matter. Since at low enough density the nucleon matter is quite asymmetric, the small percentage of protons feel a deep s.p. potential, and therefore it is energetically convenient to create a Σ^- hyperon by converting a neutron into a proton. The depth of the proton potential is mainly determined by the nuclear matter symmetry energy. Furthermore, the potentials felt by the hyperons can shift substantially the threshold density at which each hyperon sets in.

In Fig. 8 we show the chemical composition of the resulting beta-stable and asymmetric nuclear matter containing hyperons. We observe rather low hyperon onset densities of about 2-3 times normal nuclear matter density for the appearance of the Σ^- and Λ hyperons, while other hyperons do not appear in the matter. Moreover, an almost equal percentage of nucleons and hyperons are present in the stellar core at high densities. A strong deleptonization of matter takes place, since it is energetically convenient to maintain charge neutrality through hyperon formation rather than beta-decay. This can have far reaching consequences for the onset of kaon condensation [39].

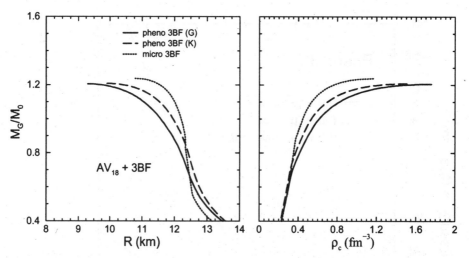

Figure 10: Neutron star gravitational mass versus radius (left panel) and the central baryon density ρ_c (right panel). Calculations involving different nucleonic TBF are compared.

The resulting EOS is displayed in Fig. 9. The upper curve shows the EOS when stellar matter is composed only of nucleons and leptons. The inclusion of hyperons (lower curves) produces a much softer EOS, which turns out to be very similar to the one obtained without TBF. This is quite astonishing, because in the pure nucleon case the repulsive character of TBF at high density increases dramatically the stiffness of the EOS. However, when hyperons are included, the presence of TBF among nucleons enhances the population of Σ^- and Λ because of the increased nucleon chemical potentials with respect to the case without TBF, thus decreasing the nucleon population. Of course, this scenario could partly change if hyperon-hyperon interactions were known or if TBF would be included also for hyperons, but this is beyond our current knowledge of the strong interaction.

The consequences for the structure of the neutron stars are illustrated in Fig. 10, where we display the resulting neutron star mass-radius curves, comparing now results obtained with different nucleonic TBF, in analogy to Fig. 6. One notes that while in Fig. 6 the different TBF still yield quite different maximum masses, the presence of hyperons equalizes the results, leading now to a maximum mass of less than 1.3 solar masses for all the nuclear TBF.

Finally, Fig. 11 shows the internal structure (energy density, pressure, and individual baryon densities) of a $M = 1.2\,M_\odot$ hyperonic neutron star. One notes that in the core the hyperon fraction reaches nearly 50% of the baryons.

The results shown so far were all obtained with the NSC89 hyperon-nucleon potentials. Fig. 12 compares the mass-radius relations using the NSC89 and the NSC96e potentials [15]. Also in the latter case the maximum mass remains below

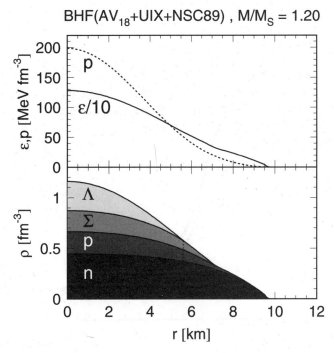

Figure 11: Structure of a $M = 1.2\ M_\odot$ neutron star.

1.4 solar masses, although the internal composition (individual particle fractions) of the star turns out quite different, which is reflected in the different radii.

These remarkable results are due to the strong softening of the hadronic EOS when including hyperons as additional degrees of freedom, and we do not expect substantial changes when introducing refinements of the theoretical framework, such as hyperon-hyperon potentials, hyperonic TBF, relativistic corrections, etc. The only remaining possibility in order to reach significantly larger maximum masses appears to be the transition to another phase of dense (quark) matter inside the star. This is indeed a reasonable assumption, since already geometrically the concept of distinguishable baryons breaks down at the densities encountered in the interior of a neutron star. This will be discussed in the following.

6 Quark matter

The results obtained with a purely hadronic EOS call for an estimate of the effects due to the hypothetical presence of quark matter in the interior of the neutron star. Unfortunately, the current theoretical description of quark matter is burdened with large uncertainties, seriously limiting the predictive power of any theoretical approach at high baryonic density. For the time being we can therefore only resort to phenomenological models for the quark matter EOS and try to constrain them as well as possible by the few experimental information on high-density baryonic matter.

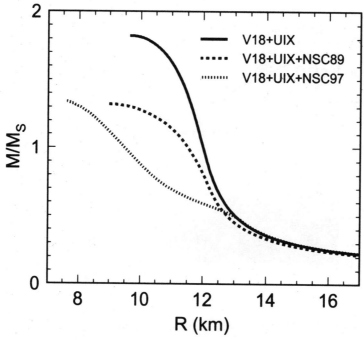

Figure 12: Mass-radius relations of neutron stars without hyperons (solid curve) and using the NSC89 (dashed curve) or the NSC97 (dotted curve) hyperon-nucleon potentials.

One important condition is due to the fact that certainly in symmetric nuclear matter no phase transition is observed below $\rho_B \approx 3\rho_0$. In fact some theoretical interpretation of the heavy ion experiments performed at the CERN SPS [40] points to a possible phase transition at a critical density $\rho_c \approx 6\rho_0 \approx 1/\text{fm}^3$. We will in the following take this value for granted and use an extended MIT bag model [41] (requiring a density-dependent bag "constant") that is compatible with this condition.

We first review briefly the description of the bulk properties of uniform quark matter, deconfined from the beta-stable hadronic matter mentioned in the previous section, within the MIT bag model [41]. The energy density of the $f = u, d, s$ quark gas can be expressed as a sum of the kinetic term and the one-gluon-exchange term [42, 43] proportional to the QCD fine structure constant α_s,

$$\epsilon_Q = B + \sum_f \epsilon_f, \tag{26}$$

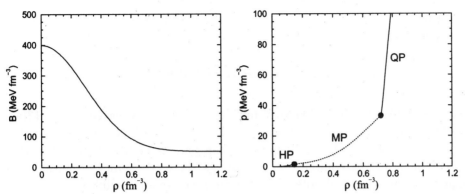

Figure 13: (a) Bag constant B versus baryon number density. (b) EOS including both hadronic and quark components. The mixed phase region of quarks and hadrons (MP) is bordered by two dots, while HP and QP label the pure hadron and quark phases.

$$
\begin{aligned}
\epsilon_f(\rho_f) = &\ \frac{3m_f^4}{8\pi^2}\left[\left(2x_f^3 + x_f\right)\sqrt{1 + x_f^2} - \operatorname{arsinh} x_f\right] \\
&- \alpha_s \frac{m_f^4}{\pi^3}\left[x_f^4 - \frac{3}{2}\left(x_f\sqrt{1 + x_f^2} - \operatorname{arsinh} x_f\right)^2\right],
\end{aligned} \tag{27}
$$

where m_f is the f current quark mass, $x_f = k_F^{(f)}/m_f$, the baryon density of f quarks is $\rho_f = k_F^{(f)3}/3\pi^2$, and B is the energy density difference between the perturbative vacuum and the true vacuum, i.e., the bag "constant."

In the original MIT bag model the bag constant $B \approx 55$ MeV fm^{-3} was used, while values of $B \approx 210$ MeV fm^{-3} are estimated from lattice calculations [44]. In this sense B can be considered as a free parameter. It has been found [30, 45], however, that within the MIT bag model (without color superconductivity) with a density-independent bag constant B, the maximum mass of a NS cannot exceed a value of about 1.6 solar masses. Indeed, the maximum mass increases as the value of B decreases, but too small values of B are incompatible with a hadron-quark transition density $\rho > 2$–$3 \; \rho_0$ in nearly symmetric nuclear matter, as demanded by heavy-ion collision phenomenology. Values of $B \gtrsim 150$ MeV/fm^3 can also be excluded within our model, since we do not obtain any more a phase transition in beta-stable matter in combination with our hadronic EOS [30].

In order to overcome these restrictions of the model, one can introduce a density-dependent bag parameter $B(\rho)$, and this approach was followed in Ref. [30]. This allows one to lower the value of B at large density, providing a stiffer QM EOS and increasing the value of the maximum mass, while at the same time still fulfilling the condition of no phase transition below $\rho \approx 3\rho_0$ in symmetric nuclear matter. In the following we present results based on the MIT model using both a constant value of the bag parameter, $B = 90$ MeV/fm^3, and a gaussian parametrization for the density dependence,

$$B(\rho) = B_\infty + (B_0 - B_\infty) \exp\left[-\beta\left(\frac{\rho}{\rho_0}\right)^2\right] \qquad (28)$$

with $B_\infty = 50$ MeV/fm^3, $B_0 = 400$ MeV/fm^3, and $\beta = 0.17$, displayed in Fig. 13(a). For a more extensive discussion of this topic, the reader is referred to Refs. [30–32].

The introduction of a density-dependent bag parameter has to be taken into account correctly for the computation of various thermodynamical quantities; in particular the quark chemical potentials and the pressure are modified as

$$\mu_q \;\rightarrow\; \mu_q + \frac{1}{3}\frac{dB(\rho)}{d\rho}\,, \qquad (29)$$

$$p \;\rightarrow\; p + \rho\frac{dB(\rho)}{d\rho}\,. \qquad (30)$$

Nevertheless, due to a cancelation of the second term in Eq. (29), occurring in relations (31) for the beta-equilibrium, the composition at a given total baryon density ρ remains unaffected by this term (and is in fact independent of B). At this stage of investigation, we disregard possible dependencies of the bag parameter on the individual quark densities.

In the beta-stable pure quark phase, the individual quark chemical potentials are fixed by Eq. (16) with $b_q = 1/3$, which implies

$$\mu_s = \mu_d = \mu_u + \mu_l\,. \qquad (31)$$

The charge neutrality condition and the total baryon number conservation read

$$0 = 2\rho_u - \rho_d - \rho_s - \rho_l\,, \qquad (32)$$
$$\rho = \rho_u + \rho_d + \rho_s\,. \qquad (33)$$

These equations determine the composition $\rho_f(\rho)$ and the pressure of the QM phase,

$$p_Q(\rho) = \rho^2\frac{d(\epsilon_Q/\rho)}{d\rho}\,, \qquad (34)$$

which can be used together with the pressure of the hadronic phase to determine the phase transition using a Maxwell construction.

However, a more realistic model for the phase transition between hadronic and quark phase inside the star is the Glendenning construction [46], which determines the range of baryon density where both phases coexist. The essential point of this procedure is that both the hadron and the quark phase are allowed to be separately charged, still preserving the total charge neutrality. This implies that neutron star

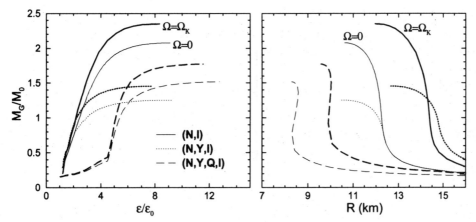

Figure 14: The gravitational mass (in units of the solar mass $M_\odot = 1.98 \times 10^{33}$g) versus the normalized central energy density ($\epsilon_0 = 156$ MeV fm^{-3}) (left panel) and versus the equatorial radius (right panel). The thin lines represent static equilibrium configurations, whereas the thick lines display configurations rotating at their respective Kepler frequencies. Several different stellar matter compositions are considered.

matter can be treated as a two-component system, and therefore can be parametrized by two independent chemical potentials like electron and baryon chemical potentials μ_e and μ_n. The chemical potentials of the different species are related to each other satisfying chemical and beta stability, Eq. (16), while the pressure is the same in the two phases to ensure mechanical stability,

$$p_H(\mu_e, \mu_n) = p_Q(\mu_e, \mu_n) = p_M(\mu_n) . \tag{35}$$

The equation for global charge neutrality,

$$0 = \chi q_Q + (1 - \chi)q_H , \tag{36}$$

with the charge densities q_H and q_Q determines the additional parameter χ, the volume fraction occupied by quark matter in the mixed phase. From this, the baryon density ρ_M and the energy density ϵ_M of the mixed phase can be calculated as

$$\rho_M = \chi\rho_Q + (1 - \chi)\rho_H , \tag{37}$$

$$\epsilon_M = \chi\epsilon_Q + (1 - \chi)\epsilon_H . \tag{38}$$

The EOS resulting from this procedure is shown in Fig. 13(b), where the pure hadron, mixed, and pure quark matter portions are indicated. The mixed phase begins actually at a quite low density around ρ_0. Clearly the outcome of the mixed

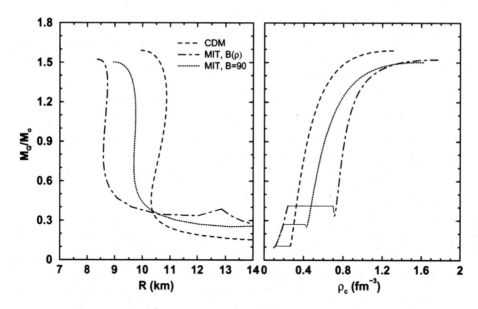

Figure 15: The NS mass as function of radius (left panel) and central density (right panel), using different QM EOS.

phase construction might be substantially changed, if surface and Coulomb energies were taken into account [47]. For the time being these are, however, unknown and have been neglected.

The final result for the structure of hybrid neutron stars is shown in Fig. 14, displaying mass-radius and mass-central density relations. It is evident that the most striking effect of the inclusion of quark matter is the increase of the maximum mass, now reaching about 1.5 $M_\odot$. At the same time, the typical neutron star radius is reduced by about 3 km to typically 9 km. Hybrid neutron stars are thus more compact than purely hadronic ones and their central energy density is larger. For completeness, the figure shows besides static neutron star configurations also those rotating at the maximum (Kepler) frequency [48]. In that case one observes a further enhancement of the maximum mass to about 1.8 $M_\odot$, and an increase of the typical equatorial radius by about 1 km.

The sensitivity to changes of the QM EOS is explored in Fig. 15, comparing results obtained using the MIT model with a bag constant $B = 90$ MeV/fm^3 and with $B(\rho)$, and also the more sophisticated color dielectric model (CDM) [32, 49]. We note that in all cases the maximum mass lies between 1.5 and 1.6 solar masses, while there are larger differences for the radii.

Finally, in Fig. 16 we display the Kepler periods P_K $(= 2\pi/\Omega_K)$ versus the rotational star mass for several different stellar sequences based on different EOS. Purely hadronic stars, shown by the dotted and long-dashed lines respectively, show

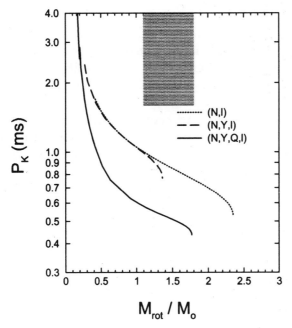

Figure 16: Kepler period versus the rotational mass for purely hadronic stars as well as hybrid stars. The following core compositions are considered: q i) nucleons and leptons (dotted line); ii) nucleons, hyperons, and leptons (dashed line); iii) hadrons, quarks, and leptons (solid line). The shaded area represents the current range of observed data.

instability against mass shedding first, because of their relatively large equatorial radii. Their limiting mass configurations are characterized by values of the Kepler period larger than half a millisecond, in agreement with results usually found in the literature [50]. In contrast, the more compact hybrid stars can reach stable periods smaller than half a millisecond.

7 Conclusions

In this contribution we reported the theoretical description of nuclear matter in the BHF approach and its various refinements, with the application to neutron star structure calculation. We pointed out the important role of TBF at high density, which is, however, strongly compensated by the inclusion of hyperons. The resulting hadronic neutron star configurations have maximum masses of less than 1.4 $M_\odot$, and the presence of quark matter inside the star is required in order to reach larger values.

Concerning the treatment of quark matter in the standard MIT bag model, we found that a density-dependent bag parameter $B(\rho)$ is necessary in order to be compatible with the CERN-SPS findings on the phase transition from hadronic to quark matter. Joining the corresponding EOS with the hadronic one, maximum masses of about 1.6 $M_\odot$ are reached, in line with other recent calculations of neutron star properties employing various phenomenological RMF nuclear EOS together with either effective mass bag model [51] or Nambu-Jona-Lasinio model [52] EOS for quark matter.

The value of the maximum mass of neutron stars obtained according to our analysis appears rather robust with respect to the uncertainties of the nuclear and the quark matter EOS. Therefore, the experimental observation of a very heavy ($M \gtrsim 1.6 \ M_\odot$) neutron star, as claimed recently by some groups [53] ($M \approx 2.2 \ M_\odot$), if confirmed, would suggest that either serious problems are present for the current theoretical modelling of the high-density phase of nuclear matter, or that the assumptions about the phase transition between hadron and quark phase are substantially wrong. In both cases, one can expect a well defined hint on the high-density nuclear matter EOS.

Acknowledgments

We would like to thank our collaborators J. Cugnon, A. Lejeune, Z.H. Li, U. Lombardo, F. Mathiot, P.K. Sahu, F. Weber, X.R. Zhou, and W. Zuo.

References

[1] P. Bonche, E. Chabanat, P. Haensel, J. Meyer, and R. Schaeffer, Nucl. Phys. **A635**, 231 (1998); **A643**, 441 (1998).

[2] B. D. Serot and J. D. Walecka, Adv. Nucl. Phys. **16**, 1 (1986).

[3] J. P. Jeukenne, A. Lejeune, and C. Mahaux, Phys. Rep. **25C**, 83 (1976); M. Baldo, in *Nuclear Methods and the Nuclear Equation of State*, International Review of Nuclear Physics, Vol.8, (World Scientific, Singapore, 1999).

[4] R. Machleidt, Adv. Nucl. Phys. **19**, 189 (1989); R. Brockmann and R. Machleidt, Phys. Rev. **C42**, 1965 (1990); G. Q. Li, R. Machleidt, and R. Brockmann, Phys. Rev. **C45**, 2782 (1992); D. Alonso and F. Sammarruca, Phys. Rev. **C67**, 054301 (2003); P. G. Krastev and F. Sammarruca, Phys. Rev. **C74**, 025808 (2006).

[5] A. Akmal and V. R. Pandharipande, Phys. Rev. **C56**, 2261 (1997); A. Akmal, V. R. Pandharipande, and D. G. Ravenhall, Phys. Rev. **C58**, 1804 (1998); J. Morales, V. R. Pandharipande, and D. G. Ravenhall, Phys. Rev. **C66**, 054308 (2002).

[6] N. Kaiser, S. Fritsch, and W. Weise, Nucl. Phys. **A697**, 255 (2002); D. R. Entem and R. Machleidt, Phys. Lett. **B524**, 93 (2002); Phys. Rev. **C68**, 041001(R) (2003).

[7] N. K. Glendenning, Nucl. Phys. **A493**, 521 (1989); *Compact Stars, Nuclear Physics, Particle Physics, and General Relativity*, 2nd ed., (Springer, New York, 2000).

[8] B. D. Day, Phys. Rev. **C24**, 1203 (1981); H. Q. Song, M. Baldo, G. Giansiracusa, and U. Lombardo, Phys. Rev. Lett. **81**, 1584 (1998); M. Baldo, A. Fiasconaro, H. Q. Song, G. Giansiracusa, and U. Lombardo, Phys. Rev. **C65**, 017303 (2002); R. Sartor, Phys. Rev. **C73**, 034307 (2006).

[9] M. Martzolff, B. Loiseau, and P. Grangé, Phys. Lett. **B92**, 46 (1980); P. Grangé, A. Lejeune, M. Martzolff, and J.-F. Mathiot, Phys. Rev. **C40**, 1040 (1989).

[10] S. A. Coon, M. D. Scadron, P. C. McNamee, B. R. Barrett, D. W. E. Blatt, and B. H. J. McKellar, Nucl. Phys. **A317**, 242 (1979); R. G. Ellis, S. A. Coon, and B. H. J. McKellar, Nucl. Phys. **A438**, 631 (1985).

[11] A. Lejeune, U. Lombardo, and W. Zuo, Phys. Lett. **B477**, 45 (2000); W. Zuo, A. Lejeune, U. Lombardo, and J.-F. Mathiot, Nucl. Phys. **A706**, 418 (2002); Eur. Phys. J. **A14**, 469 (2002); Prog. Theor. Phys. Suppl. **146**, 478 (2002).

[12] H.-J. Schulze, A. Lejeune, J. Cugnon, M. Baldo, and U. Lombardo, Phys. Lett. **B355**, 21 (1995); H.-J. Schulze, M. Baldo, U. Lombardo, J. Cugnon, and A. Lejeune, Phys. Rev. **C57**, 704 (1998).

[13] J. Cugnon, A. Lejeune, and H.-J. Schulze, Phys. Rev. **C62**, 064308 (2000); I. Vidaña, A. Polls, A. Ramos, and H.-J. Schulze, Phys. Rev. **C64**, 044301 (2001).

[14] M. Baldo, G. F. Burgio, and H.-J. Schulze, Phys. Rev. **C58**, 3688 (1998); Phys. Rev. **C61**, 055801 (2000).

[15] H.-J. Schulze, A. Polls, A. Ramos, and I. Vidaña, Phys. Rev. **C73**, 058801 (2006).

[16] G. E. Brown et al., Comm. Nucl. Part. Phys. **17**, 39 (1987).

[17] Z. H. Li, U. Lombardo, H.-J. Schulze, W. Zuo, L. W. Chen, and H. R. Ma, Phys. Rev. **C74**, 047304 (2006).

[18] F. Coester, S. Cohen, B. Day, and C. M. Vincent, Phys. Rev. **C1**, 769 (1970); F. Coester, B. Day, and A. Goodman, Phys. Rev. **C5**, 1135 (1972); B. D. Day, Phys. Rev. Lett. **47**, 226 (1981).

[19] W. Zuo, Caiwan Shen, and U. Lombardo, Phys. Rev. **C67**, 037301 (2003).

[20] B. S. Pudliner, V. R. Pandharipande, J. Carlson, and R. B. Wiringa, Phys. Rev. Lett. **74**, 4396 (1995); B. S. Pudliner, V. R. Pandharipande, J. Carlson, S. C. Pieper, and R. B. Wiringa, Phys. Rev. **C56**, 1720 (1997); R. B. Wiringa, S. C. Pieper, J. Carlson, and V. R. Pandharipande, Phys. Rev. **C62**, 014001 (2000); J. Carlson, J. Morales, V. R. Pandharipande, and D. G. Ravenhall, Phys. Rev. **C68**, 025802 (2003).

[21] J. Carlson, V. R. Pandharipande, and R. B. Wiringa, Nucl. Phys. **A401**, 59 (1983).

[22] R. B. Wiringa, V. G. J. Stoks, and R. Schiavilla, Phys. Rev. **C51**, 38 (1995).

[23] S. C. Pieper, V. R. Pandharipande, R. B. Wiringa, and J. Carlson, Phys. Rev. **C64**, 014001 (2001).

[24] A. Lejeune, P. Grangé, M. Martzolff, and J. Cugnon, Nucl. Phys. **A453**, 189 (1986); I. Bombaci and U. Lombardo, Phys. Rev. **C44**, 1892 (1991); W. Zuo, I. Bombaci, and U. Lombardo, Phys. Rev. **C60**, 024605 (1999).

[25] S. Shapiro and S. A. Teukolsky, *Black Holes, White Dwarfs, and Neutron Stars*, (John Wiley & Sons, New York, 1983)

[26] M. Baldo, I. Bombaci, and G. F. Burgio, Astron. Astroph. **328**, 274 (1997).

[27] R. Feynman, F. Metropolis, and E. Teller, Phys. Rev. **C75**, 1561 (1949).

[28] G. Baym, C. Pethick, and D. Sutherland, Astrophys. J. **170**, 299 (1971).

[29] J. W. Negele and D. Vautherin, Nucl. Phys. **A207**, 298 (1973).

[30] G. F. Burgio, M. Baldo, P. K. Sahu, and H.-J. Schulze, Phys. Rev. **C66**, 025802 (2002).

[31] M. Baldo, M. Buballa, G. F. Burgio, F. Neumann, M. Oertel, and H.-J. Schulze, Phys. Lett. **B562**, 153 (2003).

[32] C. Maieron, M. Baldo, G. F. Burgio, and H.-J. Schulze, Phys. Rev. **D70**, 043010 (2004).

[33] P. M. M. Maessen, Th. A. Rijken, and J. J. de Swart, Phys. Rev. **C40**, 2226 (1989).

[34] V. G. J. Stoks and Th. A. Rijken, Phys. Rev. **C59**, 3009 (1999).

[35] Th. A. Rijken, V. G. J. Stoks, and Y. Yamamoto, Phys. Rev. **C59**, 21 (1999).

[36] J. Cugnon, A. Lejeune, and H.-J. Schulze, Phys. Rev. **C62**, 064308 (2000); I. Vidaña, A. Polls, A. Ramos, and H.-J. Schulze, Phys. Rev. **C64**, 044301 (2001).

[37] H. Heiselberg and M. Hjorth-Jensen, Phys. Rep. **328**, 237 (2000).

[38] I. Vidaña, A. Polls, A. Ramos, M. Hjorth-Jensen, and V. G. J. Stoks, Phys. Rev. **C61**, 025802 (2000).

[39] A. Li, G. F. Burgio, U. Lombardo, and W. Zuo, Phys. Rev. **C74**, 055801 (2006).

[40] U. Heinz and M. Jacobs, nucl-th/0002042; U. Heinz, Nucl. Phys. **A685**, 414 (2001).

[41] A. Chodos, R. L. Jaffe, K. Johnson, C. B. Thorn, and V. F. Weisskopf, Phys. Rev. **D9**, 3471 (1974).

[42] E. Witten, Phys. Rev. **D30**, 272 (1984); G. Baym, E. W. Kolb, L. McLerran, T. P. Walker, and R. L. Jaffe, Phys. Lett. **B160**, 181 (1985); N. K. Glendenning, Mod. Phys. Lett. **A5**, 2197 (1990).

[43] E. Fahri and R. L. Jaffe, Phys. Rev. **D30**, 2379 (1984).

[44] H. Satz, Ann. Rev. Nucl. Part. Sc., **35**, 245 (1987).

[45] M. Alford and S. Reddy, Phys. Rev. **D67**, 074024 (2003).

[46] N. K. Glendenning, Phys. Rev. **D46**, 1274 (1992).

[47] H. Heiselberg, C. J. Pethick, and E. F. Staubo, Phys. Rev. Lett. **70**, 1355 (1993); H. Heiselberg, Phys. Rev. **D48**, 1418 (1993); N. K. Glendenning and S. Pei, Phys. Rev. **C52**, 2250 (1995); M. B. Christiansen and N. K. Glendenning, Phys. Rev. **C56**, 2858 (1997); N. K. Glendenning, Phys. Rep. **342**, 393 (2001); D. N. Voskresensky, M. Yasuhira, and T. Tatsumi, Phys. Lett. **B541**, 93 (2002); Nucl. Phys. **A718**, 359 (2003); Nucl. Phys. **A723**, 291 (2003); T. Endo, T. Maruyama, S. Chiba, and T. Tatsumi, Nucl. Phys. **A749**, 333 (2005); Prog. Theor. Phys. **115**, 337 (2006).

[48] G. F. Burgio, H.-J. Schulze, and F. Weber, Astron. Astrophys. **408**, 675 (2003).

[49] H. J. Pirner, G. Chanfray, and O. Nachtmann, Phys. Lett. **B147**, 249 (1984); A. Drago, U. Tambini, and M. Hjorth-Jensen, Phys. Lett. **B380**, 13 (1996).

[50] F. Weber, *Pulsars as Astrophysical Laboratories for Nuclear and Particle Physics*, (IOP Publishing, Bristol and Philadelphia, 1999); F. Weber, J. Phys. **G25**, R195 (1999).

[51] K. Schertler, C. Greiner, P. K. Sahu, and M. H. Thoma, Nucl. Phys. **A637**, 451 (1998); K. Schertler, C. Greiner, J. Schaffner-Bielich, and M. H. Thoma, Nucl. Phys. **A677**, 463 (2000).

[52] K. Schertler, S. Leupold, and J. Schaffner-Bielich, Phys. Rev. **C60**, 025801 (1999).

[53] P. Kaaret, E. Ford, and K. Chen, Astrophys. J. Lett. **480**, L27 (1997); W. Zhang, A. P. Smale, T. E. Strohmayer, and J. H. Swank, Astrophys. J. Lett. **500**, L171 (1998).

Physics and Astrophysics of Hadrons and Hardronic Matter
Editor: A. B. Santra

Towards a Chiral Effective Field Theory of Nuclear Matter

S. Mallik

Saha Institute of Nuclear Physics
1/AF, Bidhannagar, Kolkata-700064, India
email: mallik@theory.saha.ernet.in

Abstract

As a preliminary attempt to formulate an effective theory of nuclear matter, we undertake to calculate the effective pole parameters of nucleon in such a medium. We begin with the virial expansion of these parameters to leading order in nucleon number density in terms of the on-shell NN scattering amplitude. We then proceed to calculate the same parameters in the effective theory, getting a formula for the nucleon mass-shift to leading order, that was known already to give too large a value to be acceptable at normal nuclear density. At this point the virial expansion suggests a modification of this formula, which we carry out following Weinberg's method for the two-nucleon system in the effective theory. The results are encouraging enough to attempt a complete, next-to-leading order calculation of the off-shell nucleon spectral function in nuclear medium.

1 Introduction

A fundamental problem of nuclear physics is to explain the properties of nuclear matter (and finite nuclei) in terms of an effective field theory at low energy based only on the chiral symmetry of QCD and its assumed spontaneous breaking. While such a theory, called chiral perturbation theory, is eminently successful in dealing with pionic processes [1, 2], and also to a great extent, with those of pions and a single nucleon [3], its application to the NN system, that is called for in dealing with problems of nuclear matter, is not straightforward [4]. As first pointed out by Weinberg [5], the loop graphs for such multinucleon processes do not obey the usual power counting rule, a problem that is related to the presence of shallow bound states, such as the deuteron and the virtual bound state in the two-nucleon system.

To deal with this difficulty of the effective theory, Weinberg [5] proposes a two-step solution, which, in the context of the old fashioned perturbation theory, may be stated as follows: First, one has to construct the effective potential from those connected graphs of the T-matrix, that do not contain any pure-nucleon intermediate state and so obey the power counting rule. Then the remaining graphs with pure-nucleon intermediate states violating this rule are summed over by the Lippmann-Schwinger integral equation for the T-matrix with this effective potential.

In this work we wish to apply this procedure to find the mass and the width of a nucleon at rest in nuclear matter. For the purpose of comparison and interpretation, however, we first obtain a semi-phenomenological formula given by the first order virial expansion of the nucleon self-energy in terms of the spin-averaged NN scattering amplitude in the forward direction, to be denoted below by $\overline{M}$.

Then we turn to the effective theory. Here the one-loop formula was known earlier [6], giving the mass-shift as the nucleon number density times a constant, depending on the parameters of the effective Lagrangian. Expressing the constant in terms of scattering lengths obtainable from experiment, one finds this mass-shift at normal nuclear density to be too large to be acceptable.

It is not difficult to identify the source of the trouble with the one-loop formula. Comparing it with the virial formula, we see that the one-loop formula approximates the amplitude $\overline{M}$ within the momentum integral by its value at threshold. Clearly this is too drastic an approximation to be realistic, in view of the nearby bound states poles in the amplitude. To improve this formula within the effective theory, the Weinberg analysis suggests that we regard the constant amplitude as the effective potential (to leading order) and solve the Lippmann-Schwinger equation for the complete amplitude. Replacing the constant in the one-loop formula by this solution, we get a greatly reduced value for the effective nucleon mass. We also find the effect of including phenomenologically the effective ranges in the formula [7].

In Sec.II we obtain the virial expansion of the self-energy to leading order in nucleon density. From then onwards, we devote ourselves to the effective field theory. In Sec.III we obtain the pieces of the effective Lagrangian, that will be needed in evaluating the one-loop graphs. We use the Dirac, rather than the Pauli, spinor for the nucleon. The advantage of this relativistic treatment is the clear separation of the vacuum and the density dependent parts of the nucleon propagator, the latter part containing the on-shell delta-function. In Sec.IV we evaluate the graphs to obtain the leading terms for the effective parameters of nucleon in nuclear matter. We modify the mass shift formula in Sec. V. Finally Sec.VI concludes with a summary of our work.

2   Virial formula

Here we obtain the leading term in the virial expansion for the nucleon self-energy in nuclear medium [8, 9]. The derivation starts by considering the nucleon self-energy in *vacuum*, which may be expressed as an S-matrix element,

$$-i(2\pi)^4\delta^4(p_1' - p_1)\overline{u}(\boldsymbol{p}_1', \sigma_1')\Sigma^{(0)}(p)u(\boldsymbol{p}_1, \sigma_1)$$
$$= \langle 0|b(\boldsymbol{p}_1', \sigma_1')\,(S - 1)\,b^\dagger(\boldsymbol{p}_1, \sigma_1)|0\rangle\,, \qquad (1)$$

where S is the familiar scattering matrix operator, $S = Te^{i\int \mathcal{L}_{int}(x)\,d^4x}$ for an interaction Lagrangian $\mathcal{L}_{int}$. The subscript 1 on the variables of the particle anticipates another one with which it will interact in the medium. In fact, we shall express below the nucleon self-energy in *nuclear medium* in terms of the *NN* scattering amplitude in *vacuum*, defined as usual by

$$\langle p_1', \sigma_1'; p_2', \sigma_2' | S - 1 | p_1, \sigma_1; p_2, \sigma_2 \rangle$$
$$= i(2\pi)^4 \delta^4(p_1' + p_2' - p_1 - p_2) M(p_1, \sigma_1; p_2, \sigma_2 \to p_1', \sigma_1', ; p_2', \sigma_2'), \quad (2)$$

where M stands for the scattering matrix sandwiched between spinors corresponding to the final and the initial states of the two nucleons.

Below we shall meet the spin averaged amplitude in the forward direction,

$$\overline{M}(p_1, p_2 \to p_1, p_2)) = \frac{1}{4} \sum_{\sigma_1, \sigma_2} M(p_1, \sigma_1; p_2, \sigma_2 \to p_1, \sigma_1; p_2, \sigma_2), \quad (3)$$

With our normalization of states, the amplitude $\overline{M}$ is Lorentz invariant.

To obtain the self-energy in nuclear medium, we have to replace the vacuum expectation value in Eq. (2.1) by an appropriate one. Although we specialize later to zero temperature, we take here the most general average over an ensemble of systems maintained at temperature $T(= 1/\beta)$ with nucleon chemical potential μ. Thus the in-medium self-energy Σ is given by

$$-i(2\pi)^4 \delta^4(p_1' - p_1)\overline{u}(p_1', \sigma_1')\Sigma(p)u(p_1, \sigma_1) = \langle b(p_1', \sigma_1')(S - 1)b(p_1, \sigma_1) \rangle, \quad (4)$$

where for any operator O,

$$\langle O \rangle = Tr[e^{-\beta(H - \mu\mathcal{N})}O]/Tr e^{-\beta(H - \mu\mathcal{N})}.$$

Here H is the Hamiltonian and $\mathcal{N}$ the nucleon number operator with chemical potential μ. Clearly this form of the Boltzmann weight breaks explicit Lorentz invariance and singles out the rest frame of the medium.

We now make use of the virial expansion to first order in density. For an operator O, the ensemble average in nuclear medium can be expanded as

$$\langle O \rangle = \langle 0|O|0 \rangle + \sum_{\sigma_2} \int \frac{d^3 p_2}{(2\pi)^3 2E_{p_2}} n^-(E_{p_2}) \langle p_2, \sigma_2 | O | p_2, \sigma_2 \rangle + \cdots,$$

where $n^-(E_p)$ is the nucleon distribution function, $n^-(E_p) = 1/[e^{\beta(E_p - \mu)} + 1]$, $E_p = \sqrt{m^2 + p^2}$. Applying it to the left hand side of Eq. (2.4), we get for the difference $\Sigma^{(n)}(p) = \Sigma(p) - \Sigma^{(0)}(p)$,

$$-i(2\pi)^4 \delta^4(p_1' - p_1)\overline{u}(p_1', \sigma_1')\Sigma^{(n)}(p_1)u(p_1, \sigma_1) = \sum_{\sigma_2} \int \frac{d^3 p_2}{(2\pi)^3 2E_{p_2}} n^-(E_{p_2})$$

$$\times \langle p_2, \sigma_2 | b(p_1', \sigma_1')(S - 1)b^\dagger(p_1, \sigma_1) | p_2, \sigma_2 \rangle. \quad (5)$$

170

The matrix element in Eq. (2.5) will be immediately recognised to be the NN scattering amplitude defined above by Eq. (2.2). Cancelling the δ-function on both sides, we set $\sigma_1' = \sigma_1$ and sum over σ_1 also to get

$$-tr\{\Sigma^{(n)}(p_1)(\not{p}_1 + m)\} = 4 \int \frac{d^3p_2}{(2\pi)^3 2E_{p_2}} n^-(E_{p_2})\overline{M}(p_1, p_2 \to p_1, p_2), \qquad (6)$$

where the tr(ace) is over matrices in Dirac space.

So far we did not state explicitly the isospin structure of the amplitude $\overline{M}$, which is now easy to figure out. We consider symmetric nuclear matter and work in the limit of iso-spin symmetry. Let the traversing nucleon be in any one of its isospin states, say a proton. It may scatter with a proton or a neutron in the medium. The amplitude is therefore given by the sum,

$$\overline{M} = \overline{M}_{pp \to pp} + \overline{M}_{pn \to pn} . \qquad (7)$$

We now restrict to the case, where the three-momentum $\boldsymbol{p}_1$ is set equal to zero. Then the rest frame of the medium coincides with the lab frame of the scattering process. The shifted pole position is now readily obtained as [10],

$$m^* - \frac{i}{2}\gamma = m + U + V = m - \frac{1}{m} \int \frac{d^3p_2}{(2\pi)^3 2E_{p_2}} n(p_2)\overline{M}(p_2), \qquad (8)$$

where m^* is the effective mass of the nucleon and γ gives the damping rate of nucleonic excitations. The numerical evaluation of this formula using the phase shift analysis of the Nijmegen group [11] will be shown in Figs.3 and 4, along with those of the effective theory in Sec. V.

3 Chiral lagrangian

We now study nucleon propagation in nuclear matter in the effective theory by analysing the two-point function of the nucleon current $\eta(x)$ [12], built out of three quark fields so as to have the quantum numbers of the nucleon. In this work we do not need to spell out the form of this current; it suffices only to note that its tranformation under the chiral symmetry group $SU(2)_R \times SU(2)_L$ is the same as that of the quark doublet itself. To get the low energy structure of vertices with the nucleon current, we extend the original QCD Lagrangian by including terms that couple it to an external (spinor) field $f(x)$.

For the pion triplet, one defines as usual three different quantities, namely $U, u(u^2 = U)$ and $u_\mu = iu^\dagger(\partial_\mu u)u^\dagger$. Knowing their transformation rules, as well as those of the nucleon field and the external spinor field, we can construct the effective Lagrangian for the nucleon-nucleon system including pions in presence of the external field f. Its terms may be put into three groups,

$$\mathcal{L}_{eff} = \mathcal{L}_{\pi\psi} + \mathcal{L}_{\psi^4} + \mathcal{L}_f, \tag{9}$$

which we write below one by one. $\mathcal{L}_{\pi\psi}$ is given by the familiar terms,

$$\mathcal{L}_{\pi\psi} = \frac{F_\pi^2}{4}\partial_\mu U \partial^\mu U^\dagger + \bar{\psi}(i\slashed{\partial} - m)\psi + \frac{1}{2}g_A \bar{\psi}\slashed{\psi}\gamma_5\psi. \tag{10}$$

We choose the explicit representation $U = \exp(i\phi^a\tau^a/F_\pi)$, where ϕ^a are the hermitian pion fields ($a = 1, 2, 3$), τ^a, the Pauli matrices, F_π, the pion decay constant ($F_\pi = 93$ MeV) and g_A, the neutron decay constant ($g_A = 1.26$). In the following we do not need vertices with pion fields only.

The piece $\mathcal{L}_{\psi^4}$ giving the leading quartic interaction of nucleons has been written by Weinberg in terms of the Pauli spinor (and iso-spinor) for the nucleon field [5]. As we intend to work with the Dirac spinor $\psi(x)$, we rewrite it as

$$\mathcal{L}_{\psi^4} = -\frac{C_S}{8}\{\bar{\psi}(1+\gamma_0)\psi\}^2 - \frac{C_T}{8}\{\bar{\psi}(1+\gamma_0)\vec{\gamma}\gamma_5\psi\}^2 + \cdots. \tag{11}$$

where C_S and C_T are constants.

In the piece $\mathcal{L}_f$ dependent on the external field, we need two couplings, namely, that are linear and cubic in ψ. Using the above transformation rules, the piece linear in ψ may be written as

$$\begin{aligned}
\mathcal{L}_{f\psi} &= \lambda\overline{(uf_R)}\psi + \lambda'\overline{(u^\dagger f_L)}\psi + h.c.\\
&= \frac{\lambda}{2}\bar{f}(1-\gamma_5)u^\dagger\psi + \frac{\lambda'}{2}\bar{f}(1+\gamma_5)u\psi + h.c.,
\end{aligned} \tag{12}$$

where invariance under parity requires $\lambda = \lambda'$. Following the construction of $\mathcal{L}_{\psi^4}$, we write the other piece as

$$\mathcal{L}_{f\psi^3} = \frac{A_S}{4}\bar{\psi}(1+\gamma_0)\psi\bar{f}(1+\gamma_0)\psi + \frac{A_T}{4}\bar{\psi}(1+\gamma_0)\vec{\gamma}\gamma_5\psi\bar{f}(1+\gamma_0)\vec{\gamma}\gamma_5\psi + h.c. \tag{13}$$

where A_S and A_T are again two constants.

4 One-loop formula

We start with the ensemble average of the two-point function of nucleon currents in symmetric nuclear matter,

$$\Pi(q) = i \int d^4 x e^{iqx} \langle T\eta(x)\bar{\eta}(0)\rangle, \tag{14}$$

in the notation introduced already in (2.4).

In the real time version of quantum field theory in a medium, any two point function assumes the form of a 2×2 matrix. But the dynamics is given essentially by a single analytic function, that is determined by the 11-component itself. As the simplest example, consider the nucleon propagator. Its 11-component is given by [13],

$$\frac{1}{i}S(p)_{11} \equiv \int d^4x e^{ipx} \langle T\psi(x)\bar{\psi}(0)\rangle_{11} = (\not{p}+m) \left[\frac{i}{p^2 - m^2 + i\epsilon} \right.$$
$$\left. - \left\{ n^-(E_p)\theta(p_0) + n^+(E_p)\theta(-p_0) \right\} 2\pi\delta(p^2 - m^2) \right], \quad (15)$$

where $n^{\mp}$ are the distribution functions for nucleons and antinucleons. The corresponding analytic function $S(p)$ is identical to the free propagator in vacuum and so independent of $n^{\mp}$.

In the following we shall work in the limit of zero temperature, when the distribution functions become, $n^- \to \theta(\mu - E_p)$, $n^+ \to 0$. Then the nucleon number density, $\bar{n}$ is given by

$$\bar{n} = 4 \int \frac{d^3p}{(2\pi)^3} \theta(\mu - E_p) = \frac{2p_F^3}{3\pi^2},$$

where p_F is the Fermi momentum related to the chemical potential μ by $\mu^2 = m^2 + p_F^2$. We want to calculate the density dependent part of $\Pi(q)$ in the low energy region to first order in $\bar{n}$. To this end we draw all the Feynman graphs with one loop containing a nucleon line. In addition to the single nucleon line forming a loop, we include also the loop containing an additional pion line to account for singularities with the lowest threshold. They are depicted in Fig. 1 along with the free propagator graph (a).

We shall work at $\vec{q} = 0$, when the free propagation graph (a) takes the form

$$\Pi(E)_{(a)} = \lambda^2 S(E) = -\frac{\lambda^2}{E - m + i\epsilon} \frac{1}{2}(1 + \gamma_0), \quad (16)$$

in the vicinity of the nucleon pole. Collecting the results for the simple and the double poles at $E = m$ from the rest of the graphs, we find the vacuum pole (3.3) to be modified in nuclear medium to

$$-\frac{\lambda^{*2}}{E - m^* + i\epsilon} \frac{1}{2}(1 + \gamma_0),$$

where

$$\lambda^* = \lambda \left\{ 1 - \left(\frac{3g_A}{8mF_\pi^2} - 2\zeta \right) \bar{n} \right\}, \quad \zeta = \frac{3}{8\lambda}(A_S - A_T) \quad (17)$$

173

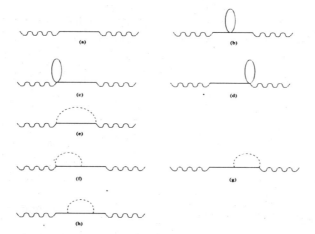

Figure 1: Free propagator and one loop Feynman graphs for the two point function in the low energy region. The wavy line represents the nucleon current, while the continuous and the dashed internal lines are for nucleon and pion propagation.

$$m^* = m + \frac{3}{4}\left(C_S - C_T + \frac{g_A^2}{4F_\pi^2}\right)\bar{n} \tag{18}$$

The coupling constant ζ is not known at present. As is known already [6] and we shall also see below, the formula for m^* is unacceptable at normal nuclear density.

5 Modified Formula

If we note that the nucleon mass-shift (3.5) is given by graphs (b) and (h) of Fig.1 and recall the presence of the mass-shell delta-function in the density dependent part of the nucleon propagator (3.2), it is easy to guess that the constant coefficient of $\bar{n}$ in this shift must be related to some on-shell NN scattering amplitude at threshold. To obtain the actual relation, we evaluate the scattering amplitude in the same chiral Lagrangian framework as that Sect. III. In the tree approximation the contributing Feynman graphs are shown in Fig.2.

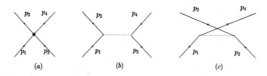

Figure 2: Nucleon-nucleon scattering amplitude in the tree approximation.

The result for the amplitude $\overline{M}$, defined by Eq.(2.3) is,

$$\overline{M} = -6m^2 \left\{ (C_S - C_T) \frac{(E_1 + m)(E_2 + m)}{4m^2} + \frac{g_A^2}{4F_\pi^2} \right\}, \tag{19}$$

E_1 and E_2 being the energies of the two nucleons. Thus at threshold $\overline{M}$ involves the same combination of constants as those in Eq.(3.5) for m^*.

The amplitude $\overline{M}$ at threshold can be expressed in terms of the s-wave spin-singlet and -triplet scattering lengths a_1 and a_3 to get

$$m^* - m = \frac{3\pi}{2m}(a_1 + a_3)\overline{n}. \tag{20}$$

With their experimental values and at normal nuclear density ($p_F = 1.36\,fm^{-1}$), it gives $m^* - m = -620$ MeV [6], which is unacceptably large.

To look for the source of the problem, we compare the one-loop formula (3.5) with the one given by the virial expansion (2.8) and note that the momentum dependent amplitude $\overline{M}$ in the latter has been approximated in the former by a constant, namely its value at threshold. Clearly we can improve this approximation in the Weinberg scheme of the effective theory by treating the constant as the effective potential (to leading order) and solving the Lippmann-Schwinger equation with this potential. As Weinberg himself shows [5], the solution in this case turns out to be the unitarized version of the potential. Accordingly we replace the scattering lengths by the corresponding momentum dependent amplitudes satisfying (elastic) unitarity,

$$-a_i \rightarrow f_i^{(1)}(k) = \left(-\frac{1}{a_i} - ik \right)^{-1}, \quad i = 1, 3. \tag{21}$$

Here k is the centre-of-mass momentum, to be distinguished from p, which is the momentum of the in-medium nucleons in the rest frame of the nucleon under consideration, the two being related by $k^2 = m(\sqrt{m^2 + p^2} - m)/2$.

An even better approximation of the scattering amplitude is the effective range approximation, that would result from including derivative terms in the effective theory [7]. The corresponding replacement would read

$$-a_i \rightarrow f_i^{(2)}(k) = \left(-\frac{1}{a_i} + \frac{1}{2}r_i k^2 - ik \right)^{-1}, \quad i = 1, 3, \tag{22}$$

where $r_{1,3}$ are the effective ranges in the spin-singlet and -triplet s waves. Without trying to relate the effective ranges to the coupling constants of the higher derivative terms in the effective Lagrangian, we shall take their values as well as those of the scattering lengths from experiment.

We thus get both the real and the imaginary parts of the pole position as,

$$m^* - \frac{i}{2}\gamma = m - \frac{6\pi}{m}\int_0^{p_F}\frac{d^3p}{(2\pi)^3}\{f_1^{(2)}(k) + f_3^{(2)}(k)\}\,. \tag{23}$$

The numerical evaluation of Eq.(4.5) is shown in Figs. 3 and 4. In normal nuclear matter, we get $\triangle m \equiv m^* - m = -33$ MeV and $\gamma/2 = 110$ MeV.

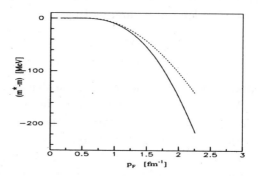

Figure 3: Shift in nucleon mass in nuclear matter as a function of Fermi momentum. The solid curve results from the virial formula, evaluated with the partial-wave analysis of the Nijmegen group and the dashed one from the modified formula of the effective theory.

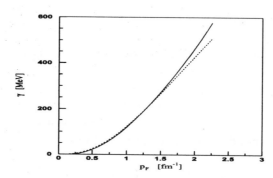

Figure 4: Damping rate of nucleonic excitation in nuclear matter as a function of Fermi momentum. The origin of solid and dashed curves are the same as in Fig.3.

Our results may be compared with those in the literature. There have been a number of calculations of the nucleon spectral function in terms of the (off-shell) NN scattering amplitude evaluated with available NN potentials. We pick out from these works the real and the imaginary parts of the on-shell nucleon self-energy in normal nuclear matter at zero three momentum. Thus Baldo et al [14] give $\triangle m = -55$ MeV,

$\gamma/2 = 63$ MeV (as reproduced in [17]). Benhar et al give $\triangle m = -68$ MeV, $\gamma/2 = 16$ MeV. Jong et al give $\triangle m = -65$ MeV, $\gamma/2 = 25$ MeV. Also a phenomenological determination of the width in terms of the differential cross-section for the NN scattering gives $\gamma/2 > 25$ MeV [17]. Finally we note the result from the virial formula [10], $\triangle m = -37$ MeV, $\gamma/2 = 112$ MeV.

It will be noted that our mass-shift and width formulae do not involve any Pauli blocking effect, our improved scattering amplitude being still in *vacuum* and not in *medium*. An estimate of this effect can be made from the work of Machleidt [18], who finds a change of about 20% in the value of the potential energy per nucleon in nuclear matter, when the Pauli projection operators are withheld from the calculation. Incidentally, the phenomenological formula for the width in Ref.[17] involves the differential cross-section in *vacuum*, the Pauli blocking factors appearing only for the final nucleons. Such factors, however, do not appear in our formula, as it depends on the forward scattering amplitude, where the final particles occupy the states vacated by the initial ones.

6 Summary

The spectral function of nucleon in nuclear matter depends on the interaction in a two-nucleon system. To treat this interaction in the framework of effective chiral field theory [5], one has to derive first the effective potential from the effective Lagrangian and then solve the dynamical equation with this potential. In this way one can restore the otherwise invalid power counting rule and accomodate at the same time the singularities of the scattering amplitude in the low energy region.

The failure of the earlier calculation of the effective nucleon mass [6] may now be understood in this framework as due to representing the two-nucleon scattering amplitude by the (leading term of the) effective potential itself. In this work we proceed further to carry out the next dynamical step, replacing the potential by the unitarised scattering amplitude. By including the effective range terms in the s-wave amplitudes, we actually include the next-to-leading order terms also in the effective potential.

The results obtained here to leading order encourages us to calculate the momentum dependent nucleon spectral function, including the density dependent terms in the effective potential.

References

[1] S. Weinberg, Physica, **96A**, 327 (1979).

[2] J. Gasser and H. Leutwyler, Ann. Phys. **158**, 142 (1984); Nucl. Phys. **B 250**, 465 (1985).

[3] T. Becher and H. Leutwyler, JHEP 106, **17** (2001), hep-ph/0103263.

[4] For a recent review see, E. Epelbaum, Prog. Part. Nucl. Phys., 57, **654** (2006), nucl-th/0509032.

[5] S. Weinberg, Phys. Lett. **B 251**, 288 (1990); Nucl. Phys. **B 363**, 3 (1991).

[6] D. Montano, H.D. Politzer and M.B. Wise, Nucl. Phys. **B 375**, 507 (1992).

[7] D.B. Kaplan, M.J. Savage and M.B. Wise, Nucl. Phys. **B 478**, 629 (1996), ibid. **B 534**, 329 (1998).

[8] H. Leutwyler and A.V. Smilga, Nucl. Phys. **B342**, 302 (1990)

[9] S. Mallik, Eur. Phys. J. **C 24**, 143 (2002).

[10] S. Mallik, A. Nyffeler, M.C.M. Rentmeester and S. Sarkar, Eur. J. Phys. **A 22**, 371 (2004).

[11] V.G.J. Stoks, R.A.M. Klomp, M.C.M. Rentmeester and J.J. de Swart, Phys. Rev. C **48**, 792, 1993

[12] B.L. Ioffe, Nucl. Phys. **B 188**, 317 (1981); Y. Chung, H.G. Dosch, M. Kremer and D. Schall, Nucl. Phys. **B 197**, 55 (1982).

[13] R.L. Kobes, G.W. Semenoff and N. Weiss, Z. Phys. **C 29**, 371 (1985).

[14] M. Baldo, I. Bombaci, G. Giansiracusa, U. Lombardo, C. Mahoux, R. Sartor, Nucl. Phys. A545, 741 (1992)

[15] O. Benhar, A. Fabrocini, S. Fantoni, Nucl. Phys. A550, 201 (1992)

[16] F. de Jong, H. Lenske, Phys. Rev. C 56, 154 (1997).

[17] J. Lehr and U. Mosel, Phys. Rev., **C 69**, 024603 (2004).

[18] R. Machleidt, Adv. in Nuc. Phys. vol 19, p.189 (Plenum Press, New York, 1989).

Physics and Astrophysics of Hadrons and Hardronic Matter
Editor: A. B. Santra

Kaons and Antikaons in Dense Hadronic Matter

Amruta Mishra

Department of Physics, I.I.T. Delhi
Hauz Khas, New Delhi 110 016, India
email: amruta@physics.iitd.ac.in

Abstract

The properties of mesons and baryons are modified in an hadronic medium. These are related to the experimental observables of the ultra relativistic heavy ion collision experiments, like the particle spectra and yield as well as their collective flow. We investigate the medium modifications of the kaons and antikaons using a chiral SU(3) model. The isospin effects on the optical potentials of kaons and antikaons are also be investigated. These can be particularly relevant at the future accelerator facility FAIR at GSI, where isospin asymmetric heavy ion collision experiments, using more neutron rich beams are planned to be adopted to study the compressed baryonic matter.

1 Introduction

The study of properties of hadrons under extreme conditions of temperature and density is an important topic in strong interaction physics. The subject has direct implications in heavy-ion collision experiments, in the study of astrophysical compact objects like neutron stars as well as in early universe. The in-medium properties of kaons have been investigated particularly because of their relevance in neutron star phenomenology as well as relativistic heavy-ion collisions. For example, in the interior of a neutron star the attractive kaon nucleon interaction might lead to kaon condensation as originally suggested by Kaplan and Nelson [1]. The in-medium modification of kaon/antikaon properties can be observed experimentally primarily in relativistic nuclear collisions, in the collective flow pattern as well as the particle yield and spectra.

The theoretical research work on the topic of medium modification of hadron properties was initiated by Brown and Rho [2] who suggested that the modifications of hadron masses should scale with the scalar quark condensate $\langle q\bar{q} \rangle$ at finite baryon density. The first attempts to extract the antikaon-nucleus potential from the analysis of kaonic-atom data were in favor of very strong attractive potentials of the order of -150 to -200 MeV at normal nuclear matter density ρ_0 [3, 4]. However, more recent self-consistent calculations based on a chiral Lagrangian [5–8] or coupled-channel G-matrix theory (within meson-exchange potentials) [9] only predicted moderate attraction with potential depths of -50 to -80 MeV at density ρ_0.

The problem with the antikaon potential at finite baryon density is that the antikaon-nucleon amplitude in the isospin channel $I = 0$ is dominated by the $\Lambda(1405)$

resonant structure, which in free space is only 27 MeV below the $\bar{K}N$ threshold. It is presently not clear if this physical resonance is a real excited state of a 'strange' baryon or it is some short lived molecular intermediate state which can be generated dynamically in a coupled channel T-matrix scattering equation using a suitable meson-baryon potential. Additionally, the coupling between the $\bar{K}N$ and πY ($Y = \Lambda, \Sigma$) channels is essential to get the proper dynamical behavior in free space. Correspondingly, the in-medium properties of the $\Lambda(1405)$, such as its pole position and its width, which in turn influence strongly the antikaon-nucleus optical potential, are very sensitive to the many-body treatment of the medium effects. Previous works have shown that a self-consistent treatment of the $\bar{K}$ self energy has a strong impact on the scattering amplitudes [5, 7–11] and thus on the in-medium properties of the antikaon. Due to the complexity of this many-body problem the actual kaon and antikaon self energies (or potentials) are still a matter of debate.

The topic of isospin effects in asymmetric nuclear matter has gained interest in the recent past [12]. The isospin effects are important in isospin asymmetric heavy ion collision experiments. Within the UrQMD model the density dependence of the symmetry potential has been studied by investigating observables like the π^-/π^+ ratio, the n/p ratio [13], the Δ^-/Δ^{++} ratio as well as the effects on the production of K^0 and K^+ [14] and on pion flow [15] for neutron rich heavy ion collisions. Recently, the isospin dependence of the in-medium NN cross section [16] has also been investigated.

In the present investigation we will use a chiral SU(3) model for the description of hadrons in the medium [17]. The nucleons – as modified in the hot hyperonic matter – have been studied within this model [18] previously. Furthermore, the properties of vector mesons [18, 19] – due to their interactions with nucleons in the medium – have also been examined and have been found to have appreciable modifications due to Dirac sea polarization effects. The chiral SU(3)$_{flavor}$ model was also generalized to SU(4)$_{flavor}$ to study the mass modification of D-mesons arising from their interactions with the light hadrons in hot hadronic matter in [20]. The energies of kaons (antikaons) at zero momentum, as modified in the medium due to their interaction with nucleons, consistent with the low energy KN scattering data [21], were also studied within this framework [22]. In the present work, we study the effect of isospin asymmetry on the kaon and antikaon optical potentials in the asymmetric nuclear matter.

2   The hadronic chiral *SU (3)* × *SU (3)* model

In this section the various terms of the effective hadronic Lagrangian used

$$\mathcal{L} = \mathcal{L}_{kin} + \sum_{W=X,Y,V,\mathcal{A},u} \mathcal{L}_{BW} + \mathcal{L}_{VP} + \mathcal{L}_{vec} + \mathcal{L}_0 + \mathcal{L}_{SB} \tag{1}$$

are discussed. Eq. (1) corresponds to a relativistic quantum field theoretical model of baryons and mesons built on a nonlinear realization of chiral symmetry and broken scale invariance (for details see [17–19]) to describe strongly interacting nuclear matter. The model was used successfully to describe nuclear matter, finite nuclei, hypernuclei and neutron stars. The Lagrangian contains the baryon octet, the spin-0 and spin-1 meson multiplets as the elementary degrees of freedom. In Eq. (1), $\mathcal{L}_{kin}$ is the kinetic energy term, $\mathcal{L}_{BW}$ contains the baryon-meson interactions in which the baryon-spin-0 meson interaction terms generate the baryon masses. $\mathcal{L}_{VP}$ describes the interactions of vector mesons with the pseudoscalar mesons (and with photons). $\mathcal{L}_{vec}$ describes the dynamical mass generation of the vector mesons via couplings to the scalar mesons and contains additionally quartic self-interactions of the vector fields. $\mathcal{L}_0$ contains the meson-meson interaction terms inducing the spontaneous breaking of chiral symmetry as well as a scale invariance breaking logarithmic potential. $\mathcal{L}_{SB}$ describes the explicit chiral symmetry breaking.

We proceed to study the hadronic properties in the chiral SU(3) model. The Lagrangian density in the mean field approximation is given as

$$\mathcal{L}_{BX} + \mathcal{L}_{BV} = -\sum_i \overline{\psi_i} \left[g_{i\omega}\gamma_0\omega + g_{i\phi}\gamma_0\phi + m_i^* \right]\psi_i \tag{2}$$

$$\mathcal{L}_{vec} = \frac{1}{2}m_\omega^2 \frac{\chi^2}{\chi_0^2}\omega^2 + g_4^4\omega^4 + \frac{1}{2}m_\phi^2 \frac{\chi^2}{\chi_0^2}\phi^2 + g_4^4\left(\frac{Z_\phi}{Z_\omega}\right)^2\phi^4 \tag{3}$$

$$\mathcal{V}_0 = \frac{1}{2}k_0\chi^2(\sigma^2 + \zeta^2) - k_1(\sigma^2 + \zeta^2)^2 - k_2(\frac{\sigma^4}{2} + \zeta^4) - k_3\chi\sigma^2\zeta$$
$$+ \ k_4\chi^4 + \frac{1}{4}\chi^4\ln\frac{\chi^4}{\chi_0^4} - \frac{\delta}{3}\chi^4\ln\frac{\sigma^2\zeta}{\sigma_0^2\zeta_0} \tag{4}$$

$$\mathcal{V}_{SB} = \left(\frac{\chi}{\chi_0}\right)^2\left[m_\pi^2 f_\pi\sigma + (\sqrt{2}m_K^2 f_K - \frac{1}{\sqrt{2}}m_\pi^2 f_\pi)\zeta\right], \tag{5}$$

where $m_i^* = -g_{\sigma i}\sigma - g_{\zeta i}\zeta$ is the effective mass of the baryon of type i ($i = N, \Sigma, \Lambda, \Xi$). In the above, $g_4 = \sqrt{Z_\omega}\tilde{g}_4$ is the renormalized coupling for ω-field. The thermodynamical potential of the grand canonical ensemble Ω per unit volume V at given chemical potential μ and temperature T can be written as

$$\frac{\Omega}{V} = -\mathcal{L}_{vec} - \mathcal{L}_0 - \mathcal{L}_{SB} - \mathcal{V}_{vac} + \sum_i \frac{\gamma_i}{(2\pi)^3}\int d^3k\, E_i^*(k)\left(f_i(k) + \bar{f}_i(k)\right)$$
$$- \sum_i \frac{\gamma_i}{(2\pi)^3}\mu_i^*\int d^3k\left(f_i(k) - \bar{f}_i(k)\right). \tag{6}$$

Here the the potential at $\rho = 0$ has been subtracted in order to get a vanishing vacuum energy. In (6) γ_i are the spin-isospin degeneracy factors. The f_i and $\bar{f}_i$ are thermal distribution functions for the baryon of species i, given in terms of the effective single particle energy, E_i^*, and chemical potential, μ_i^*, as

$$f_i(k) \;=\; \frac{1}{e^{\beta(E_i^*(k)-\mu_i^*)}+1} \;\;,\;\; \bar{f}_i(k) = \frac{1}{e^{\beta(E_i^*(k)+\mu_i^*)}+1},$$

with $E_i^*(k) = \sqrt{k_i^2 + m_i^{*2}}$ and $\mu_i^* = \mu_i - g_{i\omega}\omega$. The mesonic field equations are determined by minimizing the thermodynamical potential [18, 19]. They depend on the scalar and vector densities for the baryons at finite temperature

$$\rho_i^s = \gamma_i \int \frac{d^3k}{(2\pi)^3} \frac{m_i^*}{E_i^*} \left(f_i(k) + \bar{f}_i(k)\right) \;;\;\; \rho_i = \gamma_i \int \frac{d^3k}{(2\pi)^3} \left(f_i(k) - \bar{f}_i(k)\right). \tag{7}$$

The energy density and the pressure are given as, $\epsilon = \Omega/V + \mu_i\rho_i + TS$ and $p = -\Omega/V$.

3 Kaon (antikaon) interactions in the chiral SU(3) model

In this section, we derive the disperson relations for the $K(\bar{K})$ and calculate their optical potentials in the asymmetric nuclear matter. The medium modified energies of the kaons and antikaons arise from their interactions with the nucleons, vector mesons and scalar mesons within the chiral SU(3) model.

In this model the interactions to the scalar fields (non-strange, σ and strange, ζ), scalar–isovector field δ as well as a vectorial interaction and the vector meson (ω and ρ) - exchange terms modify the energies for $K(\bar{K})$ mesons in the medium.

The scalar meson multiplet has the expectation value $\langle X \rangle = \mathrm{diag}((\sigma+\delta)/\sqrt{2}, (\sigma-\delta)/\sqrt{2}, \zeta)$, with σ ans ζ corresponding to the non-strange and strange scalar condensates, and δ is the third isospin component of the scalar-isovector field, $\vec{\delta}$. The pseudoscalar meson field P can be written as,

$$P = \begin{pmatrix} \pi^0/\sqrt{2} & \pi^+ & \frac{2K^+}{1+w} \\ \pi^- & -\pi^0/\sqrt{2} & \frac{2K^0}{1+w} \\ \frac{2K^-}{1+w} & \frac{2\bar{K}^0}{1+w} & 0 \end{pmatrix}, \tag{8}$$

where $w = \sqrt{2}\zeta/\sigma$ and we have written down the terms that are relevant for the present investigation. From PCAC one gets the decay constants for the pseudoscalar mesons as $f_\pi = -\sigma$ and $f_K = -(\sigma + \sqrt{2}\zeta)/2$.

The interaction Lagrangian modifying the energies of the $K(\bar{K})$-mesons can be written as [23]

$$\mathcal{L}_{KN} \;=\; -\frac{i}{8f_K^2}\Big[3(\bar{N}\gamma^\mu N)(\bar{K}(\partial_\mu K) - (\partial_\mu \bar{K})K) + (\bar{N}\gamma^\mu \tau^a N)(\bar{K}\tau^a(\partial_\mu K)$$

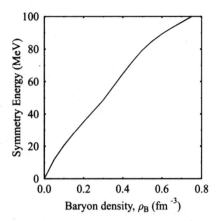

Figure 1: The symmetry energy in MeV plotted as a function of the baryon density, ρ_B (in fm $^{-3}$).

$$
\begin{aligned}
- \quad & (\partial_\mu \bar{K}) \tau^a K) \Big] + \frac{m_K^2}{2 f_K} \Big[(\sigma + \sqrt{2}\zeta)(\bar{K} K) + \delta^a (\bar{K} \tau^a K) \Big] \\
- \quad & i g_{\omega K} \Big[(\bar{K}(\partial_\mu K) - (\partial_\mu \bar{K}) K) \omega^\mu + (\bar{K} \tau^a (\partial_\mu K) - (\partial_\mu \bar{K}) \tau^a K) \rho^{\mu a} \Big] \\
- \quad & \frac{1}{f_K} \Big[(\sigma + \sqrt{2}\zeta)(\partial_\mu \bar{K})(\partial^\mu K) + (\partial_\mu \bar{K}) \tau^a (\partial^\mu K) \delta^a \Big] \\
+ \quad & \frac{d_1}{2 f_K^2} (\bar{N} N)(\partial_\mu \bar{K})(\partial^\mu K).
\end{aligned}
\tag{9}
$$

In the above, K and $\bar{K}$ are the kaon and antikaon doublets. In (9) the first line is the vectorial interaction term (Weinberg-Tomozawa term). The second term, which gives an attractive interaction for the K-mesons, is obtained from the explicit symmetry breaking term. The third term refers to the interaction in terms of the ω-meson and ρ-meson exchanges. The fourth term arises within the present chiral model from the kinetic term of the pseudoscalar mesons. The fifth term in (9) for the KN interactions arises from the term

$$
\mathcal{L}^{BM} = d_1 Tr(u_\mu u^\mu \bar{B} B),
\tag{10}
$$

in the SU(3) chiral model [22]. The last two terms in (9) represent the range term in the chiral model. The Fourier transformation of the equation-of-motion for kaons (antikaons) lead to the dispersion relations,

$$
-\omega^2 + \vec{k}^2 + m_K^2 - \Pi_K(\omega, |\vec{k}|, \rho) = 0,
$$

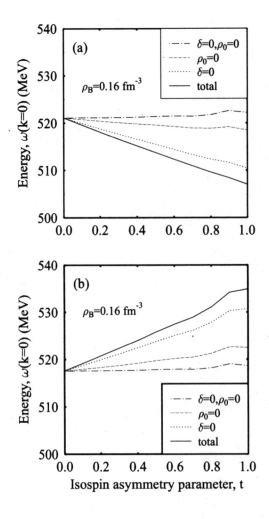

Figure 2: (Color online) The energies of the kaons, K^+ and K^0, at zero momentum and for $\rho_B=0.16$ fm^{-3}, are plotted as functions of the isospin asymmetry parameter, t in (a) and (b). The medium modifications to the energies are also shown for the situations when either the isospin asymmetric contribution from the ρ-meson or δ meson, or, both, are not taken into account. The solid line shows the total contribution.

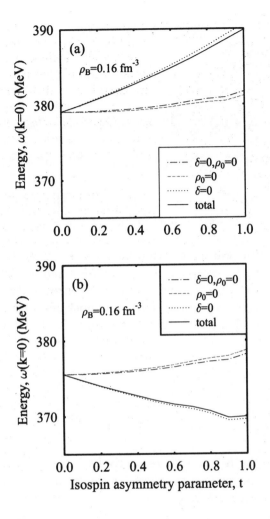

Figure 3: (Color online) The energies of the antikaons, K^- and $\bar{K}^0$, at zero momentum and for ρ_B=0.16 fm^{-3} are plotted as functions of the the isospin asymmetry parameter, t in (a) and (b). The medium modifications to the energies are also shown for the situations when the isospin asymmetric contribution from the ρ-meson or δ meson, or, both, are not taken into account. The solid line shows the total contribution. The symmetry energy in MeV plotted as a function of the baryon density, ρ_B (in fm^{-3}).

where Π_K denotes the kaon (antikaon) self energy in the medium.

Explicitly, the self energy $\Pi(\omega, |\vec{k}|)$ for the kaon doublet arising from the interaction (9) is given as

$$
\begin{aligned}
\Pi(\omega, |\vec{k}|) &= -\frac{3}{4f_K^2}(\rho_p + \rho_n)\omega + \frac{m_K^2}{2f_K}(\sigma' + \sqrt{2}\zeta' \pm \delta') - 2g_{\omega K}\omega(\omega_0 \pm \rho_0) \\
&\quad + \left[-\frac{1}{f_K}(\sigma' + \sqrt{2}\zeta' \pm \delta') + \frac{d_1}{2f_K^2}(\rho_s^p + \rho_s^n)\right](\omega^2 - \vec{k}^2),
\end{aligned}
\tag{11}
$$

where the $\pm$ signs refer to the K^+ and K^0 respectively. In the above, $\sigma'(= \sigma - \sigma_0)$, $\zeta'(= \zeta - \zeta_0)$ and $\delta'(= \delta - \delta_0)$, are the fluctuations of the scalar-isoscalar fields σ and ζ, and the third component of the scalar-isovector field, δ, from their vacuum expectation values. The vacuum expectation value of δ is zero ($\delta_0=0$), since a nonzero value for it will break the isospin symmetry of the vacuum (the small isospin breaking effect coming from the mass and charge difference of the up and down quarks has been neglected here). ρ_p and ρ_n are the number densities, and ρ_s^p for the proton and the neutron and ρ_s^n are their scalar densities.

Similarly, for the antikaon doublet, the self-energy is calculated as

$$
\begin{aligned}
\Pi(\omega, |\vec{k}|) &= \frac{3}{4f_K^2}(\rho_p + \rho_n)\omega + \frac{m_K^2}{2f_K}(\sigma' + \sqrt{2}\zeta' \pm \delta') + 2g_{\omega K}\omega(\omega_0 \pm \rho_0) \\
&\quad + \left[-\frac{1}{f_K}(\sigma' + \sqrt{2}\zeta' \pm \delta') + \frac{d_1}{2f_K^2}(\rho_s^p + \rho_s^n)\right](\omega^2 - \vec{k}^2),
\end{aligned}
\tag{12}
$$

where the $\pm$ signs refer to the K^- and $\bar{K}^0$ respectively.

After solving the above dispersion relations for the kaons and antikaons, their optical potentials can be calculated from

$$
U(\omega, k) = \omega(k) - \sqrt{k^2 + m_K^2},
\tag{13}
$$

where m_K is the vacuum mass for the kaon (antikaon).

The parameter d_1 is calculated from the the empirical value of the isospin averaged KN scattering length [21, 24, 25] taken to be

$$
\bar{a}_{KN} \approx -0.255 \text{ fm}.
\tag{14}
$$

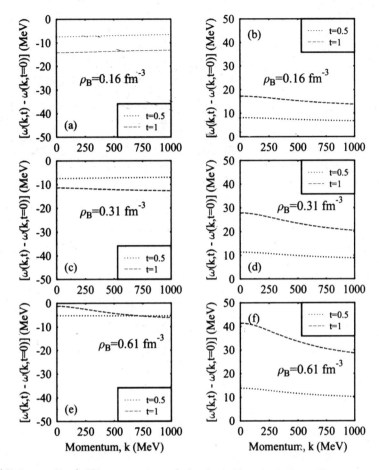

Figure 4: (Color online) The energies of the kaons, relative to the t=0 values, are plotted as functions of the momentum for t=0.5 and 1. The subplots (a),(c) and (e) refer to K^+ and, (b), (d) and (f) refer to K^0 at different densities.

4   Results and Discussions

The present calculations uses the model parameters from [17]. The values, $g_{\sigma N} = 10.623$, and $g_{\zeta N} = -0.4894$ are determined by fitting vacuum baryon masses. The other parameters as fitted to the nuclear matter saturation properties in the mean field approximation are: $g_{\omega N}$=13.606, g_4=61.466, m_ζ =1038.5 MeV, m_σ= 474.3 MeV. The coefficient d_1, calculated from the empirical value of the isospin averaged scattering length (14), is $5.196/m_K$. Using these parameters, the symmetry energy defined as

$$a_4 = \frac{1}{2}\frac{d^2E}{dt^2}|_{t=0} \tag{15}$$

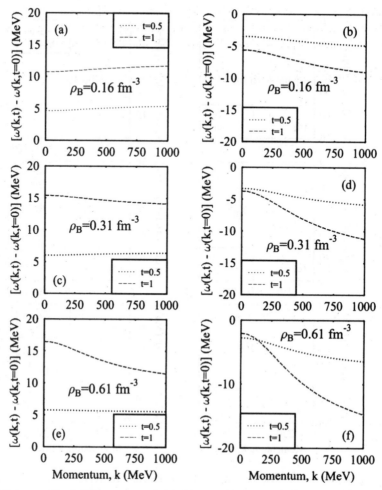

Figure 5: (Color online) The energies of the antikaons, relative to the t=0 values, are plotted as functions of the momentum for t=0.5 and 1. The subplots (a),(c) and (e) refer to K^- and, (b), (d) and (f) refer to $\bar{K}^0$ at different densities.

with the asymmetry parameter $t = (\rho_p - \rho_n)/\rho_B$, has a value of $a_4 = 28.4$ MeV at saturation nuclear matter density of ρ_0=0.15 fm $^{-3}$. Figure 1 shows the density dependence of the symmetry energy, which increases with density similar to previous calculations [26].

The kaon and antikaon properties were studied in the isospin symmetric hadronic matter within the chiral SU(3) model in ref. [22]. The contributions from the vector interaction as well as the vector meson ω- exchange terms lead to a drop for the antikaons energy, whereas they are repulsive for the kaons. The scalar meson exchange term arising from the scalar-isoscalar fields (σ and ζ) is attractive for both K and $\bar{K}$. The first term of the range term of eq. (9) is repulsive whereas the second term has an attractive contribution for the isospin symmetric matter [22] for both kaons and antikaons.

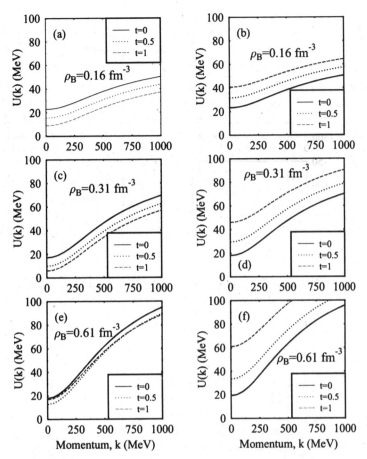

Figure 6: (Color online) The optical potentials of the kaons are plotted as functions of the momentum for various values of the isospin asymmetry parameter t. The subplots (a),(c) and (e) refer to K^+ and, (b), (d) and (f) refer to K^0 at different densities.

The contributions due to the scalar-isovector, δ-field as well as the vector-isovector ρ-meson, introduce isotopic asymmetry in the K and $\bar{\text{K}}$-energies. For $\rho_n > \rho_p$, in the kaon sector, K^+ (K^0) has negative (positive) contributions from both δ and ρ mesons. The δ contribution from the scalar exchange term is positive (negative) for K^+ (K^0), whereas that arising from the range term has the opposite sign and dominates over the former contribution. The contribution from the ρ to the kaon (antikaon) masses at ρ_B=0.16 fm^{-3} is large as compared to that of the δ contribution, as can be seen from the figures 2 and 3. When we do not account for the isospin asymmetry effects arising due to the ρ and δ fields, then the masses of kaons and antikaons stay almost constant. The small fluctuations are reflections of the deviation of the scalar density occurring in the last term of equation (11)((12)) for the kaon (antikaon) self energy from the baryon density, ρ_B.

189

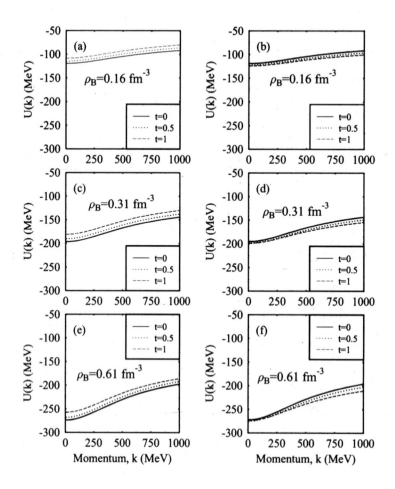

Figure 7: (Color online) The optical potentials of the antikaons are plotted as functions of the momentum for various values of the isospin asymmetry parameter, t. The subplots (a),(c) and (e) refer to K^- and, (b), (d) and (f) refer to $\bar{K}^0$ at different densities.

In figure 4, the energies of the K^+ and K^0, relative to the t=0 values, are plotted for different values of the isospin asymmetry parameter, t, at various densities. For ρ_B=0.16 fm^{-3}, the energy of K^+ is seen to drop by about 15 MeV at zero momentum when t changes from 0 to 1. On the other hand, the K^0 energy is seen to increase by a similar amount for t=1, from the isospin symmetric case of t=0. The energies of the kaons are also plotted for densities ρ_B=0.31 fm^{-3} and ρ_B=0.61 fm^{-3} in the same figure. For K^+, the t-dependence of the energy is seen to be less sensitive at higher densities, whereas the energy of K^0 is seen to have a larger drop from the t=0 case, as we increase the density. The reason for this opposite behavior for the K^+ and K^0 on the isospin asymmetry comes from the relative vector-meson contributions. For K^+, the isospin asymmetric effect to the energy arising from the ρ meson (which dominates over that from the δ meson) is opposite in sign to that of the ω meson, whereas for K^0, it has the same sign as that of the ω meson.

For the antikaons, the K^-($\bar{K^0}$) energy is seen to increase (drop) with t, as seen in figure 5. The sensitivity of the isospin asymmetry dependence of the energies is seen to be larger for K^- with density, whereas it becomes smaller for $\bar{K^0}$ at high densities. However, $\bar{K^0}$ shows an appreciable drop as we increase the momentum, whereas the value for K^- is not as sensitive to momentum change as $\bar{K^0}$.

The qualitative behavior of the isospin asymmetry dependencies of the energies of the kaons and antikaons are also reflected in their optical potentials plotted in figure 6 for the kaons, and in figure 7, for the antikaons, at selected densities. The different behavior of the K^+ and K^0, as well as for the K^- and $\bar{K^0}$ optical potentials in the dense asymmetric nuclear matter should be seen in their production as well as propagation in isospin asymmetric heavy ion collisions. The effects of the isospin asymmetric optical potentials could thus be observed in nuclear collisions at the CBM experiment at the proposed project FAIR at GSI, where experiments with neutron rich beams are planned to be adopted.

5 Summary

To summarize, we have investigated, within a chiral SU(3) model, the density dependence of the $K, \bar{K}$-meson optical potentials in asymmetric nuclear matter, arising from the interactions with nucleons and scalar and vector mesons. The properties of the light hadrons – as studied in a SU(3) chiral model – modify the $K(\bar{K})$-meson properties in the hadronic medium. The model with parameters fixed from the properties of hadron masses, nuclei and KN scattering data, takes into account all terms up to the next to leading order arising in chiral perturbative expansion for the interactions of $K(\bar{K})$-mesons with baryons. There is seen to be significant density dependence of the isospin asymmetry on the optical potentials of the kaons and antikaons. The results can be used in heavy-ion simulations that include mean fields for the propagation of mesons [22]. The different potentials of kaons and antikaons can be particularly relevant for neutron-rich beams at the CBM experiment at the proposed project FAIR at GSI, Germany, as well as at the experiments at the Rare Isotope Accelerator (RIA) laboratory, USA.

This work has been done in collaboration with Stefan Schramm.

References

[1] D. B. Kaplan and A. E. Nelson, Phys. Lett. B **175**, 57 (1986); A. E. Nelson and D. B. Kaplan, *ibid*, 192, 193 (1987).

[2] G. E. Brown and M. Rho, Phys. Rev. Lett. **66**, 2720 (1991).

[3] E. Friedman, A. Gal, and C.J. Batty, Nucl. Phys. A **579**, 518 (1994).

[4] A. Gal, Nucl. Phys. A **691**, 268 (2001).

[5] M. Lutz, Phys. Lett. B **426**, 12 (1998).

[6] M. Lutz and E. E. Kolomeitsev, Nucl. Phys. A **700**, 193 (2002).

[7] M. Lutz and C. L. Korpa, Nucl. Phys. A **700**, 309 (2002).

[8] A. Ramos and E. Oset, Nucl. Phys. A **671**, 481 (2000).

[9] L. Tolós, A. Ramos, and A. Polls, Phys. Rev. C **65**, 054907 (2002).

[10] L. Tolós, A. Ramos, A. Polls, and T.T.S. Kuo, Nucl. Phys. A **690**, 547 (2001).

[11] J. Schaffner-Bielich, V. Koch, and M. Effenberger, Nucl. Phys. A **669**, 153 (2000).

[12] B.A. Li, Phys. Rev. Lett. 88, 192701 (2002); B.A.Li, G.C. Yong and W. Zuo, Phys. Rev. C **71**, 014608 (2005); L. W. Chen, C. M. Ko and B. A. Li, Phys. Rev. C **72**, 064606 (2005). T. Gaitanos, M. Di Toro, S. Typee, V. Baran, C. Fuchs, V. Greco, H.H.Wolter, Nucl. Phys. A **732**, 24 (2004); T. Gaitanos, M. Di Toro, G. Ferini, M. Colonna, H. H. Wolter, nucl-th/0402041; Q. Li, Z. Li, E. Zhao, R. K. Gupta, Phys. Rev. C **71**, 054907 (2005); A. Akmal and V. R. Pandharipandhe, Phys. Rev. C **56**, 2261 (1997); H. Heiselberg and H. Hjorth-Jensen, Phys. Rep. **328**, 237 (2000).

[13] Q. Li, Z. Li, S. Soff, M. Bleicher, H. Stöcker, Phys. Rev. C **72**, 034613 (2005)

[14] Q. Li, Z. Li, S. Soff, R. K. Gupta, M. Bleicher, H. Stöcker, J. Phys. G **31**, 1359 (2005).

[15] Q. Li, Z. Li, S. Soff, M. Bleicher, H. Stöcker, J. Phys. G **32**, 151 (2006).

[16] Q. Li, Z. Li, S. Soff, M. Bleicher, H. Stöcker, J. Phys. G **32**, 407 (2006).

[17] P. Papazoglou, D. Zschiesche, S. Schramm, J. Schaffner-Bielich, H. Stöcker, and W. Greiner, Phys. Rev. C **59**, 411 (1999).

[18] A. Mishra, K. Balazs, D. Zschiesche, S. Schramm, H. Stöcker, and W. Greiner, nucl-th/0308064; Phys. Rev. C **69**, 024903 (2004).

[19] D. Zschiesche, A. Mishra, S. Schramm, H. Stöcker, and W. Greiner, Phys. Rev. C **70**, 045202 (2004).

[20] A. Mishra, E. Bratkovskaya, J. Schaffner-Bielich, S. Schramm, and H. Stöcker, Phys. Rev. C **69**, 015202 (2004).

[21] J. Schaffner-Bielich, I. N. Mishustin, J. Bondorf, Nucl. Phys. A **625**, 325 (1997).

[22] A. Mishra, E. L. Bratkovskaya, J. Schaffner-Bielich, S. Schramm and H. Stöcker, Phys. Rev. C **70**, 044904 (2004).

[23] Amruta Mishra and Stefan Schramm, Phys. Rev. C **74**, 064904, 2006.

[24] G. E. Brown, C.-H. Lee, M. Rho, and V. Thorsson, Nucl. Phys. A **567**, 937 (1994).

[25] T. Barnes and E. S. Swanson, Phys. Rev. C **49**, 1166 (1994).

[26] G-C Yong, B. A. Li and L. W. Chen, nucl-th/0606003.

Physics and Astrophysics of Hadrons and Hardronic Matter
Editor: A. B. Santra

Equation of State with Short Range Correlations

Prafulla K. Panda[a,b]*, C. Providência[b], J. Da Providência[b]

a) Institute of Physics, Bhubaneswar, India - 751005
b) Departamento de Física, Universidade de Coimbra,
P-3004-516 Coimbra, Portugal
* email: panda@iopb.res.in

Abstract

Short range correlations are introduced in a relativistic approach to the equation of state for the infinite nuclear matter in the frame work of Hartree-Fock approximation using an effective Hamiltonian derived from scalar and vector mesons. The unitary correlation method is used to introduce short range correlations. The effects of the correlations on the ground state properties of nuclear matter are studied. It is shown that the correlations give rise to an extra node in the ground state wave function in the nucleons, contrary to what happens in non-relativistic calculations with a hard core. The nucleon effective mass and equation of state (EOS) are very sensitive to short range correlations.

1 Introduction

This chapter will deal with the role of short range correlations in a relativistic approach to the description of nuclear matter. Short range correlations may be introduced into the model wave function by several procedures which are related with the so called e^S method or coupled-cluster expansion, introduced by Coester and Kümmel [1], and extensively developed by R. Bishop [2]. In this note, we consider the unitary operator method as proposed by Villars [3], which automatically guarantees that the correlated state is normalized. The general idea of introducing short range correlations in systems with short range interactions exists for a long time [4, 5] but has not been pursued for the relativistic case. .

Non-relativistic calculations based on realistic NN potentials predict equilibrium points which do not reproduce simultaneously the binding energy and saturation density. Either the saturation density is reproduced but the binding energy is too small, or the binding energy is reproduced at too high a density [6]. In order to solve this problem, the existence of a repulsive potential or density-dependent repulsive mechanism [7] is usually assumed. Due to Lorentz covariance and self-consistency, relativistic mean field theories [8] include automatically contributions which are equivalent to n-body repulsive potentials in non-relativistic approaches. The relativistic quenching of the scalar field provides a mechanism for saturation, though, by itself, it may lead to too small an effective mass and too large incompressibility of nuclear matter, a situation which is encountered in the Walecka model [8].

In non-relativistic models the saturation arises from the interplay between a long range attraction and a short range repulsion, so strong that it is indispensable to take short range correlations into account. In relativistic mean field models, the parameters are phenomenologically fitted to the saturation properties of nuclear matter. Although in this approach short range correlation effects may be accounted for, to some extent, by the model parameters, it is our aim to study explicitly the consequences of actual short range correlations. In a previous publication [9–11] we have discussed the effect of the correlations in the ground state properties of nuclear matter in the framework of the Hartree-Fock approximation using an effective Hamiltonian derived from the $\sigma - \omega$ Walecka model. We have shown, for interactions mediated only by sigma and omega mesons, that the equation of state (EOS) becomes considerably softer when correlations are taken into account, provided the correlation function is treated variationally, always paying careful attention to the constraint imposed by the "healing distance" requirement. In the present chapter we will use the same approach and will include also the exchange of pions and ρ-mesons.

2 Effective relativistic model Hamiltonian

We consider the effective Hamiltonian

$$
\begin{aligned}
H &= \int \psi_\alpha^\dagger(\vec{x}) \, (-i\vec{\alpha} \cdot \vec{\nabla} + \beta M)_{\alpha\beta} \, \psi_\beta(\vec{x}) \, d\vec{x} \\
&+ \frac{1}{2} \int \psi_\alpha^\dagger(\vec{x}) \, \psi_\gamma^\dagger(\vec{y}) V_{\alpha\beta,\gamma\delta}(|\vec{x} - \vec{y}|)\psi_\delta(\vec{y}) \, \psi_\beta(\vec{x}) \, d\vec{x} \, d\vec{y},
\end{aligned} \tag{1}
$$

where the exchange of σ, ω and ρ mesons is taken into account through the interaction

$$
V_{\alpha\beta,\gamma\delta}(r) = \sum_{i=\sigma,\omega,\rho} V^i_{\alpha\beta,\gamma\delta}(r) \tag{2}
$$

with

$$
V^\sigma_{\alpha\beta,\gamma\delta}(r) = -\frac{g_\sigma^2}{4\pi}(\beta)_{\alpha\beta}(\beta)_{\gamma\delta}\frac{e^{-m_\sigma r}}{r}, \qquad V^\omega_{\alpha\beta,\gamma\delta}(r) = \frac{g_\omega^2}{4\pi}\left(\delta_{\alpha\beta}\delta_{\gamma\delta} - \vec{\alpha}_{\alpha\beta} \cdot \vec{\alpha}_{\gamma\delta}\right)\frac{e^{-m_\omega r}}{r}
$$

$$
\text{and} \quad V^\rho_{\alpha\beta,\gamma\delta}(r) = \frac{g_\rho^2}{4\pi}\left(\delta_{\alpha\beta}\delta_{\gamma\delta} - \vec{\alpha}_{\alpha\beta} \cdot \vec{\alpha}_{\gamma\delta}\right)\vec{\tau}_1 \cdot \vec{\tau}_2\frac{e^{-m_\rho r}}{r} \, .
$$

The exchange of π mesons is described by an interaction of the form [12]

$$
V^\pi_{\alpha\beta,\gamma\delta}(\vec{r}) = \frac{G_{\pi NN}^2}{4\pi 4M^2}(\vec{\tau}_1 \cdot \vec{\tau}_2)(\vec{\sigma}_1 \cdot \vec{\nabla})_{\alpha\beta}(\vec{\sigma}_2 \cdot \vec{\nabla})_{\gamma\delta}\frac{e^{-m_\pi r}}{r} \tag{3}
$$

where $\vec{\sigma}_i = \vec{\alpha}_i \gamma_5$, M is the nucleon mass and $m_\sigma, m_\omega, m_\rho, m_\pi$ are the meson masses. This can be rewritten in momentum space as

$$V^\pi_{\alpha\beta\gamma\delta}(\vec{q}) = \frac{G^2_{\pi NN}}{4\pi 4 M^2}(\vec{\tau}_1 \cdot \vec{\tau}_2)(\vec{\sigma}_1 \cdot \vec{q})_{\alpha\beta}(\vec{\sigma}_2 \cdot \vec{q})_{\gamma\delta}\frac{1}{q^2 + m^2_\pi}. \tag{4}$$

As explained in the paper by Bouyssy $et\ al$ [12], eq. (3) may be approximated as

$$V^\pi_{\alpha\beta,\gamma\delta}(\vec{r}) \approx \frac{1}{3}\left[\frac{f_\pi}{m_\pi}\right]^2 \frac{G^2_{\pi NN}}{4\pi 4 M^2}(\vec{\tau}_1 \cdot \vec{\tau}_2)((\vec{\sigma})_{\alpha\beta} \cdot (\vec{\sigma})_{\gamma\delta})\frac{m^2_\pi}{3}\left[\frac{4\pi}{m^2_\pi}\delta(\vec{r}) - \frac{e^{-m_\pi r}}{r}\right]. \tag{5}$$

The first term in the above is the repulsive contact interaction and the second term is an attractive Yukawa potential. In equation (1), ψ is the nucleon field and $\vec{\alpha}$, β are the Dirac-matrices. The equal time quantization conditions $[\psi_\alpha(\vec{x},t),\psi_\beta(\vec{y},t)^\dagger]_+ = \delta_{\alpha\beta}\delta(\vec{x}-\vec{y})$ are satisfied. At time $t = 0$, the field ψ may be expanded as [13]

$$\psi(\vec{x}) = \frac{1}{\sqrt{V}}\sum_{r,k}\left[U_r(\vec{k})c_{r,\vec{k}} + V_r(-\vec{k})\tilde{c}^\dagger_{r,-\vec{k}}\right]e^{i\vec{k}\cdot\vec{x}},$$

where U_r and V_r are given by

$$U_r(\vec{k}) = \begin{pmatrix} \cos\frac{\chi(\vec{k})}{2} \\ \vec{\sigma}\cdot\hat{k}\sin\frac{\chi(\vec{k})}{2} \end{pmatrix}u_r; \quad V_r(-\vec{k}) = \begin{pmatrix} -\vec{\sigma}\cdot\hat{k}\sin\frac{\chi(\vec{k})}{2} \\ \cos\frac{\chi(\vec{k})}{2} \end{pmatrix}v_r.$$

For free fields, we have $\cos\chi(\vec{k}) = M/\epsilon(\vec{k})$, $\sin\chi(\vec{k}) = |\vec{k}|/\epsilon(\vec{k})$ with $\epsilon(\vec{k}) = \sqrt{\vec{k}^2 + M^2}$, while, for interacting fields, $\cos\chi(\vec{k}) = M^*(\vec{k})/\epsilon^*(\vec{k})$, $\sin\chi(\vec{k}) = |\vec{k}^*|/\epsilon^*(\vec{k})$, with $\epsilon^*(\vec{k}) = \sqrt{\vec{k}^{*2} + M^{*2}(\vec{k})}$, where $\vec{k}^*$ and $M^*(\vec{k})$ denote, respectively, the effective momentum and mass, determined self-consistently by the Hartree-Fock (HF) prescription. The equal time anti-commutation conditions $[c_{r,\vec{k}},c^\dagger_{s,\vec{k}'}]_+ = \delta_{rs}\delta_{\vec{k},\vec{k}'} = [\tilde{c}_{r,\vec{k}},\tilde{c}^\dagger_{s,\vec{k}'}]_+$ are satisfied. We wish to describe hadronic matter in terms of the occupation of positive energy states, leaving empty the negative energy states, so that the vacuum denoted by $\mid 0\rangle$ is defined by $c_{r,\vec{k}}\mid 0\rangle = \tilde{c}^\dagger_{r,\vec{k}}\mid 0\rangle = 0$. One-particle states are written as $|\vec{k},r\rangle = c^\dagger_{r,\vec{k}}\mid 0\rangle$. Two-particle and three-particle uncorrelated states are written as $|\vec{k},r;\vec{k}',r'\rangle = c^\dagger_{r,\vec{k}}c^\dagger_{r',\vec{k}'}\mid 0\rangle$, and $|\vec{k},r;\vec{k}',r';\vec{k}'',r''\rangle = c^\dagger_{r,\vec{k}}c^\dagger_{r',\vec{k}'}c^\dagger_{r'',\vec{k}''}\mid 0\rangle$, respectively; and so on. Considering the energy expectation value $E = \langle\Phi|H|\Phi\rangle/\langle\Phi|\Phi\rangle$, the HF method determines the Slater determinant $|\Phi\rangle$, which is obtained by filling up all positive energy states up to the Fermi momentum. We take into account the effect of short range correlations through the unitary operator method. The correlated wave function [14] is $|\Psi\rangle = e^{iS}|\Phi\rangle$ where S is, in general, a n-body Hermitian operator, splitting into a 2-body part, a 3-body part, etc.. The expectation value of H becomes

$$E = \frac{\langle \Psi | H | \Psi \rangle}{\langle \Psi | \Psi \rangle} = \frac{\langle \Phi | e^{-iS} \, H \, e^{iS} | \Phi \rangle}{\langle \Phi | \Phi \rangle}. \tag{6}$$

In the present calculation, we only take into account two-body correlations. Let us denote the two-body correlated wave function by $\overline{|\vec{k}, r; \vec{k}', r'\rangle} = e^{iS} |\vec{k}, r; \vec{k}', r'\rangle \approx f_{12} |\vec{k}, r; \vec{k}', r'\rangle$ where f_{12} is the short range correlation factor, the so-called Jastrow factor [15]. For simplicity, we consider $f_{12} = f(|\vec{r}_{12}|)$, $\vec{r}_{12} = \vec{r}_1 - \vec{r}_2$, and $f(r) = 1 - (\alpha + \beta r) \, e^{-\gamma r}$ where α, β and γ are parameters. The choice of a real function to describe the effect of the unitary operator has to be supplemented by a normalization condition

$$\int \left(f^2(r) - 1 \right) d^3 r = 0, \tag{7}$$

which assures unitarity to leading cluster order. In the present case the function $f(r)$ introduces a node in the wave function.

The important effect of the short range correlations is the expression for the correlated ground-state energy. Here, in the leading order of the cluster expansion, the interaction matrix element $\langle \vec{k}, r; \vec{k}', r' | V_{12} | \vec{k}, r; \vec{k}', r' \rangle$ of the HF expression is replaced by $\langle \overline{\vec{k}, r; \vec{k}', r'} | V_{12} + t_1 + t_2 | \overline{\vec{k}, r; \vec{k}', r'} \rangle - \langle \vec{k}, r; \vec{k}', r' | t_1 + t_2 | \vec{k}, r; \vec{k}', r' \rangle$, where t_i is the kinetic energy operator of particle i. As argued by Moszkowski [16] and Bethe [17], it is expected that the true ground-state wave function of the nucleus, containing correlations, coincides with the independent particle, or HF wave function, for inter particle distances $r \geq r_h$, where $r_h \approx 1$ fm is the so-called "healing distance". This behavior is a consequence of the restrictions imposed by the Pauli Principle. A natural consequence of having the correlations introduced by a unitary operator is the normalization constraint on $f(r)$ expressed by eq. (7). The correlated ground state energy of symmetric nuclear matter with σ, ω and ρ potentials reads

$$\mathcal{E}_{\sigma, \omega, \rho} = \mathcal{T}_c^d + \mathcal{T}_c^e + \mathcal{V}_c^d + \mathcal{V}_c^e$$

where the first terms result from the kinetic contribution, including the direct $(\mathcal{T}_c^d)$ and exchange $(\mathcal{T}_c^e)$ correlation contribution from the kinetic energy,

$$\mathcal{T}_c^d = \frac{\nu(1 + \mathcal{C}\rho_B)}{\pi^2} \int_0^{k_F} k^2 dk \left[|k| \sin \chi(k) + M \cos \chi(k) \right]$$

$$\mathcal{T}_c^e = -\frac{1}{4\pi^4} \int_0^{k_f} k^2 dk \, k'^2 dk' \left\{ \left[|k| \sin \chi(k) + 2M \cos \chi(k) \right] I(k, k') \right.$$
$$\left. + |k| \sin \chi(k') J(k, k') \right\}$$

197

with $C = \int (f^2(r) - 1)d^3r = 0$ from equation (7). The last terms are the direct and exchange contributions from the potential energy with correlations

$$V_c^d = \frac{\tilde{F}_\sigma(0)}{2}\rho_s^2 + \frac{\tilde{F}_\omega(0)}{2}\rho_B^2,$$

$$V_c^e = \frac{1}{(2\pi)^4}\int_0^{k_f} k\, dk\, k'\, dk' \Big[\sum_{i=\sigma,\omega,\rho} A_i(k,k') + \cos\chi(k)\cos\chi(k') \sum_{i=\sigma,\omega,\rho} B_i(k,k')$$
$$+ \sin\chi(k)\sin\chi(k') \sum_{i=\sigma,\omega\rho} C_i(k,k') \Big].$$

In the last expressions A_i, B_i, C_i , I and J are exchange integrals defined in reference [9]. The baryon and the scalar densities are, respectively, ρ_B and ρ_s.

We next describe in more detail the treatment of pionic exchange. Starting with eq. (3), the one pion exchange potential may also be expressed as

$$V^\pi(r) = \frac{G_{\pi NN}^2}{16\pi M^2}(\vec{\tau}_1 \cdot \vec{\tau}_2) \Big[\quad (\vec{\sigma}_1 \cdot \hat{r})(\vec{\sigma}_2 \cdot \hat{r})(m_\pi^2 r^2 + 3m_\pi r + 3)$$
$$- \vec{\sigma}_1 \cdot \vec{\sigma}_2(m_\pi r + 1)\Big]\frac{e^{-m_\pi r}}{r^3}. \qquad (8)$$

In the above, $\vec{r} = \vec{x} - \vec{y}$ and $r = |\vec{r}|$. Thus, the pion contribution to H is [18]

$$\mathcal{H}_\pi = \frac{G_{\pi NN}^2 m_\pi^2}{48\pi M^2}\int dy\Big[\psi(\vec{x})^\dagger\sigma_i\tau_a\psi(\vec{x})\psi(\vec{y})^\dagger\sigma_i\tau_a\psi(\vec{y})\,(y_0(r) - y_2(r))$$
$$+3y_2(r)\,\psi(\vec{x})^\dagger(\sigma\cdot\hat{r})\tau_a\psi(\vec{x})\psi(\vec{y})^\dagger(\sigma\cdot\hat{r})\tau_a\psi(\vec{y})\Big]. \qquad (9)$$

where $y_0(r)$ and $y_2(r)$ are given by

$$y_0(r) = -\frac{e^{-m_\pi r}}{r}, \quad y_2(r) = \left(1 + \frac{3}{m_\pi r} + \frac{3}{(m_\pi r)^2}\right)\frac{e^{-m_\pi r}}{r}. \qquad (10)$$

In order to introduce correlations we are led to replace the potential $g(r) = \frac{e^{-m_\pi r}}{r}$ by the potential $g_{corr}(r)$, solution of the differential equation:

$$\left(\frac{1}{r}\frac{d^2}{dr^2} - \frac{1}{r^2}\frac{d}{dr}\right)g_{corr}(r) = \frac{1}{r}U(r)f^2(r), \qquad (11)$$

where $U(r)$ is given by

$$\frac{1}{r}U(r) = \left(\frac{1}{r}\frac{d^2}{dr^2} - \frac{1}{r^2}\frac{d}{dr}\right)g(r) \qquad (12)$$

and $f(r)$ is the Jastrow factor. There remains one extra correction term, which may be easily determined. From (8) we have

$$f^2 V^\pi(r) = \frac{G_{\pi NN}^2}{16\pi M^2} (\vec{\tau}_1 \cdot \vec{\tau}_2) \left[V_1(r) + V_2(r) \right],$$ (13)

where

$$V_1(r) = (\vec{\sigma}_1 \cdot \hat{r})(\vec{\sigma}_2 \cdot \hat{r}) \left(\frac{1}{r} \frac{d^2}{dr^2} - \frac{1}{r^2} \frac{d}{dr} \right) g_{corr}(r) + \vec{\sigma}_1 \cdot \vec{\sigma}_2 \left(\frac{1}{r} \frac{d}{dr} \right) g_{corr}(r),$$

$$= (\vec{\sigma}_1 \cdot \vec{\nabla})(\vec{\sigma}_2 \cdot \vec{\nabla}) g_{corr}(r)$$ (14)

and

$$V_2(r) = \vec{\sigma}_1 \cdot \vec{\sigma}_2 \, h(r), \quad h(r) = f^2 \left(\frac{1}{r} \frac{d}{dr} \right) g(r) - \left(\frac{1}{r} \frac{d}{dr} \right) g_{corr}(r).$$ (15)

In momentum space:

$$V_1(q) = (\vec{\sigma}_1 \cdot \vec{q})(\vec{\sigma}_2 \cdot \vec{q}) \tilde{g}_{corr}(q), \quad V_2(q) = \vec{\sigma}_1 \cdot \vec{\sigma}_2 \tilde{h}(q).$$ (16)

The function $g_{corr}(r)$ is the solution of the differential equation

$$\left(\frac{d^2}{dr^2} - \frac{1}{r} \frac{d}{dr} \right) g_{corr}(r) = f^2 \left(3 + 3 m_\pi r + m_\pi^2 r^2 \right) \frac{e^{-m_\pi r}}{r^3}.$$ (17)

The contribution of the pion exchange potential to the correlated Hamiltonian becomes:

$$\begin{aligned}
\tilde{\mathcal{H}}_\pi = {}& \frac{G_{\pi NN}^2}{16\pi M^2} \int dy \Big[\psi(\vec{x})^\dagger \, (\vec{\sigma} \cdot \vec{\nabla}) \tau_a \, \psi(\vec{x}) \, \psi(\vec{y})^\dagger (\vec{\sigma} \cdot \vec{\nabla}) \tau_a \psi(\vec{y}) \, g_{corr}(r) \\
& + \psi(\vec{x})^\dagger \, \sigma_i \, \tau_a \psi(\vec{x}) \, \psi(\vec{y})^\dagger \sigma_i \tau_a \psi(\vec{y}) \, h(r) \Big].
\end{aligned}$$ (18)

The direct term of the expectation value of $\tilde{\mathcal{H}}_\pi$ vanishes. The exchange term splits into to terms: $I_\pi^e = I_\pi^{e1} + I_\pi^{e2}$. The first one reads

$$\begin{aligned}
I_\pi^{e1} = {}& \frac{G_{\pi NN}^2}{4 M^2} \frac{3\nu}{2\,V^2} \sum_{\vec{k},\vec{k}'} \tilde{g}_{corr}(\vec{k} - \vec{k}') \Big[(\vec{k} - \vec{k}')^2 \{ 1 + \cos\chi(\vec{k}) \cos\chi(\vec{k}') \} \\
& - (2kk' - \hat{k} \cdot \hat{k}'(k^2 + k'^2)) \sin\chi(\vec{k}) \sin\chi(\vec{k}') \Big].
\end{aligned}$$ (19)

In the above the factor 3 comes from isospin. The second term reads

$$\begin{aligned}
I_\pi^{e2} = {}& \frac{G_{\pi NN}^2}{4 M^2} \frac{3\nu}{2\,V^2} \sum_{\vec{k},\vec{k}'} \tilde{h}(\vec{k} - \vec{k}') \Big[1 + \cos\chi(\vec{k}) \cos\chi(\vec{k}') \\
& - \hat{k} \cdot \hat{k}' \sin\chi(\vec{k}) \sin\chi(\vec{k}') \Big].
\end{aligned}$$ (20)

Finally the total energy contribution is $\mathcal{E}_{total} = \mathcal{E}_{\sigma,\omega,\rho} + I_\pi^e$. The reported calculation involves the following approximation. The interaction $V_1(q)$, eq. (16), may be separated into a tensor term and a central term,

$$V_1(q) = \left((\vec{\sigma}_1 \cdot \vec{q})(\vec{\sigma}_2 \cdot \vec{q})\tilde{g}_{corr}(q) - \frac{1}{3}(\vec{\sigma}_1 \cdot \vec{\sigma}_2)q^2\tilde{g}_{corr}(q) \right) + \frac{1}{3}(\vec{\sigma}_1 \cdot \vec{\sigma}_2)q^2\tilde{g}_{corr}(q).$$

We have replaced $V_1(q)$ by its central part, $\frac{1}{3}(\vec{\sigma}_1 \cdot \vec{\sigma}_2)q^2\tilde{g}_{corr}(q)$. This means replacing I_π^e by

$$I_\pi^e = \frac{G_{\pi NN}^2}{4M^2} \frac{3\nu}{2\,V^2} \sum_{\vec{k},\vec{k}'} (\tilde{h}(\vec{k} - \vec{k}') + \frac{1}{3}(\vec{k} - \vec{k}')^2\tilde{g}_{corr}(\vec{k} - \vec{k}'))$$

$$\times \left[1 + \cos\chi(\vec{k})\cos\chi(\vec{k}') - \hat{k} \cdot \hat{k}' \sin\chi(\vec{k})\sin\chi(\vec{k}') \right].$$

Moreover, we have treated this term within two distinct schemes: leaving the term as it stands, and replacing in it $(\vec{k} - \vec{k}')^2\tilde{g}_{corr}(\vec{k} - \vec{k}')$ by $((\vec{k} - \vec{k}')^2\tilde{g}_{corr}(\vec{k} - \vec{k}') - \lim_{q \to 0} q^2\tilde{g}_{corr}(q))$. The second scheme amounts to neglecting the contact term in the pion interaction.

3   Results and discussions

The models contain the following parameters: the couplings g_σ, g_ω, g_ρ, g_π, the meson masses, m_i, $i = \sigma, \omega, \rho, \pi$ and the three parameters specifying the short range correlation function, α, β and γ. The couplings g_σ and g_ω are chosen so as to reproduce the ground state properties of nuclear matter. For the ρ and π-meson couplings we take the usual values $g_\rho^2/4\pi = 0.55$ and $f_\pi^2/4\pi = 0.08$, where $f_\pi = G_{\pi NN}m_\pi/(2M)$. We choose $m_\sigma = 550$ MeV, $m_\omega = 783$ MeV, $m_\rho = 770$ MeV and $m_\pi = 138$ MeV. The normalization condition (7) determines β. We fix α variationally, by minimizing the energy. The parameter γ is such that it reproduces a reasonable healing distance, assuming that this quantity decreases as k_F increases. Therefore, we assume that γ depends on k_F according to $\gamma = m_\omega + a_1\,k_F/k_{F0}$, where the parameters $a_1 = 150$ MeV is conveniently chosen. The behavior of the EOS with the exchange of sigma, omega, rho mesons and pions taken into account is shown in Fig. 1 (left panel) and the corresponding model parameters are displayed in Table I.

We found that short-range correlations have important consequences and should not be neglected. From Figs. 1 (right panel), it may be seen that their main effect is the tendency to considerably soften the EOS. Moreover, as seen in Tables II, the phenomenological values of the couplings g_σ and g_ω are strongly dependent on the correlations, being considerably reduced by their presence.

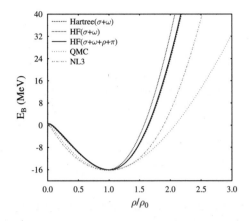 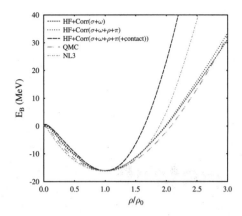

Figure 1: Left panel: The EoS for the pure HF approach (excludingcorrelations). For comparison, the EOS obtained within the QMC [19] and the NL3 [20] parametrizations of the $\sigma - \omega$ model are also shown. Right panel: EOS with short range correlations when rho and pion exchange is taken into account. The stiffness of the long-dashed curve is due to the insufficient flexibility of the correlation factor. For comparison, a typical EOS in the absence of π and ρ mesons exchange is also shown.

When the contact term in the pion exchange interaction is included, the EOS appears to show a stiffer behavior, even when correlations are taken into account, but this is due to the very simplified parametrization of the adopted correlation function which was not flexible enough to respond to both repulsive and attractive components of the interactions. In Fig. 2 (left panel), we plot the correlation function. Short range correlations give rise to an extra node in the dependence of the ground state wave-function on the relative coordinate, contrary to what generally happens in non-relativistic calculations with hard core, when the wave function acquires a wound. The present parametrization of $f(r)$ is not able to respond adequately to the

Table 1: Coupling constants for HF calculations without correlations. The parameters where obtained for a binding energy of -16.07 MeV at the saturation density $\rho_0 = 0.148 \text{fm}^{-3}$

	g_σ	g_ω	M^*/M	K (MeV)
Hartree $\sigma\omega$	8.8839	13.8330	0.538	555
HF $\sigma\omega$	8.49	12.6566	0.513	680
HF $\sigma\omega\rho\pi$(total)	10.2192	12.0809	0.515	587

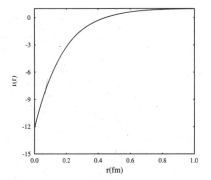
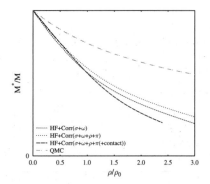

Figure 2: Left panel: The correlation function for the $\sigma - \omega$ calculation. Right panel: Behavior of the effective mass. When the contact term (of eq. (5)) is omitted, the decrease is less pronounced than for the corresponding HF result. The lowest curve is the one labeled HF-corr ($\sigma + \omega + \rho + \pi$) (with cont. term), but we believe that the presence of the contact term may obscure the results and requires a special interpretation, as explained in the text.

very short range repulsion associated with the contact term. A richer prescription [21] seems to be required when the contact term is included.

As shown in Fig. 2(right panel), if short-range two-body correlations are included and the contact term (c. t.) in the pion exchange interaction is neglected, the effective mass does not decrease so fast with the increase of density as in the Hartree and HF calculations, which explains the softer behavior of the EOS with correlations. The faster decrease of the mass found when the c.t. is not neglected, may be attributed to an insufficient flexibility of the correlation factor. Correlations increase the kinetic energy to the order of magnitude one expects from non-relativistic approaches. This is an important effect since the Hartree and, even more dramatically,

Table 2: A $\equiv \sigma + \omega$, B $\equiv \sigma + \omega + \rho + \pi$, C $\equiv \sigma + \omega + \rho + \pi$ (+contact). Model parameters with short range correlations of nuclear matter at saturation density are given. These results were obtained with fixed: $M = 939$ MeV $m_\sigma = 550$ MeV, $m_\omega = 783$ MeV, $m_\rho = 770$ MeV, $m_\pi = 138$ MeV, $g_\rho^2/4\pi = 0.55$ and $f_\pi^2/4\pi = 0.08$ at $k_{F0} = 1.3$ fm^{-1} with binding energy $E_B = \varepsilon/\rho - M = -16.07$ MeV. Here, $\mathcal{T}'$ stands for $\mathcal{T} - M\rho_B$.

HF+Corr	g_σ	g_ω	α (MeV)	β (MeV)	γ	$\mathcal{T}'/\rho_B$ (MeV)	$\mathcal{V}_d/\rho_B$ (MeV)	$\mathcal{V}_e/\rho_B$ (MeV)	$\mathcal{T}_c^e/\rho_B$ (MeV)
A	4.645	3.779	13.1529	-2225.6839	933.	14.366	-70.87	10.87	29.55
B	3.310	3.085	13.3282	-2201.2453	933	15.57	-19.86	-40.13	28.34
C	11.588	13.432	2.0939	-621.443	933	11.58	-64.71	6.61	4.99

Table 3: Ground state properties of nuclear matter with correlations at saturation density.

HF+Corr	M^*/M	K(MeV)	r_h
$\sigma + \omega$	0.571	392	0.286
$\sigma + \omega + \rho + \pi$	0.602	384	0.288
$\sigma + \omega + \rho + \pi$ (+contact)	0.568	602	0.254

the HF approximation, predict an unrealistically low value of the kinetic contribution to the total energy at saturation, as shown in Table II.

Acknowledgments

I would like to thank Dr. A.B. Santra and the organizers for giving me the oporunity to participate in this conference. This work was partially supported by FCT (Portugal) under the projects POCI/FP/63918/2005 and POCI/FP/63419/2005.

References

[1] F. Coester, Nucl. Phys. **7**, 421 (1958); F. Coester and H. Kümmel, Nucl. Phys. **17**, 477 (1960).

[2] R. F. Bishop and H. Kümmel, Phys. Today **40(3)**, 95 (1991); R. F. Bishop, Theor. Chem. Acta **80**, 95 (1991).

[3] F. Villars, "Proceedings of the International School of Physics, 'Enrico Fermi'- Course 23, (1961)." Academic Press, New York, 1963; J.S. Bell, " Lectures on the Many-Body Problem, First Bergen International School of Physics." Benjamin, New York, (1962).

[4] J. da Providência and C.M. Shakin, Ann. Phys.(NY) **30**, 95 (1964).

[5] H. Feldmeier, T. Neff, R. Roth, and J. Schnack, Nucl. Phys. **A 632**, 61 (1998); T. Neff and H. Feldmeier, Nucl. Phys. **713**, 311 (2003).

[6] F. Coester, S. Cohen, B.D. Day, and C.M. Vincent, Phys. Rev. **C 1**, 769 (1970); R. Brockmann and R. Machleidt, Phys. Rev. **C 42**, 1965 (1990).

[7] R.B. Wiringa, V. Fiks and A. Fabrocini, Phys. Rev. **C 38**, 1010 (1988); W. Zuo, A. Lejeune, U. Lombardo and J.-F. Mathiot, Nucl. Phys. **A 706**, 418 (2002).

[8] B.D. Serot, J.D. Walecka, Int. J. Mod. Phys. **E6**, 515 (1997).

[9] P.K. Panda, D.P. Menezes, C. Providência and J. da Providência, Phys. Rev. **C 71**, 015801 (2005); P.K. Panda, D.P. Menezes, C. Providência and J. da Providência, Braz. J. Phys **35**, 873 (2005).

[10] P.K. Panda, C. Providência and J. da Providência, Phys. Rev. **C 73**, 035805 (2006).

[11] P.K. Panda, J. da Providência, C. Providência and D.P. Menezes, Proceedings online of the Conference on Microscopic Approaches to Many-Body Theory (MAMBT), in honor of Ray Bishop, Manchester, UK, http://www.qmbt.org/MAMBT/pdf/Providencia.pdf, 2005.

[12] A. Bouyssy, J.-F. Mathiot and N.Van Giai and S. Marcos, Phys. Rev. C **36**, 380 (1987).

[13] A. Mishra, P.K. Panda, S. Schramm, J. Reinhardt and W. Greiner, Phys. Rev. C **56**, 1380 (1997).

[14] J. da Providência and C. M. Shakin, Phys. Rev C **4**, 1560 (1971); C. M. Shakin, Phys. Rev. C **4**, 684 (1971).

[15] R. Jastrow, Phys. Rev. **98**, 1479 (1955).

[16] S.A. Moszkowski and B.L. Scott, Ann. Phys. (N.Y.) i **11**, 65 (1960).

[17] H. Bethe, Ann. Rev. Nucl. Sci. **21**, 93 (1971).

[18] P.K. Panda, R. Sahu and S.P. Misra, Phy. Rev. C **45** (1992) 2079.

[19] P. A. M. Guichon, Phys. Lett. B **200**, 235 (1988); K. Saito and A.W. Thomas, Phys. Lett. B **327**, 9 (1994); P.K. Panda, A. Mishra, J.M. Eisenberg, W. Greiner, Phys. Rev. C **56**, 3134 (1997).

[20] G.A. Lalazissis, J. König, P. Ring, Phys. Rev. **C55**, 540 (1997).

[21] P.K. Panda, C. Providência, J. da Providência, submitted for publication.

Physics and Astrophysics of Hadrons and Hardronic Matter
Editor: A. B. Santra

Strangeness in Nucleon

S. K. Singh[a], H. Arenhovel[b]

a) Department of Physics, Aligarh Muslim University
Aligarh-202 002, India.
email: pht13sks@rediffmail.com

b) Institut fur Kernphysik, Johannes Gutenberg-Universitat
D-55099 Mainz, Germany

Abstract

There are many evidences to suggest that strangeness content of nucleon is nonzero. The recent advances in doing electron scattering experiment with polarized electrons and polarized targets have made it possible to determine the strangeness form factors of nucleon from experimental observation of parity violating asymmetry in polarized electron scattering. The present status of the field and future program of research in this field is presented.

1   Introduction

The constituent quark model is one of the most successful models to describe the structure of nucleons. In this model the nucleon is described as bound state of three valence quarks moving in a potential. However, an important question in understanding the quark structure of nucleons involves the role of sea quarks. There are evidences, suggesting that there is sizable contribution of the sea quarks to the quark momentum distribution coming from the analysis of quark distribution functions in deep inelastic scattering of electrons from nucleons. The analysis of the sigma term from the low energy pion nucleon scattering gives evidence for the presence of $s\bar{s}$ contribution to nucleon mass. In addition, the analysis of spin dependent parton distribution functions from the polarized electron nucleon scattering and the study of axial form factor in the neutral current neutrino-nucleon scattering gives evidence for the presence of sea quarks in nucleon spin [1].

Kaplan and Manohar [2] suggested that the contributions of sea quarks to the ground state nucleon properties such as spin, charge, the study of magnetic moment and axial coupling could be learnt through the weak neutral current probes of the nucleons in electron and neutrino nucleon scattering from nucleons and nuclei. From this point of view, the neutral current induced neutrino scattering and the neutral current induced parity violating effects in polarized electron scattering experiments have attracted much attention in last 10-15 years. The experimental and theoretical investigations have been presently carried out mainly in the area of parity violating polarized electron scattering from nucleons and nuclei where sea quark content, specially the $s\bar{s}$ content has been studied. There are many theoretical studies made in the area of neutral current neutrino scattering from nucleons and nuclei. The specially the $s\bar{s}$ content has been studied. There are many theoretical studies made in the area of neutral current neutrino scattering from nucleons and nuclei. The future experiments planned with intermediate energy neutrinos for neutrino oscillation studies, can be well designed to look for sea quark degrees of freedom, specially the strangeness content of the nucleon.

In section-2, we describe the basic formalism for studying the strangeness content of the nucleon from polarized electron scattering from nucleons. In section-3, we describe the formalism to study parity violating effects in electron scattering from nucleons and nuclei, and present some experimental results in section-4. In section-5, we present summary.

2   Lagrangian and Matrix element in Standard Model

In Standard Model, the neutral current Lagrangian is written as:

$$\mathcal{L}_{int}^{NC} = \frac{G_F}{\sqrt{2}} J_{e,NC}^{\mu} J_{\mu}^{had,NC}, \qquad e-q \text{ interaction} \tag{1}$$

$$= \frac{G_F}{\sqrt{2}} J_{\nu,NC}^{\mu} J_{\mu}^{had,NC}, \qquad \nu-q \text{ interaction} \tag{2}$$

where $G_F = \frac{g^2}{4\sqrt{2}M_W^2 \sin^2\theta}$ and

$$J_{e,NC}^{\mu} = \frac{1}{2}\bar{\psi}_e\gamma^{\mu}\left(4\sin^2\theta_W - 1 + \gamma^5\right)\psi_e, \quad J_{\nu,NC}^{\mu} = \frac{1}{2}\bar{\psi}_\nu\gamma^{\mu}\left(1 - \gamma^5\right)\psi_\nu \tag{3}$$

$$J_{\mu}^{had,NC} = V_{\mu}^{had,NC} - A_{\mu}^{had,NC} \tag{4}$$

where

$$V_{\mu}^{had,NC} = \frac{1}{2}\left(1 - \frac{8}{3}\sin^2\theta_W\right)\bar{\psi}_u\gamma_\mu\psi_u - \frac{1}{2}\left(1 - \frac{4}{3}\sin^2\theta_W\right)\left(\bar{\psi}_d\gamma_\mu\psi_d + \bar{\psi}_s\gamma_\mu\psi_s\right) \tag{5}$$

$$A_{\mu}^{had,NC} = \frac{1}{2}\left(\bar{\psi}_u\gamma_\mu\gamma_5\psi_u - \bar{\psi}_d\gamma_\mu\gamma_5\psi_d - \bar{\psi}_s\gamma_\mu\gamma_5\psi_s\right) \tag{6}$$

Similarly, the electromagnetic Lagrangian in Standard model is written as:

$$\mathcal{L}_{int}^{em} = eJ_e^{\mu}J_{\mu}^{had} \tag{7}$$

where

$$J_e^{\mu} = \bar{\psi}_e\gamma^{\mu}\psi_e \tag{8}$$

$$J_{\mu}^{had} = \frac{2}{3}\bar{\psi}_u\gamma_\mu\psi_u - \frac{1}{3}\bar{\psi}_d\gamma_\mu\psi_d - \frac{1}{3}\bar{\psi}_s\gamma_\mu\psi_s \tag{9}$$

The matrix element of electromagnetic J_μ^γ and weak neutral current J_μ^Z are defined as:

$$\langle p'|J_\mu^Z|p\rangle = \bar{u}(p')\left[F_1^Z(Q^2)\gamma_\mu + iF_2^Z(Q^2)\sigma_{\mu\nu}\frac{q^\nu}{2M} + G_A^Z(Q^2)\gamma_\mu\gamma_5\right]u(p) \tag{10}$$

$$\langle p'|J_\mu^\gamma|p\rangle = \bar{u}(p')\left[F_1^\gamma(Q^2)\gamma_\mu + iF_2^\gamma(Q^2)\frac{\sigma_{\mu\nu}}{2M}q^\nu\right]u(p) \cdot \tag{11}$$

where $F_1^{\gamma,Z}(Q^2)$ and $F_2^{\gamma,Z}(Q^2)$ are Dirac form factors for electromagnetic current and weak neutral current and are given in terms of Sachs form factors $G_{E,M}^{\gamma,Z}(Q^2)$ as:

$$G_E^{\gamma,Z} = F_1^{\gamma,Z} - \tau F_2^{\gamma,Z} \tag{12}$$

$$G_M^{\gamma,Z} = F_1^{\gamma,Z} + F_2^{\gamma,Z}, \quad \tau = \frac{Q^2}{4M^2} \tag{13}$$

where electromagnetic nucleon form factors are:

$$G_E^{\gamma,p} = \frac{2}{3}G_{E,M}^u - \frac{1}{3}G_{E,M}^d - \frac{1}{3}G_{E,M}^s \tag{14}$$

$$G_E^{\gamma,n} = \frac{2}{3}G_{E,M}^d - \frac{1}{3}G_{E,M}^u - \frac{1}{3}G_{E,M}^s \tag{15}$$

and weak nucleon form factors are:

$$G_E^{Z,p} = \frac{1}{2}\left(1 - \frac{8}{3}\sin^2\theta\right)G_{E,M}^u - \frac{1}{2}\left(1 - \frac{4}{3}\sin^2\theta\right)\left(G_{E,M}^d + G_{E,M}^s\right) \tag{16}$$

$$G_E^{Z,n} = \frac{1}{2}\left(1 - \frac{8}{3}\sin^2\theta\right)G_{E,M}^d - \frac{1}{2}\left(1 - \frac{4}{3}\sin^2\theta\right)\left(G_{E,M}^u + G_{E,M}^s\right) \tag{17}$$

Using the electromagnetic form factors $G_{E,M}^{\gamma,p}(Q^2)$ and $G_{E,M}^{\gamma,n}(Q^2)$ to eliminate the weak form factors, the neutral current form factors are given as:

$$G_{E,M}^{Z,(p,n)} = \frac{1}{2}\left(1 - 4\sin^2\theta_W\right)G_{E,M}^{p,n} - \frac{1}{2}G_{E,M}^{p,n} - \frac{1}{2}G_{E,M}^s \tag{18}$$

Similarly axial vector form factors are given as

$$G_A^Z = -\frac{1}{2}G_A\tau_3 + \frac{1}{2}G_s \tag{19}$$

2.1 Electromagnetic and Weak Form Factors [3]

The neutral current nucleon form factors which are completely determined in terms of electromagnetic and weak from factors $G_{E,M}^{p,n}(Q^2)$ and $G_A(Q^2)$, which are parametri-zed as dipoles in the following way:

207

$$G_E^p(q^2) = \left(1 - \frac{q^2}{M_V^2}\right)^{-2}, \quad G_E^n(q^2) = \left(\frac{q^2}{4M^2}\right)\mu_n G_E^p(q^2)\xi_n \tag{20}$$

$$G_M^p(q^2) = (1 + \mu_p)G_E^p(q^2), \quad G_M^n(q^2) = \mu_n G_E^p(q^2) \tag{21}$$

$$\xi_n = \frac{1}{1 - \lambda_n \frac{q^2}{4M^2}}, \quad \mu_p = 1.792847, \quad \mu_n = -1.913043, \quad \lambda_n = 5.6.$$

$$G_A(q^2) = G_A(0)\left(1 - \frac{q^2}{M_A^2}\right)^{-2}, \quad G_A(0) = 1.25 \tag{22}$$

The numerical value of the vector dipole mass $M_V= 0.84$ GeV is taken from experimental data on electron proton scattering, and axial dipole mass $M_A=1.05$ GeV from neutrino scattering from proton and deuteron.

2.2 Strangeness Form Factors

Strangeness form factors G_E^s, G_M^s and G_A^s can be obtained experimentally from the determination of the neutral current form factors $G_{E,M}^Z(Q^2)$ and $G_A^Z(Q^2)$ from polarized electron scattering and neutrino scattering. Theoretically there are many models, which make predictions for the strangeness form factors. Some of them are as follows.

2.2.1 Pole Model [4]

In this model, the strangeness form factors are determined using vector meson dominance through following diagrams: Based on pole fit to the isoscalar electromagnetic

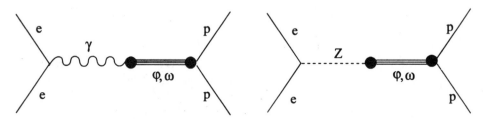

Figure 1: Pole model for strangeness form factors

nucleon form factors, which are dominated by $\omega(780)$, $\varphi(1020)$ and a phenomenological third resonance $\omega(1600)$, the strangeness form factors $F_1^S(Q^2)$ and $F_2^S(Q^2)$ are given by:

$$F_1^S(Q^2) = \sum_{i=1}^{3} a_i^S \frac{Q^2}{m_i^2 + Q^2}, \quad F_2^S(Q^2) = \sum_{i=1}^{3} b_i^S \frac{Q^2}{m_i^2 + Q^2} \tag{23}$$

where a_i^S and b_i^S are determined from fitting the data on nucleon form factors.

2.2.2 Meson Cloud Model [5],[6]

These models give quite different results depending upon how many particle contributions are calculated in the loop diagrams shown in Fig.2.

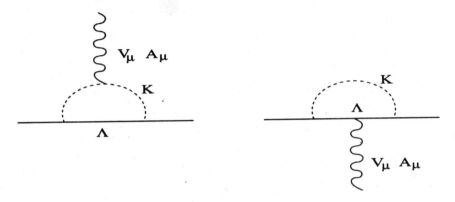

Figure 2: Virtual meson models for strangeness form factors

Standard perturbation theory using effective Lagrangian is used to calculate these diagrams. These calculations make definite predictions, but suffer from the following problems:

(i) Convergence is a problem, and higher order diagrams are not well calculated.

(ii) Q^2 dependence of the form factors are model dependent.

2.2.3 Phenomenological Models

Most of the calculations on parity violating electron scattering experiments make use of phenomenological models which are parametrized as dipole and are give by:

$$F_1^S(Q^2) = \frac{1}{6} \frac{r_S^2 Q^2}{\left(1 + \frac{Q^2}{M_1^2}\right)^2}, \quad F_2^S(Q^2) = \frac{\mu_S^2}{\left(1 + \frac{Q^2}{M_2^2}\right)^2}, \quad G_A^S(Q^2) = \frac{g_A^S}{\left(1 + \frac{Q^2}{M_A^2}\right)^2} \quad (24)$$

where $Q^2 = -q^2$.

2.3 Corrections to the Electromagnetic and Weak Form factors

The determination of strangeness form factors from the analysis of parity violating electron scattering experiments involves observation of very small effects which are generally of the order of $10^{-4}(q^2/M^2)$. It is important to know the small corrections which should be applied to the known electromagnetic and weak form factors. These are:

2.3.1 Radiative corrections to G_E^Z, G_M^Z and G_A^Z [1],[7]

These corrections come from following diagrams which lead to following corrections

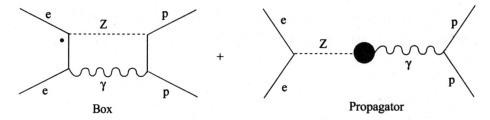

Figure 3: Box and propagator diagrams for radiative corrections

$$G_A^Z = -\frac{1}{2}\left(1 + R_A^1\right)G_A\tau_3 + \frac{1}{2}R_A^0 + \frac{1}{2}G_A^s \tag{25}$$

$$G_{E,M}^{Z,(p,n)} = \left(1 - 4\sin^2\theta_W\right)\left(1 + R_V^{p,n}\right)G_E^{p,n}\left(1 + R_V^n\right)G_{E,M}^{p,n} - G_{E,M}^s \tag{26}$$

$$R_A^1 = -0.41 \pm 0.24, \qquad R_A^0 = -0.6 \pm 0.14 \tag{27}$$

$$R_V^p = -0.054 \pm 0.033, \qquad R_V^n = -0.0143 \pm 0.0004 \tag{28}$$

2.3.2 Anapole Moments [8]

These are defined as the effective parity violating coupling of photon to nucleon and appear when parity violation induced by weak interaction of nucleons through weak nucleon-nucleon potential is included.

Its coupling to real photon is zero as $q^2 = 0, q \cdot \epsilon = 0$ and they contribute only to the electron scattering processes which are equivalent to virtual photon coupling to the hadronic current. Nonrelativistically it can be generated through the spin dependent potential containing momentum term in the parity violating nucleon-nucleon potential. For example, nonrelativistically

$$\vec{J} \propto q^2 \langle\vec{\sigma}\rangle - \vec{q}\langle q \cdot \sigma\rangle \tag{29}$$

which mimics an axial coupling. The model calculations give ($Q^2 = -q^2$)

$$F_A = \frac{Q^2}{2M^2}\left[a_S(Q^2) + a_V(Q^2)\tau_3\right] \tag{30}$$

where

$$a_{S,V}(0) = \frac{f_{\pi NN}g_{\pi NN}}{4\sqrt{2}\pi^2}\alpha_{S,V}(0); \quad \alpha_S = 1.6, \quad \alpha_V = 0.4 \tag{31}$$

Thus, the effective axial coupling $G_A^Z(Q^2)$ is given by:

$$G_A^Z = -\frac{1}{2}G_A\left(1 + R_A^1\right) + \frac{1}{2}G_A^S + \frac{1}{2}R_A^0 + F_A \tag{32}$$

3 Parity Violating Electron Scattering From Nucleons and Nuclei

The parity violating longitudinal asymmetry in the polarized electron scattering arises due to the interference of the photon(γ) and Z exchange diagrams shown in Fig.4.

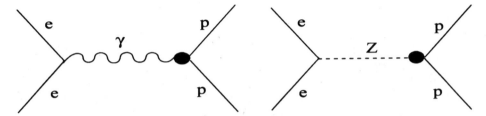

Figure 4: Photon and Z exchnage diagrams for ep scattering

The asymmetry $\mathcal{A}$ is defined as

$$\mathcal{A} = \frac{d\sigma_R - d\sigma_L}{d\sigma_R + d\sigma_L} \tag{33}$$

where $d\sigma_R$ and $d\sigma_L$ are the cross sections for electron nucleon and electron nucleus scattering for right handed and left handed electrons.

3.1 Nucleon Targets [9]

In the case of proton targets, this asymmetry can be calculated using the matrix elements of neutral current Lagrangian and the electromagnetic Lagrangian described in section-2 and using the matrix elements in terms of the neutral current and electromagnetic form factors. The expressions for the asymmetry is given as:

$$\mathcal{A} = \frac{d\sigma_R - d\sigma_L}{d\sigma_R + d\sigma_L} = -\frac{G_F Q^2}{2\sqrt{2}\pi\alpha}\left[\frac{\mathcal{A}_E + \mathcal{A}_M + \mathcal{A}_A}{D}\right] \tag{34}$$

where

$$D = \epsilon\, G_E^{\gamma^2} + \tau\, G_M^{\gamma^2} \tag{35}$$

$$\tau = \frac{Q^2}{4M^2}, \quad \epsilon = \left[1 + 2\left(1 + \tau\right)\tan^2\frac{\theta}{2}\right]^{-1} \tag{36}$$

211

$$\mathcal{A}_E = \epsilon\, G_E^\gamma\, G_E^Z, \quad \mathcal{A}_M = \tau\, G_M^\gamma\, G_M^Z \tag{37}$$

$$\mathcal{A}_A = \left(1 - 4\sin^2\theta_W\right)\epsilon'\, G_M^\gamma\, G_A^Z, \quad \epsilon' = \sqrt{\tau\left(1+\tau\right)\left(1-\epsilon^2\right)} \tag{38}$$

It can be seen that this asymmetry is not very sensitive to G_A^Z at low values of q^2/M^2 for electron scattering experiments with electron energies in the MeV region.

3.2 Nuclear Targets

In case of nuclear targets, there will be additional contribution to the parity violating asymmetry in addition to the $\gamma - Z$ interference term described above. These come mainly from two effects described below.

3.2.1 Hadronic Parity Violating Effects [10]

It is well known that weak interaction between two nucleons gives rise to parity violating potential which are generated by W and Z exchanges. There are quite a few calculations which make use of the quark model and Standard model of the electroweak interactions to generate this parity violating weak nucleon-nucleon potential. These are described in one boson exchange(OBE) model in analogy with the strong nucleon-nucleon potential and is parametrized in terms of the parity violating πNN, ρNN and ωNN couplings in addition to the parity conserving πNN, ρNN and ωNN couplings. This potential is given as

$$
\begin{aligned}
V_{PNC} =\ & i\frac{h_\pi^1 g_A m_N}{\sqrt{2}F_\pi}\left(\frac{\tau_1 \times \tau_2}{2}\right)_3 (\vec{\sigma}_1 + \vec{\sigma}_2)\left[\frac{\vec{p}_1 - \vec{p}_2}{2m_N}, \omega_\pi(r)\right] \\
& - g_\rho\left(h_\rho^0 \tau_1 \cdot \tau_2 + h_\rho^1\left(\frac{\tau_1 + \tau_2}{2}\right)_3 + h_\rho^2\frac{(3\tau_1^3\tau_2^3 - \tau_1 \cdot \tau_2)}{2\sqrt{6}}\right) \\
& \left((\vec{\sigma}_1 - \vec{\sigma}_2)\cdot\left\{\frac{\vec{p}_1 - \vec{p}_2}{2m_N}, \omega_\rho(r)\right\} + i\left(1 + \chi_\rho\right)\vec{\sigma}_1 \times \vec{\sigma}_2 \right. \\
& \left. \cdot\left[\frac{\vec{p}_1 - \vec{p}_2}{2m_N}, \omega_\rho(r)\right]\right) - g_\omega\left(h_\omega^0 + h_\omega^1\left(\frac{\tau_1 + \tau_2}{2}\right)_3\right) \\
& \left((\vec{\sigma}_1 - \vec{\sigma}_2)\cdot\left\{\frac{\vec{p}_1 - \vec{p}_2}{2m_N}, \omega_\omega(r)\right\} + i\left(1 + \chi_\omega\right)\vec{\sigma}_1 \times \vec{\sigma}_2 \right. \\
& \left. \cdot\left[\frac{\vec{p}_1 - \vec{p}_2}{2m_N}, \omega_\omega(r)\right]\right) - \left(g_\omega h_\omega^1 - g_\rho h_\rho^1\right)\left(\frac{\tau_1 - \tau_2}{2}\right)_3 \\
& (\vec{\sigma}_1 + \vec{\sigma}_2)\cdot\left\{\frac{\vec{p}_1 - \vec{p}_2}{2m_N}, \omega_\rho(r)\right\} - g_\rho h_\rho'^1 i\left(\frac{\tau_1 \times \tau_2}{2}\right)_3 \\
& (\vec{\sigma}_1 + \vec{\sigma}_2)\cdot\left[\frac{\vec{p}_1 - \vec{p}_2}{2m_N}, \omega_\rho(r)\right]
\end{aligned}
\tag{39}
$$

where the best values of the parameters entering in the potential are given in table-2 corresponding to two models given by Desplanques et al. [11] and Kaiser and Meissner [12]. This potential give rise to mixing of odd parity states in the wave function, leading to additional parity violating effects. Such calculations have been done extensively for deuteron.

Table 1: Numerical values of pnc coupling constants in units of 10^{-7} [11],[12].

Coupling Constants	DDH	KM	Coupling Constants	DDH	KM
h_π^1	4.6	0.2	$h_\rho^{1'}$	0.0	-2.2
h_ρ^0	-11.4	-3.7	h_ω^0	-1.9	-6.2
h_ρ^1	-0.2	-0.1	h_ω^1	-1.1	-1.1
h_ρ^2	-9.5	-3.3	-	-	-

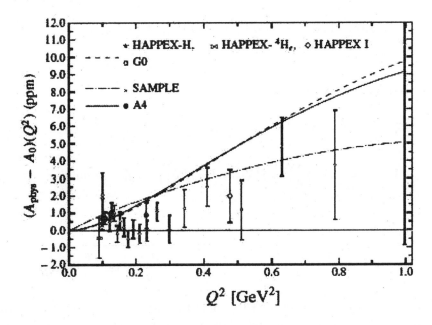

Figure 5: Experimental asymmetry vs Q^2.

3.2.2 Exchange Current Effects [13]

In analogy with the exchange current effect in the electromagnetic interactions which are well known in processes like $ed \rightarrow enp$, $ed \rightarrow ed$, $e\,^3\text{He} \rightarrow e\,^3\text{He}$ and $e\,^3\text{H} \rightarrow e\,^3\text{H}$, the parity violating exchange currents may exist, which are generated by the parity violating nucleon-nucleon potential in a potential model approach. There are only a few calculations of such effects and they are done in the case of deuteron.

3.3 Deuteron Targets

The parity violating effects in $ed \rightarrow enp$ have been calculated by many authors [13],[14], but the calculation for elastic scattering are not many. Here we present the theoretical results for the parity violating asymmetry in elastic deuteron scattering, which is given as [15]:

$$\mathcal{A}(Q^2) = \frac{G_F Q^2}{4\sqrt{2}\pi\alpha} \left(\frac{W^{PV}(\gamma Z)}{W^{EM}} \right) \tag{40}$$

where

$$W^{EM} = \sum_J v_L F_{CJ}^2 + \sum v_T \left(F_{EJ}^2 + F_{MJ}^2 \right) \tag{41}$$

$$W^{PV}(\gamma Z) = \left[v_L W_{AV}^L + v_T W_{AV}^T + v_{T'} W_{AV}^{T'} \right] \tag{42}$$

and

$$W_{AV}^L = \sum_J F_{CJ} \tilde{F}_{CJ} \tag{43}$$

$$W_{AV}^T = g_A^e \sum_J \left(F_{EJ} \tilde{F}_{EJ} + F_{MJ} \tilde{F}_{MJ} \right). \quad g_A^e = \frac{1}{2} \tag{44}$$

$$W_{AV}^{T'} = g_V^e \sum_J \left(F_{EJ} \tilde{F}_{MJ}^5 + F_{MJ} \tilde{F}_{EJ}^5 \right), \quad g_V^e = 4\sin^2\theta - 1 \tag{45}$$

$$v_L = \left(\frac{Q^2}{q^2} \right)^2 \tag{46}$$

$$v_T = \frac{1}{2} \left(\frac{Q^2}{q^2} \right)^2 + \tan^2\frac{\theta}{2}, \quad v_{T'} = \sqrt{\frac{Q^2}{q^2} + \tan^2\frac{\theta}{2}} \, \tan\frac{\theta}{2} \tag{47}$$

F_{CJ}, F_{EJ} and F_{MJ} are the standard Coulomb, Electric and Magnetic multipoles. $\tilde{F}_{CJ}$, $\tilde{F}_{EJ}$ and $\tilde{F}_{MJ}$ are the Coulomb, Electric and Magnetic multipoles calculated with the parity admixture wave functions. $\tilde{F}_{EJ}^5$ and $\tilde{F}_{MJ}^5$ are the Electric and Magnetic multipoles calculated with the axial vector current. The strangeness form factors enter through the form factors $G_E^{\gamma,Z}$, $G_M^{\gamma,Z}$ and G_A^Z in the definitions of the multipoles $\tilde{F}_{CJ}$, $\tilde{F}_{EJ}$, $\tilde{F}_{MJ}$ and $\tilde{F}_{EJ}^5$ and $\tilde{F}_{MJ}^5$.

4 Experimental Studies

The following experiments on parity violating electron scattering experiments have been already done or being done at various laboratories.

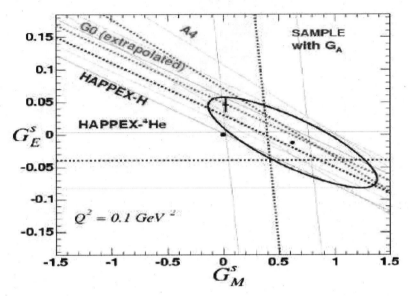

Figure 6: Limits on G_E^s and G_M^s at Q^2=0.1 Gev^2 for experimental observations of the asymmetry

(i) **MIT-BATES [16]:** The parity violating electron scattering experiments at MIT-BATES are SAMPLE-I, SAMPLE-II and SAMPLE-III which use 200MeV electron beam to do experiments on proton and deuteron targets. The Q^2 for SAMPLE-I is $Q^2 = 0.091$ GeV^2/c^2 while it is $Q^2 = 0.038$ GeV^2/c^2 for SAMPLE-II. The results have been reported and a reanalysis of SAMPLE-II have also been done to explain away some discrepancies in SAMPLE-III.

(ii) **TJNAF Experiment [17]:** The parity violating electron scattering experiments at Jefferson National Accelerator Facility are being done by the G0 and HAPPEX collaboration. These experiments use the high energy electron of $E_e = 3.3$ GeV and $E_e = 3.356$ GeV to perform experiments at $\theta = 10^o$ and $\theta = 8^o$ corresponding to $Q^2 = 0.48$ GeV^2/c^2. The targets being used are proton and ^{4}He. There are plans to use deuteron targets as well in future to do these experiments, with various energies of electron varying from $E_e = ($ 0.29, 0.37, 0.445, 0.535, 0.641, 0.765, 0.900) MeV. There are also plans to do $e + p \rightarrow e + \Delta$ experiment on proton and deuteron targets. (ii) **MAMI [18]:** The parity violating electron scattering experiments at MAINZ, MICROTRON, Germany is being done by A4 collaboration. The experiments on proton target have been completed at $E_e = 855$ MeV and

215

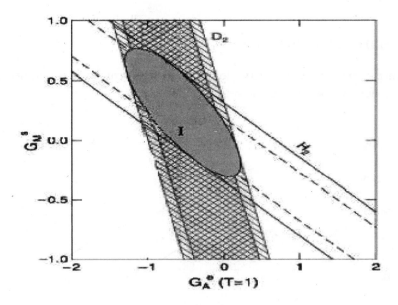

Figure 7: Limits on G_M^s and $G_A^e (= (1 + R_A^{(1)})G_A^3 + R_A^0 + G_A^s)$ for experimental observations of the asymmetry. See equations 24 & 25 for definition of $R_A^{(1)}$, R_A^0 and G_A^s.

$\theta = 35^o$, corresponding to $Q^2 = 0.225$ GeV^2/c^2. The experiments have been also done at $\theta = 145^o$. The experiments with deuteron targets and observation of the asymmetry with deuteron targets is planned for the future.

In Fig.5, we give the summary of the observed asymmetry from various experiments which have already been completed. In Fig.6, Fig.7 and Fig.8, we show the limits on various strangeness form factors, which are obtained from these experiments.

5 Summary

It has been shown that there is evidence that the strangeness content of the nucleon is nonvanishing. There are many theoretical models which give nonzero values of strangeness form factors $G_E^s(Q^2)$, $G_M^s(Q^2)$ and $G_A^s(Q^2)$. Experimentally, there is a possibility to determine these form factors from the parity violating electron scattering as well as neutral current neutrino scattering from nucleons and nuclei.

The recent advances in performing electron scattering experiments with polarized electrons have made it possible to experimentally determine these form factors. Various experiments done at MIT, TJNAF and MAMI electron accelerator laboratories in the energy range of 200MeV-3.5 GeV are not conclusive but they indicate that $G_E^s(Q^2)$ may be very small. The determination of $G_M^s(Q^2)$ and $G_A^s(Q^2)$ is not conclusive and depends upon the small radiative corrections etc. in the various form factors. The new experiments being done at various energies for a large range of Q^2, at these laboratories will be able to determine the strangeness form factors in near future.

Acknowledgments

The work is financially supported by Alexander von Humbolt Foundation and one of us S. K. Singh would like to thank Institut fur Kernphysik, Mainz, Germany for hospitality during the tenure of the fellowship.

References

[1] M. J. Musolf et al. Phys. Rept. **239**, 1 (1994).

[2] D. B. Kaplan and A. Manohar, Nucl. Phys. **B310**, 527 (1988).

[3] W. M. Alberico, S. M. Bilenky and C. Maieron, Phys. Rept. **358**, 227 (2002).

[4] R. L. Jaffe, Phys. Lett. **B229**, 275 (1989); H. W. Hammer, U. G. Meissner and D. Drechsel, Phys. Lett. **B367**, 323 (1996).

[5] M. J. Musolf and M. Burkardt, Z. Phys. **C61**, 433 (1994); N. W. Park, J. Schechter and H. Weigel, Phys. Rev. **D43**, 869 (1991); S. Hong and B. Park, Nucl. Phys. **A561**, 525 (1993).

[6] S. C. Phatak and Sarira Sahu, Phys. Lett. **B321**, 11 (1994); H. Ito, Phys. Rev. **C52**, R1750 (1995); H. Weigel et al., Phys. Lett. **B353**, 20 (1995).

[7] M. J. Musolf and B. R. Holstein, Phys. Lett. **B242**, 461 (1990).

[8] W. C. Haxton, E. M. Henley and M. J. Musolf, Phys. Rev. Lett. **63**, 949 (1989); V. V. Flambaum and D. W. Murray, Phys. Rev. **C56**, 1641 (1997).

[9] R. D. McKeown, Phys. Lett. **B219**, 140 (1989); D. H. Beck, Phys. Rev. **D39**, 3248 (1989).

[10] B. Desplanques, Phys. Rept. **297**, 1 (1998); B. R. Holstein, e-print:nucl-th/0607038.

[11] B. Desplanques, J. F. Donoghue and B. R. Holstein, Ann. Phys. (N.Y.) **124**, 449 (1980).

[12] N. Kaiser and U. G. Meissner, Nucl. Phys. **A489**, 671 (1988); ibid. Nucl. Phys. **A510**, 759 (1990).

[13] R. Schiavilla, J. Carlson and M. Paris, Phys. Rev. **C67**, 032501(R) (2003); C. P. Liu, G. Prezeau and M. J. Ramsey-Musolf, Phys. Rev. **C67**, 035501 (2003).

[14] G. Kuster and H. Arenhovel, Nucl. Phys. **A626**, 911 (1997); B. Mosconi and P. Ricci, Phys. Rev. **C55**, 3115 (1997); W. Y. P. Hwang et al., Ann. Phys. (N.Y.) **129**, 47 (1980); ibid., Ann. Phys. (N.Y.) **137**, 378 (1981).

[15] H. Arenhovel and S. K. Singh, Eur. Phys. J. **A10**, 183 (2001) and to be published.

[16] B. Mueller et al. (SAMPLE Collaboration), Phys. Rev. Lett. **78**, 3824 (1997); D. T. Spyde et al. (SAMPLE Collaboration), Phys. Rev. Lett. **84**, 1106 (2000); Phys. Lett. **B583**, 79 (2004); R. Hasty et al. (SAMPLE Collaboration), Science **290**, 2117 (2000).

[17] K. A. Aniol et al. (HAPPEX Collaboration), Phys. Rev. Lett. **96**, 022003 (2006); Phys. Lett. **B635**, 275 (2006); D. Armstrong et al. (G0 Collaboration), Phys. Rev. Lett. **95**, 092001 (2005).

[18] F. E. Mass et al. (A4 Collaboration), Eur. Phys. J. **A17**, 339 (2003); Phys. Rev. Lett. **93**, 022002 (2004); **94**, 152001 (2005).

Physics and Astrophysics of Hadrons and Hardronic Matter
Editor: A. B. Santra

Fusion of Hadrons and Multi-quark Objects[1]

<u>B.N. Joshi</u>*, Arun K. Jain

Nuclear Physics Division
Bhabha Atomic Research Centre, Mumbai-400 085, India.
* email: bnjoshi@barc.gov.in

Abstract

There has been tremendous interest but there are only unconfirmed results of dibaryon and pentaquark resonances so far. The main hurdle one faces in the search of dibaryon and pentaquark is their identification, their signature and practically no guide to their location. The quest of where and how to look for such weak resonances is not clear yet. The transition from a bipolar to a unipolar non-strange dibaryon may possibly be seen in the $(p, 2p)$ reactions on heavy nuclei. The change of the finite size of the $p - p$ interaction vertex can be identified as a sudden change in the extracted DWIA spectroscopic factor. The DWIA anomalies are to be searched for in the existing $(p, 2p)$ reaction data for the identification of non strange dibaryons and pentaquark with knockout reaction using K^+ beam.

1 Introduction

Dibaryons, six quark objects with or without strangeness, are predicted by QCD based phenological models above 1.9 GeV and have baryon number equal to two $(B = 2)$. Such object are being searched in the past couple of decades with a view to provide additional support to the standard model. These six quark states have been theoretically modeled [1] in terms of bag model, similar to the description of baryons and mesons in this model. This of course is done without any additional parameters. This model however, predicted basically only one bound state of $s-$ wave flavor-singlet dihyperon $(H, J^P = 0^+$ at 2150 MeV)which was to decay only weakly. The other dibaryon state $(H^*, J^P = 1^+$ at 2335 MeV) was to decay strongly into $\Lambda\Lambda$ or $N\Xi$.

Experimentally, however, the situation is very confusing as regards the observation of narrow dibaryons. The narrowness arises due to the hindered decay either dynamically due to their exotic configuration or due to quantum numbers prohibiting their decay by strong interaction. Claims for observations but unconfirmed results arise due to weakness of dibaryon's signatures compared to the physical background in a given process. These resonances normally occur beyond the pion production threshold and so one has to search for them like a "Needle in Haystack" in the background. It has however been argued that the measurement of total cross section is

---
[1]Work supported by a grant from the Department of Science and Technology, Govt of India.

in a given process. These resonances normally occur beyond the pion production threshold and so one has to search for them like a "Needle in Haystack" in the background. It has however been argued that the measurement of total cross section is not a suitable characteristic to study the dibaryon resonances. There have been attempts to obtain the non-strange dibaryon characteristics from the $p - p$ total cross sections [2]. The possibility of having different analyzing powers of a resonance and its background has led Tatischeff [3] to look for dibaryons in $^1H(\vec{d}, pp)X$ reaction at around $T_d \simeq 2 GeV$. These resulted in tensor analyzing power with small oscillatory pattern around $M_{pp} \simeq 1.646\ GeV$. A reaction mechanism assuming the physics of mesons and nucleons in interaction to give rise to a continuous background and a simple spectator mechanism calculation however, do not show any oscillations. Thus the narrow structure around 1.945 GeV and lack of any signal in the corresponding calculation led them to associate it with a dibaryon.

Dibaryon searches have been made in the form of narrow isoscalar and isovector resonances which couple to the γd channel. The search for these are made both in the charged pion production channel [4,5] as well as in the neutral pion production channel [6] . However the π^0-production channel is better suited for this purpose since in the energy region of interest below Δ-resonance down to the threshold the cross section for the conventional π^0-production is much lower than the $\pi^{\pm}$-production channels. Due to this there is much less background leading to better signal to background ratio for searching the narrow structures. Here use is made of the tagged photons in the $\gamma d \longrightarrow \pi^0 X$ reaction in the energy range $E_\gamma \sim 140 - 300$ MeV. Both total $\pi^0 X$-production as well as partial integral cross sections were measured. The coherent $\pi^0 X$ -production is forward peaked whereas the incoherent channel dominates the backward angles. The dibaryon resonance being produced incoherently should be present at the backward angles and hence should be enhanced by the large angle cuts. Except for three structures the excitations were found to be rather smooth. These three structures were however found to be consistently present in all the cuts. The most prominent one was found to be at 2084 MeV. This matched closely with one of the three sharp resonances reported by Tatischeff et al [7]. After subtraction of the smooth background the fluctuations fitted with a Gausian of full width at half maximum equal to 1 MeV led to a resonance production of $\sim 2 - 5\mu b.MeV$. This therefore puts the stringent constraint on the sensitivity of the dibaryon production to the level of a few $\mu b.MeV$. Tatischeff et al [7] are more confident in their claim to have dibaryon resonances at 2050, 2122 and 2150 MeV. They corroborate their weak dibaryonic findings with the phenomenological mass formula by Mulders et al [8]. for two clusters of quarks situated at the end of a stretched bag. In the mass formula the clusters are taken as a combination of

220

$q^2 - q^4$ with two phenomenological parameters. It was found that with these two parameters all the experimental energy levels are reproduced from 2 GeV to 2.2 GeV.

If we look at Multi-quark objects as peaks in the invariant mass spectra or excitation function then it appears difficult to distinguish them or separate from the background. Thus either one has to look for improving the signal to background ratios through better technological inputs or one has to look for some other characteristics of dibaryons which are more discernible.

2 Investigating the Multi-quark Objects

If we know some other characteristics of these Multi-quark objects such as their nature in terms of fused hadrons in the form of a single bag containing all the quarks or a dumbbell shape of two hadrons close to each other. Then one can search for these objects in that domain. This has been the main point of this paper to find and discuss the other characteristics besides the energy-position and the sharpness in the energy domain. Here it means that one is looking for a longer life time of six-quark comp lex. It can be further argued that for the dibaryon one is looking for the energy at which the six-quarks are confined in a smaller volume than that in the neighboring energy region where these are more or less separated in the form of three-quark clusters of baryons. Therefore instead for looking for $(\Delta t, \Delta E)$ domain if one looks for $(\Delta x, \Delta p)$ domain then this will open up another way altogether to look for dibaryons. This aspect requires some means for looking at the reactions producing dibaryons which can measure sizes and distances, or otherwise their Fourier transforms in terms of momentum distributions. The direct nuclear reactions are normally such that they can be employed to determine the Fourier transform of the overlap integral of the initial and final bound states. In this case therefore the dibaryons have to be first produced through some reaction mechanism and then have to be detected during the life span of their existence. The width of the dibaryonic states is expected to be around few MeV which corresponds to a life time of around 10^{-21} sec. For such a short life time one therefore has to have the detector of the dibaryon to be an attachment part of the reactants. .This therefore requires the main reactant target particle to be a part of a bigger nucleus such that the residual or the spectator portion of the bigger target nucleus can act as a detecter or a probe of the dibaryon produced in its vicinity. i.e there should be a third body at the instant of their production. Three body final state reaction such as the knockout reactions are the reactions where these conditions are ideally satisfied when the target is chosen to be a medium-heavy mass nucleus. Here the residual

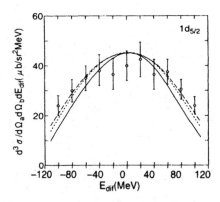

Figure 1: $^{40}Ca(p, 2p)^{39}K$ 200 MeV ZR-RDWIA

nucleus is close enough to act as a detector yet far away to act as a spectator for the purpose of the production of dibaryon. This is because when the kinematics is chosen such that the spectator residual nucleus is almost stationary in the laboratory frame then in the quasi-free approximation the reaction is extreme surface localized. The reactions $A(p, 2p)B$, $A(\pi, \pi p)B$, $A(K^+, K^+ n)B$ etc.. are such that the residual nucleus B, due to its strong optical distortion effects, can gauge the size of the $2p$, πp, $K^+ n$ complex in terms of Zero-Range or Finite-Range character of this vertex.

The reaction $A(p, 2p)B$ if performed at appropriate incident proton energy where the two protons are predicted to produce dibaryon resonance then one can look for them in the energy sharing spectrum of the final two protons. As is expected for a dibaryon to have a compact six-quark structure one should be able to observe a change of cross section in the summed energy spectrum in the region where the two protons formed the dibaryon. Infact one need not change the energy of the incident proton beam to scan the region of the dibaryon resonance. The Fermi motion of the bound struck proton will lead to a spectrum of the residual spectator momenta corresponding to a variation of the relative $p - p$ energy. If, however, the dibaryon width, ΔE is large, corresponding to a shorter life time, it will reflect in the overall reduction of the absolute cross section over the whole summed energy spectrum peak.

This behaviour has been verified in the following example. Figs. 1-2 and Figs. 3-4 [9] show knockout of proton from $1d_{5/2}$ state of $^{40}Ca(p, 2p)^{39}K$ reaction at the proton incident energies of 200 MeV and 300 MeV in the laboratory system. The scattering angles (θ_a, θ_b) of the outgoing protons are $(30^0, 30^0)$ at 200 MeV and the are $(30^0, 55^0)$ at 300 MeV. The dash, dotted and solid curves corresponds to

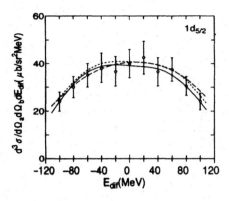

Figure 2: $^{40}Ca(p, 2p)^{39}K$ 200 MeV FR-RDWIA

the rms radii 3.282 fm, 3.479 fm and 3.721 fm respectively. In Figs 1-2 and 3-4 comparison of the zero range relativistic DWIA (ZR-RDWIA) and finite range relativistic DWIA (FR-RDWIA) of $^{40}Ca(p, 2p)^{39}K$ reaction with the data at 200 MeV and 300 MeV respectively is made. In FR-RDWIA relativistic Love-Frenny model [10] of NN t-matrix is employed. The Love-Frenny relativistic NN t-matrix, having a short range repulsion, is a phenomenological model calculation which fits the NN elastic scattering angular distributions. Observing Figs. 1-2 and Figs. 3-4 one can immediately conclude that at 200 MeV only FR-RDWIA fits the data well whereas at 300 MeV only ZR-RDWIA fits the data nicely. Since in the FR calculations Love-Franny NN interaction incorporates a short range repulsion which fits the experimental data well at 200 MeV means that at this particular energy a bipolar p-p vertex is formed. Whereas at 300 MeV it forms a monopolar p-p vertex since only the ZR fits the data well. Hence at 300 MeV there is a transition from a bipolar p-p vertex to a monopolar p-p vertex indicating the formation of a dibaryon.

Similar effects were seen in the $(\alpha, 2\alpha)$ reaction where the $\alpha - \alpha$ resonance at 19.8 MeV observed in the $\alpha - \alpha$ scattering has been very clearly seen in the $^6Li(\alpha, 2\alpha)$ reaction data at 42.8 MeV [11]. Through this experiment the authors could experimentally prove the validity of the post form prescription for the off shell behaviour of the quasi-free vertex in the knockout reactions. This is a separate point that the 19.8 MeV ^8Be resonance seen in the experiment is a resonance corresponding to a dumbbell shape of the ^8Be nucleus. In fact the aim of the experiment was not to look for the shape of the ^8Be nucleus at 19.8 MeV excitation energy but to demonstrate that resonances in the knockout reactions are representations of the resonances at the quasi-free vertex. Supplementary observation of the effect of the

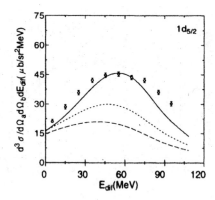

Figure 3: $^{40}Ca(p, 2p)^{39}K$ 300 MeV ZR-RDWIA

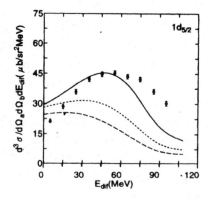

Figure 4: $^{40}Ca(p, 2p)^{39}K$ 300 MeV FR-RDWIA

sharp variation in the quasi free vertex in the knockout reaction angular correlation cross sections in the 55 MeV $^{6}Li(\alpha, 2\alpha)$ also provide further evidence of this effect.

One of the disturbing features of the knockout reactions, which has been a matter of concern for several decades is the large anomaly witnessed in the predictions and the the corresponding observations of the absolute cross sections. Infact, in the case of cluster knockout reactions on medium mass nuclei this anomaly is of few orders of magnitude too large. Moreover the DWIA analyses of the energy sharing distributions of these reactions indicate that the optical distortion effects are to be reduced substantially to fit even the energy dependent shape of these distributions. This as well as other observations can be consistently understood if the finite range nature of the quasi-free vertex (where the large energy and/or momentum is transferred) is incorporated properly. Now the distortions (confined

224

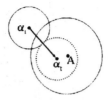

Figure 5: $\alpha - \alpha$ repulsion shown schematically.

to few fms) make orders of magnitude change in the cross sections and infact the plane wave impulse approximation (PWIA) calculations are most of the time closer to the observed cross sections. This therefore implies that one has to incorporate the finite range nature of the quasi-free vertex non-perturbatively. In this connection it is to be remarked that in the $(\alpha, 2\alpha)$ reactions the strong $\alpha - \alpha$ repulsion below $\sim 2fms$ does not allow the two α's to come closer than $\sim 2fms$, see Fig. 6.

From Fig.6, however, it can be seen that in any of the $l_{\alpha-\alpha} = 0, 2$ or 4 partial waves the two α's can not come closer than this distance due to the short range repulsion between two α particles. These three partial waves are the main partial waves contributing to the $\theta_{cm} = 90^o$ scattering region.

In the impulse approximation therefore the $r_{\alpha-\alpha}$, instead of being zero, is $\sim 2fms$. This range is however, comparable to the sizes of the nuclei under investigation. In the zero range program, for fitting the data, it would appear that the incoming α-particle meets the bound α-cluster at a distance of $\sim 2fms$ outside the nucleus than actually it was. In other words the bound state wave function has to shifted outwards by $\sim 2fms$ to reproduce the data. In fact it has been found that the $\sim 2fms$ shifted wave function overlaps the wave function generated by a bound potential of $2.52 \times A^{1/3} fm$ radius. This $\sim 2fms$ hard core of the $\alpha - \alpha$ potential therefore explains the observation of the need to use a large radius to fit the absolute cross sections. It has however been neglected that the wave function generated by the $2.52 \times A^{1/3} fm$ radius potential contains comparatively lesser amount of the larger-momentum components leading to the sharper energy sharing distributions as compared to the experimental data.

One can have another perspective from the bound $\alpha-$side also where in the zero range limit of the impulse approximation, for fitting the data, it would appear that

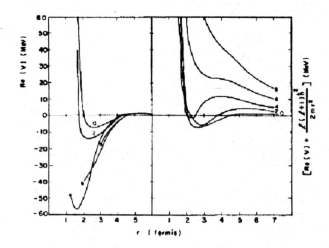

Figure 6: $\alpha - \alpha$ real potentials for various partial waves, $l = 0, 2, 4$ and 6 as obtained by Darriulat *et al.* [12]

the bound α-cluster meets the incoming α particle at a distance of $\sim 2fms$ inside the nucleus than actually it was. Or in other words effectively it corresponds to the situation where the incoming α-particle has penetrated $\sim 2fms$ more inside the α-residual nucleus potential. This gave rise to the observed phenomenological weak attenuation behaviour for the DWIA explanation of the $(\alpha, 2\alpha)$ reactions.

In the case of $(p, 2p)$ reactions on ^{12}C and ^{40}Ca, where anomalies similar to the $(\alpha, 2\alpha)$ reactions have been observed, although to a much lesser extent, one can seek similar explanation. As one knows that the hard core in the $p - p$ interaction is having a much smaller range, $\sim 0.8fm$ it is therefore expected that the influence of the hard core will be much smaller as compared to that in the $(\alpha, 2\alpha)$ reactions. However in the knockout from $l \neq 0$ case the shape of the momentum distribution will also manifest the lower attenuation characteristic as observed in the case of $^{7}Li(\alpha, 2\alpha)$ reactions [14].

The pentaquark, θ^+ itself has been observed as a weak resonance and it may be possible with the kind of arguments given here that it should be possible to see a much stronger evidence in the reaction $A(K^+, K^+n)(A-1)$ or more easily observable channel $A(K^+, K^0p)(A-1)$. The vertex $K^+ + n \rightarrow \theta^+ \rightarrow K^0 + p$ will have the same anomaly as in the case of the $(\alpha, 2\alpha)$ or the $(p, 2p)$ reactions except at the place where the resonance exists as a monopole.

3 Conclusions

From the observations and explanation of various discrepancies in the understanding of the cluster knockout reactions it can be concluded that a non-perturbative treatment of the quasi- free vertex is necessary for the proper description of the data. The phenomenological description of these knockout reactions in terms of increasing the bound state potential or reducing the optical distortions have one common feature that there exists a strong repulsion between the quasi-free scattering partners. In terms of this background it also becomes evident that the behaviour of the (p,2p) reaction on ^{40}Ca nucleus changes drastically when the quasi-free $p - p$ vertex makes a transition from normal two separated three quark clusters to a compound 6-quark bag dibaryon. This transition not only makes a change in the extracted spectroscopic factor but also changes the shape of the energy sharing spectrum. The procedure described in this article paves a new way to look at the resonances not in the time-energy domain, where the weak resonances almost merge with the background, but in the space-momentum domain where the resonance is seen to be enhanced much above the background. Similar experiment has been suggested here to look for the pentaquark state more clearly in a knockout reaction using K^+ beam.

References

[1] R L Jaffe, Phys. Rev. Lett., **38**, 195, (1977).

[2] A A Archipov, Talk at the IX*th* Int. Conf. on Hadron Spectroscopy, HADRON 2001, Protvino, Russia, August 2001.

[3] B Tatischeff *et al.*, Phys. Rev., **C45**, 2005, (1992).

[4] P E Argan *et al.*, Phys. Rev. Lett., **46**, 96, (1981).

[5] W Duhm *et al.*, Nucl. Phys., **459**, 557, (1986).

[6] U Siodlaczek *et al.*, Europ. Phys. Jour., **A9**, 309, (2000).

[7] B Tatischeff *et al.*, Phys. Rev., **C59**, 1878, (1999).

[8] P J Mulders *et al.*, Phys. Rev., **D21**, 2653, (1980); Phys. Rev. Lett., **D40**, 1543, (1978).

[9] Y. Ikebata, Phys. Rev., **C52**, 890, (1995).

[10] C.J. Horowitz, Phys. Rev., **C31**, 1340, (1985).

[11] P Gaillard *et al.*, Phys. Rev. Lett., **C25**, 593, (1970).

[12] P Darriulat *et al.*, Phys. Rev., **157**, B315, (1965).

[13] J R Pizzi *et al.*, Nucl. Phys., **A136**, 496, (1969).

[14] A K Jain and S Mythili, Phys. Rev., **C53**, 508, (1996).

Physics and Astrophysics of Hadrons and Hardronic Matter
Editor: A. B. Santra

Experimental Studies of η-Nucleus Interaction and Search of η-Nucleus Bound States

V. Jha[1], B. J. Roy[1], A. Chatterjee[1], H. Machner[2]

[1] *Nuclear Physics Division, B. A. R. C, Mumbai - 400 085*
[2] *IP, Forschungszentrum, Juelich, D-52425, Germany*
email: vjha@barc.gov.in

The preliminary results of an experiment to search for the formation of η-mesic nuclei $^{25}Mg_\eta$, using recoil-free $(p, ^3He)$ transfer reactions on ^{27}Al target, are presented here. We also describe an experiment to study the η-nucleus interaction using $p^6Li \to {}^7Be\eta$ reaction near threshold . These experiments have been performed at the cooler synchrotron accelerator COSY using Big Karl magnetic spectrometer and its associated detectors. Physics motivation, experimental details and the results obtained so far from these experiments have been presented here.

1 Introduction

The study of η meson interaction with nucleons and nuclei has evoked strong interest in recent years. Many of these studies are motivated by the theoretical predictions of the existence of a η-nucleus quasibound state, the so-called η-mesic nucleus[1]. Studies of η-mesic nucleus - a bound state of η meson in a nucleus provide a unique way of understanding the η-nucleus $(\eta - A)$ interaction, which is not understood well. Formation of such bound η-mesic nuclear states of various η-A systems will elucidate the nature of elementary η-nucleon $(\eta - N)$ interaction in the presence of medium. In the absence of any η-meson beam due to their short-lived nature, the η-nucleus interaction can only be studied through the production of η in the final state of some reaction. An interesting aspect in investigating the η meson interaction with nucleons and nuclei is the possibility to extract information on the $S_{11}(1535)N^*$ nuclear resonance. The η-nucleon (ηN) interaction at low energies is strongly influenced by the presence of the S-wave nucleon resonance N^* which lies close to the η production threshold. As a result, the interaction of nucleons with η meson in this energy region, where the S-wave contribution dominates, is very strong. The large width of this resonance ($\Gamma \sim 150$ MeV) means that it covers the entire low energy of the ηN interaction, almost exclusively. The N^* (1535) resonance hás $\eta - N$ channel as its dominant decay channel in addition to its decay in the $\pi - N$ channel. The excitation and decay of $N^*(1535)$ resonance through the $\eta - N$ channel suggests that the interaction of η meson with nucleons inside a nucleus at low energies can be considered as a series of formation and decay of the intermediate $N^*(1535)$ resonance, which can lead to the binding of η inside the nucleus if the N^* induced ηN interaction is strong enough . However, such η-nucleus binding will have finite lifetime and nonzero width due to alternate decay mode of N*(1535) in the $\pi - N$ channel. The formation of the bound state and the measurement of its binding energy and width can provide the important information on the behaviour of N*(1535) resonance in the nuclear medium. The in-medium properties of N* and its possible mass shift in nuclear matter at normal density can be related to partial restoration of chiral symmetry in the hadronic medium.

The other method to investigate these interesting aspects is through the studies of η meson production reaction on light nuclei. While the low energy η-nucleon interaction is relatively well studied, the data for η-nucleus interaction are scarce. The near threshold data for $pd \rightarrow {}^3He\eta$ and $dd \rightarrow {}^4He\eta$ reactions are remarkable for both their large strength and the rapid energy variation. Investigations on these reactions have indicated the presence of a strong final state interaction in $\eta-{}^3He$ and $\eta - {}^4He$ systems. There is practically no data existing on the η-nucleus final state interaction studies for the nuclei heavier than ${}^4He$. In this context, measurement of $p^6Li \rightarrow {}^7Be\eta$ reaction is useful to understand that how the interaction changes as the mass number increases. Apart from studying the behaviour of N*(1535) the η-nucleus interaction studies can also be used to address some detailed problems of fundamental nature. For example, the $\eta - {}^4He$ S-wave scattering length is of importance for the determination of $\pi^0 - \eta$ mixing angle. In addition to this, the strange quark content of the nucleon can be tested by comparing the scattering of η and η' meson on nucleon in a nuclear medium. The studies of η production and η-bound states in nuclei can also provide insight into the gluonic effects in the hadronic sector due to the presence of strong $\eta - \eta'$ mixing. All these features make the studies of η-nucleus interaction through the η production reactions on nuclei and the formation of η-nucleus bound states, quite fascinating.

In the present work we describe two experiments performed by GEM collaboration using the proton beam from cooler synchrotron accelerator at COSY, Juelich, Germany. First measurement strives to search for formation of the η-mesic nuclei ${}^{25}Mg_\eta$ using recoil-free transfer reactions. In the other experiment, the interaction of η meson with a ${}^7Be$ nucleus is studied using the $p^6Li \rightarrow {}^7Be\eta$ reaction at an energy just above the reaction threshold. The physics motivation and details of the experimental setup used in these two experiments have been discussed . The data analysis and the results obtained so far from these experiments are also presented.

2 Search of η-nucleus Bound states

2.1 Present status of studies of η-Mesic nuclei search

The existence of the η-nucleus bound states can be related to the positive values of real part of the η-nucleon scattering length Re($a_{\eta N}$). The size of nucleus required to form a quasi-bound system with an η-meson depends on the estimates of Re($a_{\eta N}$). It is an open and widely debated question that what might be the lightest nuclei to bind an η to form an η-mesic nucleus. A value of $\mathrm{Re}(a_{\eta N}) = 0.27$ fm derived by the initial calculations [2], predicted the existence of η-nucleus bound states for nuclei with mass number $A \geq 10$. Other analyses of the scattering length however, predict a value upto three times larger for $\mathrm{Re}(a_{\eta N})$, which has led to speculations that η-nucleus bound state may be possible for all nuclei with $A \geq 2$. It is premature to conclude which of these $a_{\eta N}$ values is a realistic one . Since the elementary $a_{\eta N}$ is not fixed any η-nucleus calculation has large uncertainties associated with calculated η-nucleus scattering lengths. For the existence of the bound states the real part of η-nucleus scattering length should be relatively large and negative. Information on the η-nucleus interaction can be obtained from the energy dependance of cross section in η production reactions such as $pd \rightarrow {}^3He\eta$, $dd \rightarrow {}^4He\eta$. However, the sign of real part of η-nucleus scattering length cannot be determined from these measurements. Such a determination can be made only after η-bound nuclei are experimentally discovered. Recent calculations using the techniques of unitarized chiral perturbation theory conclude that nuclei in the region of $A = 24$ are the most promising for the search of bound states [3]. The calculated widths of the possible η-nucleus bound states for the light nuclei become comparatively larger than the binding energies. For heavy nuclei, although there are many states possible, the separation of levels is much smaller than the half-widths of the bound states, making their experimental detection impossible. Therefore, the suitable nuclei where the bound state peaks can be experimentally observed seems to be the ones with medium mass.

Although the theoretical studies indicate the existence of η-bound states, a compelling experimental proof for this is still awaited. There have been few experimental attempts in the past to search for the formation of η-mesic nuclei. First measurements for search of bound η-mesic states using the (π^+, p) reaction on various targets were performed at Brookhaven [4]. The experiment looked for the narrow peaks of a few-MeV width in the missing mass spectrum of $\pi A \rightarrow pX$ reaction . However, the experiments couldn't confirm their existence, perhaps owing to larger width of η-mesic nuclei than was expected in experiment. The strong energy dependance of the near-threshold amplitude observed in the measurements of hadron induced η-production reactions e.g., $pd \rightarrow {}^3He\eta$, has been interpreted as the formation of the quasi-bound state [5], but more direct and unambiguous evidence is required to confirm their existence. There have been recent attempts to search for the η-mesic nuclei ${}^{11}B_\eta$ and ${}^3He_\eta$ from photon induced η production reactions. Sokol et al. [6] claimed the formation of η-mesic nuclei in the $\gamma + {}^{12}C$ reaction with the measurement of $\gamma + A \rightarrow p(n) + (A - 1)_\eta \rightarrow p(n) + (\pi^+ + n) + (A - 1)$. Here, $\pi^+ - n$ back-to-back correlation spectra was assumed to arise through stages of formation of the η-mesic nuclei and a peak in the $\pi^+ - n$ invariant mass distribution was seen. However, measurements with better statistics are required for drawing any further conclusion. In a more recent work on the photoproduction reaction $\gamma {}^3He \rightarrow \pi^o pX$ [7], an enhancement of yield observed in the π^0-p pair emitted at back-to-back angles.

The enhancement seen in the photon energy spectrum corresponding to excitation energy below the $^3He - \eta$ production threshold was construed as the observation of an $\eta - {}^3He$ quasibound state. However, it was pointed out in Ref. [8], that the data of Ref. [7] do not allow to unambiguously deduce the existence of $^3He_\eta$ bound state.

2.2   GEM Experiment at COSY for search of η-nucleus bound states

The study of $(p, {}^3He)$ reaction with the Big karl magnetic spectrometer on ^{27}Al target has been carried out by GEM collaboration. The experimental search employs the reaction $p + {}^{27}Al \rightarrow {}^3He + {}^{25}Mg_\eta$ at the recoilless conditions. Experiments have been performed using the 1.745 GeV/c proton beam from the accelerator COSY at IKP, Juelich with typical beam currents of $\sim 5 \times 10^8$ p/s incident on $1mm$ thick ^{27}Al target. Decision to use a ^{27}Al target has been based on the binding energy and width predictions given by the calculations of ref. [3]. The signature for the η-mesic nucleus production in this experiment is a peak in the spectrum of ^3He at the momentum corresponding to the production of negative energy η's in two-body kinematics. The beam momentum for the experiment is chosen to make use of recoilless kinematics so that the slow η produced in elementary reaction gets embedded inside the nuclear medium. The outgoing ^3He particles have been detected by the magnetic spectrometer Big Karl (BK) alongwith its focal plane detectors shown in Fig.1. Because the cross section is sharply peaked in the forward direction for ^3He particles [9] the major part of possible η bound state cross section can be detected within the angular acceptance of BK. Tracks of ^3He particles were measured in the focal plane of the spectrometer with two stacks of multi-wire drift chambers (MWDC) which allow measurements in both x and y direction. These MWDC stacks are followed by two layers of scintillator hodoscopes P and Q to measure energy loss and time of flight of particles. The trigger for data readout was a coincidence between P and Q hodoscope layers. Two momenta settings of spectrometer at 859 MeV/c and 897 MeV/c, covering ~ 80 MeV range each in the η-nucleus binding energy range have been used for measurements. The production of ^3He particles is primarily associated with multipion processes which act as an background to the η-nucleus bound state formation. To enhance the sensitivity of measurement and to reduce the background the decay products of the η mesic nuclei are measured in coincidence with ^3He particles. A large acceptance plastic scintillator detector ENSTAR [10] has been constructed for measuring the decay products from bound η formation. ENSTAR detector consists of three layer of scintillator pieces arranged in cylindrical geometry around the target over a carbon fiber beam pipe. ENSTAR gives the energy loss and angle information for the η bound state decay particles, namely protons and pions.

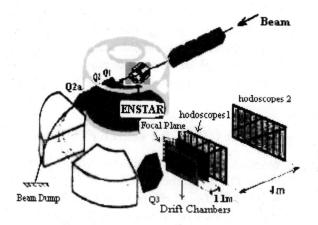

Figure 1: Experimental setup for the search of η-mesic nucleus: The Big Karl Magnetic spectrometer alongwith its focal plane detection system and ENSTAR detector placed around the target has been used for measurements.

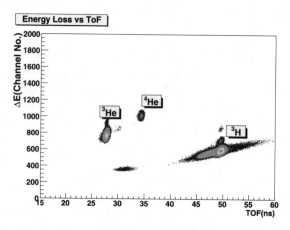

Figure 2: Energy loss versus the Time of Flight spectrum from the focal plane hodoscopes: ^{3}He particles are clearly identified.

2.3 Data Analysis and Results

The ^3He particles have been selected by applying cuts on energy loss and time of flight spectra obtained from the scintillator layers, as shown in the Fig. 2. The momenta of these particles at the target point have been determined using their measured trajectories in drift chambers and Big Karl optical matrix. The missing mass of the residual reaction products in $p^{27}Al \rightarrow {}^3HeX$ reaction thus obtained is a uniform distribution dominated by the background events. To suppress the background and select signal corresponding to η bound state formation software cuts using ENSTAR detector signal have been used. Monte Carlo (MC) simulation of the η-mesic events show that the decay particles i.e, proton and pions have mean energies of 100 and 300 MeV respectively. This implies that the most of protons stop in the second layer of ENSTAR and the pions leave a measurable signal in third layer of ENSTAR while passing through it. Events with two tracks having the above characteristics in ENSTAR were selected. Further, MC simulations predict that the outgoing decay particles, protons and pions from the $^{25}Mg_\eta$ decay come out at almost back-to-back angles. With an additional cut on the angle correlation of the two decay particles from ENSTAR,a peak-like structure, in the excitation energy spectra of the residual '$^{25}Mg + \eta$' system below the free threshold, has been observed at almost same range of values for two different spectrometer settings as shown in Fig. 3. The present low statistics data and relatively large background do not allow to infer any bound state formation, nevertheless, an upper limit to the formation cross section can be determined.

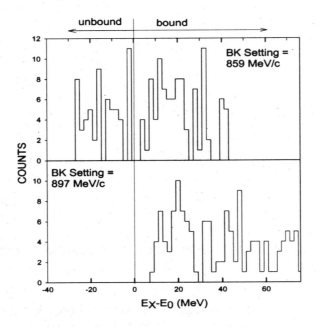

Figure 3: Excitation energy spectra of residual '25Mg+η ' nuclear system at two different Big-Karl momentum settings. Events corresponding to measured ^3He particles in coincidence with ENSTAR cuts for η-mesic states have been selected.

Figure 4: The photograph of the detector setup consisting of Multi-wire Avalanche chambers and scintillator layers placed inside a vacuum chamber. The detection setup is used for the identification and measurement of ^{7}Be particles.

3 η-Nucleus Final state Interaction Studies:

3.1 Motivation:

In addition to η-nucleus bound states searches, studies of eta-nucleus final state interaction(FSI) have been carried out using the $p^6Li \rightarrow {}^7Be\eta$ reaction near threshold. The threshold of the reaction is at a beam momentum of 1273.17 MeV/c . Investigations on $\eta - {}^3He$ and $\eta - {}^4He$ systems have indicated presence of a strong final state interaction. There is practically no data existing on the η-nucleus final state interaction studies for the nuclei heavier than ^4He. The only data available is on the coherent η production for $p^6Li \rightarrow {}^7Be\eta$ reaction at an excess energy of 19 MeV corresponding to incident proton kinetic energy of 683 MeV. The measurement, in this experiment were performed by detecting two forward emitted γ rays resulting from the decay of η-meson. A total of only such 8 events were detected within the acceptance of the detector, of which 3 were estimated to be associated with the background. The estimated c.m. cross section as an average over the range of c.m. angles $0° \leq \theta_\eta \leq 50°$ is 4.6 ±3.8 nb/sr. This number includes the sum of ^7Be states upto 10MeV in excitation energy. With the energy resolution of the experimental set-up used it was not possible to distinguish different final states of the residual nucleus in that experiment. Measurements with better resolution, where separation of L=1 states (ground and 1st excited state of ^7Be) from higher L values is possible, are required to extract more reliable information about $\eta - {}^7Be$ low energy interaction. Further, this experimental data lie outside the FSI peak and therefore the studies of this reaction closer to threshold are needed to study the $\eta - {}^7Be$ FSI.

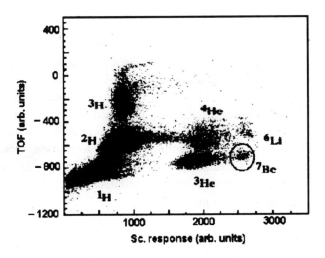

Figure 5: Time of Flight vs light output from the thin scintillator layers for identification of ^{7}Be particles.

Further, the large difference in the rigidities of beam proton and ^7Be ensures a very low background in the focal plane of BK spectrometer. The experiment was performed on 2mm ^6Li target using the 1310 MeV/c proton beam at COSY. The target choice is motivated by the fact that it is lightest solid target that can be prepared . Also, because of the $d + {}^4He$ cluster structure of ^6Li nucleus, a possible connection between the present reaction and the $pd \rightarrow {}^3He\eta$ reaction can be made. The multiwire drift chambers used in the previous experiment for measuring the particle position and direction were not suitable for such strongly ionising particles. Hence a new detection system consisting of two sets of multiwire avalanche chambers (MWACs) has been constructed. This allows us to measure both the position and the direction of the heavy recoiling nucleus. From these measured values the three momentum vector of the emitted particle can be reconstructed using the optics of the Big Karl magnetic spectrograph. These MWAC detectors are followed by the two layers of $\Delta E - E$ hodoscope system, consisting of a $0.5mm$ and $2mm$ thick plastic scintillators having a distance of $1m$ between them. These scintillators use energy loss and time of flight between them to identify various particles. Whole detection system is housed in a huge vacuum chamber. A photograph of the vacuum chamber with MWACs and hodoscopes placed inside it, is shown in Fig.4. Time-of-flight and energy loss information from the scintillator layers are used to identify various particles. Fig.5 shows the spectrum of time of flight between the $\Delta E - E$ layers and light output from the first layer in an earlier test measurement performed with 1293 MeV/c momentum proton beam. Detailed analysis of the present data using the similar identification technique and a momentum reconstruction through MWACs for getting the missing mass, is in progress.

4   Summary

A dedicated experiment to search for the signals of formation of eta-mesic nuclei has been performed at COSY, Juelich using the $p + {}^{27}Al \rightarrow {}^{3}He + {}^{25}Mg_{\eta}$ reaction. The missing mass technique supplemented with the measurement of decay products of eta mesic nucleus for background reduction has been used. The statistics from the present measurement is not sufficient to provide the clear evidence for the existence of η-mesic nuclei. However an upper limit on the formation cross section of η-nucleus bound state can be estimated. The measurement of $p^{6}Li \rightarrow {}^{7}Be\eta$ reaction for studies of η-nucleus final state interaction has also been performed. The analysis of the experimental data is in progress.

Acknowledgements

The work presented here is based on the measurements performed by the GEM collaboration. We are grateful to all its members.

References

[1] Q. Haider and L.C. Liu. Phys. Rev. C 34 1845 (1986)

[2] R.S. Bhalerao and L.C. Liu ,Phys. Rev. Lett. 54 865 (1985)

[3] C. Garcia-Recio, T Inoue, J Nieves, E Oset Phys. Lett. 550B 47 (2002)

[4] R.E. Cherien et al Phys. rev. Lett. 60 2595 (1988)

[5] C. Wilkin Phys. Rev. C 47 R938 (1993)

[6] G. A. Sokol et al. Fisika B8, 81 (1998)

[7] M. Pfeiffer et al Phys. rev. Lett. 92 252001 (2004)

[8] C. Hanhart, Comment in Phys. Rev. Lett. 94 049101 (2005)

[9] L.C.Liu Private communication.

[10] M. G. Betigeri et al to be published (2007)

[11] E. Scomparin et al J Phys. G 19 L51 (1993)

Physics and Astrophysics of Hadrons and Hardronic Matter
Editor: A. B. Santra
Copyright © 2008, Narosa Publishing House, New Delhi, India

Role of Antikaon Condensation in r-Mode Instability

Debarati Chatterjee*, Debades Bandyopadhyay

Saha Institute of Nuclear Physics
1/AF Bidhannagar, Kolkata-700064, India
* email: debarati@theory.saha.ernet.in

Abstract

We investigate the effect of antikaon condensed matter on bulk viscosity in rotating neutron stars. We use relativistic field theoretical models to construct the equation of state of neutron stars with the condensate, where, the phase transition from nucleonic to K^- condensed phase is assumed to be of first order. We calculate the coefficient of bulk viscosity due to the nonleptonic weak interaction $n \rightleftharpoons p + K^-$. The influence of antikaon bulk viscosity on the gravitational radiation reaction driven instability in the r-modes is investigated. We compare our results with the previously studied nonleptonic weak interaction $n + p \rightleftharpoons p + \Lambda$ involving hyperons on the damping of the r-mode oscillations.

We find that the bulk viscosity coefficient due to the nonleptonic weak process involving the condensate is suppressed by several orders of magnitude in comparison with the non-superfluid hyperon bulk viscosity coefficient. Consequently, the antikaon bulk viscosity may not be able to damp the r-mode instability, while hyperon bulk viscosity can effectively suppress r-mode oscillations at low temperatures. Hence neutron stars containing K^- condensate in their core could be possible sources of gravitational waves.

1 Introduction

Immense information about the internal composition of neutron stars can be obtained through the study of unstable modes of oscillations associated with rotating neutron stars. Among the various possible instabilities, inertial r-modes restored by Coriolis force are particularly interesting as they are thought to play an important role in regulating the spins of newly-born neutron stars as well as old, accreting neutron stars in low mass X-ray binaries (LMXBs). If the r-mode is unstable, a rapidly rotating neutron star could emit a significant fraction of its rotational energy and angular momentum as gravitational waves, which could be detectable by the upcoming generation of gravitational wave detectors.

Bulk viscosity in neutron star is caused by the energy dissipation due to nonleptonic weak interaction processes in pulsating dense matter. It was argued that the r-mode instability could be effectively suppressed by bulk viscosity due to exotic matter in neutron star interior. Neutron star matter spans a wide range of densities, from the density of iron nucleus at the surface of the star to several times normal nuclear matter density in the core. Since the chemical potentials of nucleons and leptons increase rapidly with density in the neutron star core, different exotic forms of matter with large strangeness fraction such as hyperons, Bose-Einstein condensates of antikaons or deconfined quarks may appear there. The coefficient of bulk viscosity due to nonleptonic weak processes involving hyperons was calculated by several authors [1–5]. Also, the impact of bulk viscosity due to unpaired and paired quark matter on the r-mode instability had been investigated extensively [5–7].

In this paper, we study the effect of the presence of antikaon condensates on the equation of state (EoS), bulk viscosity and the corresponding damping timescale. We compare the bulk viscosities generated by the nonleptonic weak interaction $n \rightleftharpoons p + K^-$ involving antikaons with that due to the nonleptonic process $n + p \rightleftharpoons p + \Lambda$ involving hyperons as previously obtained, and investigate their role in damping of r-mode oscillations in rotating neutron stars.

This paper is organized in the following way. In Sec. II, we describe the model to calculate equation of state, bulk viscosity coefficient and the corresponding timescale. The parameters of the model are listed in Sec. III. The results of our calculations are discussed in Sec. IV, and the summary and conclusions are given in Sec. V.

2 Theoretical Model

We assume a first order phase transition from nuclear to antikaon condensed matter. A relativistic field theoretical model is adopted to describe the β-equilibrated and charge neutral matter in both the phases.

2.1 Hadronic Phase

The constituents of hadronic phase are octet baryons, electrons and muons. In this model, baryon-baryon interaction is mediated by the exchange of scalar and vector mesons. For hyperon-hyperon interaction, two additional strange mesons, scalar f_0 (denoted by σ^*) and vector ϕ are incorporated. The Lagrangian density for the hadronic phase is given by

$$
\begin{aligned}
\mathcal{L}_B = & \sum_B \bar{\Psi}_B \left(i\gamma_\mu \partial^\mu - m_B + g_{\sigma B}\sigma - g_{\omega B}\gamma_\mu \omega^\mu - g_{\rho B}\gamma_\mu t_B \cdot \boldsymbol{\rho}^\mu \right) \Psi_B \\
& + \frac{1}{2} \left(\partial_\mu \sigma \partial^\mu \sigma - m_\sigma^2 \sigma^2 \right) - U(\sigma) \\
& - \frac{1}{4} \omega_{\mu\nu}\omega^{\mu\nu} + \frac{1}{2} m_\omega^2 \omega_\mu \omega^\mu - \frac{1}{4} \boldsymbol{\rho}_{\mu\nu} \cdot \boldsymbol{\rho}^{\mu\nu} + \frac{1}{2} m_\rho^2 \boldsymbol{\rho}_\mu \cdot \boldsymbol{\rho}^\mu + \mathcal{L}_{YY}.
\end{aligned}
\tag{1}
$$

The isospin multiplets for baryons B are represented by the Dirac bispinor Ψ_B with vacuum baryon mass m_B, and isospin operator t_B, and $\omega_{\mu\nu}$ and $\rho_{\mu\nu}$ are field strength tensors. The scalar self-interaction term [8]

$$U(\sigma) = \frac{1}{3}g_2\sigma^3 + \frac{1}{4}g_3\sigma^4, \tag{2}$$

is introduced to reproduce the correct compressibility of nuclear matter. We perform this calculation in the mean field approximation [9]. The Lagrangian density for hyperon-hyperon interaction is given by

$$\begin{aligned}
\mathcal{L}_{YY} &= \sum_B \bar{\Psi}_B \left(g_{\sigma^* B}\sigma^* - g_{\phi B}\gamma_\mu \phi^\mu \right) \Psi_B \\
&\quad + \frac{1}{2} \left(\partial_\mu \sigma^* \partial^\mu \sigma^* - m_\sigma^{*2}\sigma^{*2} \right) - \frac{1}{4}\phi_{\mu\nu}\phi^{\mu\nu} + \frac{1}{2}m_\phi^2 \phi_\mu \phi^\mu.
\end{aligned} \tag{3}$$

The scalar density and baryon number density are

$$n_B^S = \frac{2J_B + 1}{2\pi^2} \int_0^{k_{F_B}} \frac{m_B^*}{(k^2 + m_B^{*2})^{1/2}} k^2 \, dk , \tag{4}$$

$$n_B = (2J_B + 1)\frac{k_{F_B}^3}{6\pi^2} , \tag{5}$$

where Fermi momentum is k_{F_B}, spin is J_B, and isospin projection is I_{3B}. Effective mass and chemical potential of baryon B are $m_B^* = m_B - g_{\sigma B}\sigma - g_{\sigma^* B}\sigma^*$ and $\mu_B = (k_{F_B}^2 + m_B^{*2})^{1/2} + g_{\omega B}\omega_0 + g_{\phi B}\phi_0 + I_{3B}g_{\rho B}\rho_{03}$, respectively. Charge neutrality in the hadronic phase is imposed through the condition

$$Q = \sum_B q_B n_B - n_e - n_\mu = 0 , \tag{6}$$

where n_B is the number density of baryon B, q_B is the electric charge and n_e and n_μ are charge densities of electrons and muons respectively. The total energy density in the hadronic phase is given by

$$\begin{aligned}
\varepsilon &= \frac{1}{2}m_\sigma^2\sigma^2 + \frac{1}{3}g_2\sigma^3 + \frac{1}{4}g_3\sigma^4 + \frac{1}{2}m_{\sigma^*}^2\sigma^{*2} \\
&\quad + \frac{1}{2}m_\omega^2\omega_0^2 + \frac{1}{2}m_\phi^2\phi_0^2 + \frac{1}{2}m_\rho^2\rho_{03}^2 \\
&\quad + \sum_B \frac{2J_B + 1}{2\pi^2} \int_0^{k_{F_B}} (k^2 + m_B^{*2})^{1/2}k^2 \, dk \\
&\quad + \sum_{l=e^-,\mu^-} \frac{1}{\pi^2} \int_0^{K_{F_l}} (k^2 + m_l^2)^{1/2}k^2 \, dk,
\end{aligned} \tag{7}$$

and the pressure is

$$
\begin{aligned}
P = \ & -\frac{1}{2}m_\sigma^2\sigma^2 - \frac{1}{3}g_2\sigma^3 - \frac{1}{4}g_3\sigma^4 \\
& -\frac{1}{2}m_{\sigma^*}^2\sigma^{*2} + \frac{1}{2}m_\omega^2\omega_0^2 + \frac{1}{2}m_\phi^2\phi_0^2 + \frac{1}{2}m_\rho^2\rho_{03}^2 \\
& +\frac{1}{3}\sum_B \frac{2J_B+1}{2\pi^2}\int_0^{k_{FB}} \frac{k^4\,dk}{(k^2+m_B^{*2})^{1/2}} \\
& +\frac{1}{3}\sum_{l=e^-,\mu^-} \frac{1}{\pi^2}\int_0^{K_{Fl}} \frac{k^4\,dk}{(k^2+m_l^2)^{1/2}} \cdot
\end{aligned}
\tag{8}
$$

2.2 Antikaon condensed phase

The constituents of the pure antikaon condensed phase are baryons (neutrons, protons), leptons (electrons, muons) and antikaons, where the baryons are embedded in the condensate. The (anti)kaon-(anti)kaon interaction in the pure condensed phase is described using the relativistic field theoretical approach, through the exchange of σ, ω, ρ, σ^* and ϕ mesons. However, nucleons do not couple with the strange mesons, hence $g_{\sigma^*N} = g_{\phi N} = 0$.

The Lagrangian density for (anti)kaons in the minimal coupling scheme is,

$$
\mathcal{L}_K = D_\mu^* \bar{K} D^\mu K - m_K^{*2}\bar{K}K \,,
\tag{9}
$$

where the covariant derivative $D_\mu = \partial_\mu + ig_{\omega K}\omega_\mu + ig_{\phi K}\phi_\mu + ig_{\rho K}\mathbf{t}_K \cdot \boldsymbol{\rho}_\mu$ and the effective mass of (anti)kaons is $m_K^* = m_K - g_{\sigma K}\sigma - g_{\sigma^* K}\sigma^*$. The in-medium energies of K^- mesons for s-wave ($\vec{k}=0$) condensation is given by

$$
\mu_{K^-} = m_K^* - g_{\omega K}\omega_0 - g_{\phi K}\phi_0 - \frac{1}{2}g_{\rho K}\rho_{03} \,,
\tag{10}
$$

where the isospin projection $I_{3K^-} = -1/2$. The scalar and number density of baryon B in the antikaon condensed phase are given by

$$
n_B^{K,S} = \frac{2J_B+1}{2\pi^2}\int_0^{k_{FB}} \frac{m_B^*}{(k^2+m_B^{*2})^{1/2}}k^2\,dk \,,
\tag{11}
$$

$$
n_B^K = (2J_B+1)\frac{k_{FB}^3}{6\pi^2} \,,
\tag{12}
$$

The scalar density of K^- mesons in the condensate is given by [10]

$$
n_{K^-} = 2\left(\omega_{K^-} + g_{\omega K}\omega_0 + g_{\phi K}\phi_0 + \frac{1}{2}g_{\rho K}\rho_{03}\right)\bar{K}K = 2m_K^*\bar{K}K \,.
\tag{13}
$$

The total charge density in the antikaon condensed phase is

$$Q^K = \sum_{B=n,p} q_B n_B^K - n_{K^-} - n_e - n_\mu = 0. \tag{14}$$

The total energy density in the antikaon condensed phase is

$$\begin{aligned}
\varepsilon^K &= \frac{1}{2}m_\sigma^2\sigma^2 + \frac{1}{3}g_2\sigma^3 + \frac{1}{4}g_3\sigma^4 + \frac{1}{2}m_{\sigma^*}^2\sigma^{*2} + \frac{1}{2}m_\omega^2\omega_0^2 + \frac{1}{2}m_\phi^2\phi_0^2 + \frac{1}{2}m_\rho^2\rho_{03}^2 \\
&+ \sum_{B=n,p} \frac{2J_B+1}{2\pi^2} \int_0^{k_{FB}} (k^2 + m_B^{*2})^{1/2} k^2 \, dk + \sum_{l=e^-,\mu^-} \frac{1}{\pi^2} \int_0^{K_{Fl}} (k^2 + m_l^2)^{1/2} k^2 dk \\
&+ m_K^* n_{K^-},
\end{aligned} \tag{15}$$

where last term denotes the contribution of the K^- condensate. The pressure in this phase is

$$\begin{aligned}
P^K &= -\frac{1}{2}m_\sigma^2\sigma^2 - \frac{1}{3}g_2\sigma^3 - \frac{1}{4}g_3\sigma^4 - \frac{1}{2}m_{\sigma^*}^2\sigma^{*2} + \frac{1}{2}m_\omega^2\omega_0^2 + \frac{1}{2}m_\phi^2\phi_0^2 + \frac{1}{2}m_\rho^2\rho_{03}^2 \\
&+ \frac{1}{3}\sum_{B=n,p} \frac{2J_B+1}{2\pi^2} \int_0^{k_{FB}} \frac{k^4 \, dk}{(k^2 + m_B^{*2})^{1/2}} + \frac{1}{3}\sum_{l=e^-,\mu^-} \frac{1}{\pi^2} \int_0^{K_{Fl}} \frac{k^4 \, dk}{(k^2 + m_l^2)^{1/2}}
\end{aligned} \tag{16}$$

2.3   The Mixed Phase

The mixed phase of hadronic and K^- condensed matter is governed by the Gibbs phase equilibrium rules [10, 11],

$$P^h = P^K, \tag{17}$$

$$\mu_B^h = \mu_B^K. \tag{18}$$

where μ_B^h and μ_B^K are chemical potentials of baryons B in the pure hadronic and K^- condensed phase, respectively. The global charge neutrality and baryon number conservation laws are

$$(1 - \chi)Q^h + \chi Q^K = 0, \tag{19}$$

$$n_B = (1 - \chi)n_B^h + \chi n_B^K, \tag{20}$$

where χ is the volume fraction of K^- condensed phase in the mixed phase. The total energy density in the mixed phase is given by

$$\epsilon = (1 - \chi)\epsilon^h + \chi\epsilon^K. \tag{21}$$

2.4 Bulk Viscosity

Energy dissipation due to pressure and density variations associated with r-mode oscillation, which drive the system out of β equilibrium, gives rise to bulk viscosity. The reactions between different constituent particles try to bring the system back to an equilibrium configuration, with a delay which depends on the characteristic timescale of the interaction. Strong interaction processes are insignificant as the strong interaction equilibrium is reached so fast that these processes are considered to be in equilibrium compared to typical pulsation timescales.

As we are concerned about bulk viscosity coefficient in young neutron stars where temperature is $\sim 10^9$ - 10^{10} K, we want to find out whether non-leptonic processes involving antikaons might lead to a high value for the bulk viscosity coefficient. The relevant nonleptonic process involving antikaons is

$$n \rightleftharpoons p + K^-. \tag{22}$$

As this reaction involves variation of neutron number density (n_n) due to density perturbation, we consider neutron fraction as a primary parameter. The general expression for the real part of bulk viscosity coefficient [3, 12] is

$$Re\,\zeta = \frac{P(\gamma_\infty - \gamma_0)\tau}{1 + (\omega\tau)^2} , \tag{23}$$

where P is the pressure, τ is the net microscopic relaxation time and γ_∞ and γ_0 are 'infinite' and 'zero' frequency adiabatic indices respectively. The factor

$$\gamma_\infty - \gamma_0 = -\frac{n_b^2}{P}\frac{\partial P}{\partial n_n}\frac{d\bar{x}_n}{dn_b} , \tag{24}$$

can be determined from the EoS. Here $\bar{x}_n = \frac{n_n}{n_b}$ gives the neutron fraction in the equilibrium state and $n_b = \sum_B n_B$ is the total baryon density. In the co-rotating frame, the generic equation relating the angular velocity (ω) of (l, m) r-mode to the angular velocity of rotation of the star (Ω) is $\omega = \frac{2m}{l(l+1)}\Omega$ [13].

The relaxation time (τ) for the process is given by [3]

$$\frac{1}{\tau} = \frac{\Gamma_K}{\delta\mu}\frac{\delta\mu}{\delta n_n^K}. \tag{25}$$

Here $\delta n_n^K = n_n^K - \bar{n}_n^K$ is the departure of neutron fraction from its thermodynamic equilibrium value $\bar{n}_n^K$ in the K^- condensed phase. The reaction rate per unit volume is [39]

$$\Gamma_K = \frac{< |M_K|^2 > k_{F_n}^2 \,\delta\mu}{16\pi^3 \mu_{K^-}} , \tag{26}$$

242

where k_{F_n} is the Fermi momentum for neutrons in the condensed phase and $<|M_K|^2>$ is the squared matrix element, averaged over initial spins and summed over final spins. The in-medium energy of K^- mesons in the condensate is $E_3 = \mu_3 = \mu_{K^-}$. If the quantity $\frac{\delta\mu}{\delta n_n^K}$ is calculated numerically, as soon as we know the relaxation time, we can calculate the bulk viscosity coefficient.

2.5 Matrix element

Next, we focus on the evaluation of the matrix element for the nonleptonic process (22). In general, the matrix element for the decay of a $\frac{1}{2}^+$ baryon to another $\frac{1}{2}^+$ baryon and a 0^- meson can be written as

$$\mathcal{M} = \bar{u}(k_2)(A + B\gamma_5)u(k_1) \tag{27}$$

where $u(k_1)$ and $u(k_2)$ are the spinors of neutrons and protons respectively. Here A is the parity-violating amplitude, and B is the parity-conserving amplitude. The squared and spin averaged matrix element is given by

$$<|\mathcal{M}|^2> = 2[(k_1 \cdot k_2 + m_n^* m_p^*)|A|^2 + (k_1 \cdot k_2 - m_n^* m_p^*)|B|^2]. \tag{28}$$

For s-wave K^- condensation, $\vec{k}_3 = 0 \implies |\vec{k}_1| = |\vec{k}_2|$. As fermion momenta lie close to the Fermi surfaces, $k_1 \cdot k_2 = E_1 E_2 - |\vec{k}_1||\vec{k}_2|\cos\theta = \mu_n\mu_p - k_{F_n}k_{F_p}$. This leads to the squared matrix element

$$<|\mathcal{M}^2|> = 2[(\mu_n\mu_p - k_{F_n}k_{F_p} + m_n^* m_p^*)|A|^2 + (\mu_n\mu_p - k_{F_n}k_{F_p} - m_n^* m_p^*)|B|^2]. \tag{29}$$

Here we apply the weak SU(3) symmetry to the non-leptonic weak decay amplitudes for the process (22). The weak decays of the octet hyperons can be described by an effective SU(3) interaction with a parity violating (A) and parity conserving (B) amplitudes [14, 15]. The weak operator is proportional to Gell-Mann matrix λ_6 to ensure hypercharge violation $|\Delta Y| = 1$ and $|\Delta I| = 1/2$. Similarly, the amplitudes for (22) are extracted from experimentally known decay parameters of the weak decay of hyperons [15]. The amplitudes are $A = -1.62 \times 10^{-7}$ and $B = -7.1 \times 10^{-7}$. It is to be noted that all quantities in (29) are to be calculated in the condensed phase.

2.6 Critical Angular Velocity

The bulk viscosity damping timescale (τ_B) due to the non-leptonic process involving antikaons is given by [3, 16]

$$\frac{1}{\tau_B} = -\frac{1}{2E}\frac{dE}{dt}, \tag{30}$$

where E is the energy of the perturbation as measured in the co-rotating frame of the fluid and is expressed as

$$E = \frac{1}{2}\alpha^2\Omega^2 R^{-2} \int_0^R \epsilon(r)r^6 dr .$$ (31)

Here, α is the dimensionless amplitude of the r-mode, R is the radius of the star and $\epsilon(r)$ is the energy density profile. The derivative of the co-rotating frame energy with respect to time is

$$\frac{dE}{dt} = -4\pi \int_0^R \zeta(r) < |\vec{\nabla} \cdot \delta\vec{v}|^2 > r^2 dr,$$ (32)

where the angle average of the square of the hydrodynamic expansion [17] is

$$< |\vec{\nabla} \cdot \delta\vec{v}|^2 > = \frac{(\alpha\Omega)^2}{690} \left(\frac{r}{R}\right)^6 \left(1 + 0.86 \left(\frac{r}{R}\right)^2\right) \left(\frac{\Omega^2}{\pi G\bar{\epsilon}}\right)^2 ,$$

and $\bar{\epsilon}$ is the mean energy density of a non-rotating star. The total r-mode time scale (τ_r) is defined as

$$\frac{1}{\tau_r} = -\frac{1}{\tau_{GR}} + \frac{1}{\tau_B} + \frac{1}{\tau_U}.$$ (33)

where the time scales for gravitational radiation (τ_{GR}) and modified Urca process (τ_U) involving only nucleons have also been included. The gravitational radiation timescale is given by [18]

$$\frac{1}{\tau_{GR}} = \frac{131072\pi}{164025}\Omega^6 \int_0^R \epsilon(r)r^6 dr .$$ (34)

The time scale due to modified Urca process (τ_U) involving only nucleons is calculated from (31) using the following expression for bulk viscosity coefficient for modified Urca process [18, 19]

$$\zeta_U = 6 \times 10^{-59}\epsilon^2 T^6\omega^2 .$$ (35)

For a star of given mass, solving $\frac{1}{\tau_r} = 0$, we can obtain the critical angular velocity at each temperature above which the r-mode becomes unstable.

3   Parameters of the Theory

3.1   Nucleon-Meson coupling constants

Nucleon-meson coupling constants are determined from saturation properties of nuclear matter [20]. We used the following values: binding energy = −16.3 MeV, baryon density $n_0 = 0.153$ fm^{-3}, asymmetry energy coefficient $a_{\text{asy}} = 32.5$ MeV, incompressibility $K = 240$ MeV, and effective nucleon mass $m_N^*/m_N = 0.78$. We also studied the parameter set [21] with incompressibility $K = 300$ MeV, and effective nucleon mass $m_N^*/m_N = 0.70$.

3.2 Kaon-Meson coupling constants

According to the quark model and isospin counting rule, the vector coupling constants are given by

$$g_{\omega K} = \frac{1}{3} g_{\omega N} \quad \text{and} \quad g_{\rho K} = g_{\rho N} . \tag{36}$$

The scalar coupling constant is obtained from the real part of K^- optical potential depth at normal nuclear matter density

$$U_{\bar{K}}(n_0) = -g_{\sigma K}\sigma - g_{\omega K}\omega_0 . \tag{37}$$

It is known from K^--atomic data that antikaons experience an attractive potential in nuclear matter while kaons feel a repulsive interaction [22–27]. The strength of antikaon optical potential depth ranges from shallow attractive (-40 MeV) to strongly attractive (-180 MeV). Here we perform the calculation for antikaon optical potential depth at normal nuclear matter density $U_{\bar{K}}(n_0) = -120$ MeV. The strange meson fields couple with (anti)kaons. The σ^*-K coupling constant is $g_{\sigma^* K} = 2.65$ as determined from the decay of $f_0(925)$ meson and the vector ϕ meson coupling with (anti)kaons $\sqrt{2}g_{\phi K} = 6.04$ follows from the SU(3) relation [28].

3.3 Hyperon-Meson coupling constants

Hyperon-vector meson coupling constants are determined from SU(6) symmetry of the quark model [29–31]. The scalar σ meson coupling to hyperons is calculated from the potential depth of a hyperon (Y)

$$U_Y^N(n_0) = -g_{\sigma Y}\sigma + g_{\omega Y}\omega_0 , \tag{38}$$

in normal nuclear matter. The potential depth of Λ hyperons in normal nuclear matter $U_\Lambda^N(n_0) = -30$ MeV is obtained from the analysis of energy levels of Λ hypernuclei [30,32]. Recent Ξ-hypernuclei data from various experiments [33,34] give a relativistic potential of $U_\Xi^N(n_0) = -18$ MeV. However, the analysis of Σ^- atomic data implies a strong isoscalar repulsion for Σ^- hyperons in nuclear matter [35]. Also, recent Σ hypernuclei data indicate a repulsive Σ-nucleus potential depth [36]. Therefore, a repulsive potential depth of 30 MeV for Σ hyperons [35] is adapted.

The hyperon-σ^* coupling constants are estimated by fitting them to a potential depth, $U_Y^{(Y')}(n_0)$, for a hyperon (Y) in a hyperon (Y') matter at normal nuclear matter density obtained from double Λ hypernuclei data [29,37]. This is given by

$$U_\Xi^{(\Xi)}(n_0) = U_\Lambda^{(\Xi)}(n_0) = 2U_\Xi^{(\Lambda)}(n_0) = 2U_\Lambda^{(\Lambda)}(n_0) = -40 \ . \tag{39}$$

4   Results and Discussions

The Equation of State (pressure versus energy density) for neutron star matter with K^- condensate is plotted in Fig. 1 (solid line) for K=240 MeV. The equations

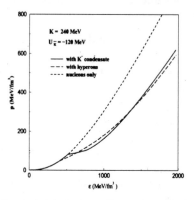

Figure 1: The equation of state (pressure P vs energy density ϵ) for matter containing nucleons only (short dashed line), with hyperons (long dashed line) and with K^- condensate (solid line) for antikaon optical potential depth at normal nuclear matter density $U_{\bar{K}} = -120$ MeV.

of state for neutron star matter containing nucleons only (short dashed line) and with hyperons (long dashed line) are superimposed on the same figure. The two kinks on the EoS involving the condensate at 3.26 n_0 and 4.62 n_0 mark the beginning and end of the mixed phase. For K^- condensed matter with K=300 MeV, similar kinks are observed at 2.23 n_0 and 3.59 n_0 defining the mixed phase. The threshold density for the appearance of the Λ hyperon is 2.6 n_0. The EoS becomes softer in presence of exotic matter (K^- condensate or hyperons) compared with that of nucleon matter. However, the EoS for antikaon condensed matter is stiffer than that for hyperon matter beyond the mixed phase.

For the calculation of damping time scale due to bulk viscosity using (30), we need the energy density profile and bulk viscosity profile of the neutron star. We choose a neutron star of gravitational mass $1.63 M_\odot$ corresponding to a central baryon density 3.94 n_0 and rotating at an angular velocity $\Omega = 1180 s^{-1}$. The neutron star is so chosen to ensure that it contains K^- condensate in its core because the central baryon density is well above the threshold of K^- condensation (3.26 n_0). The bulk viscosity profiles for K = 240 MeV and K = 300 MeV are

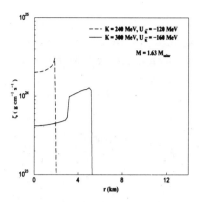

Figure 2: Bulk viscosity profile is plotted with equatorial distance for a rotating neutron star of mass 1.63 $M_\odot$ for two parameter sets.

plotted in Fig 2 as a function of equatorial distance. Here we note that the bulk viscosity profile drops to zero value beyond a certain equatorial distance, when the baryon density decreases below the threshold density of K^- condensation and the nonleptonic process in (22) ceases to occur in the star.

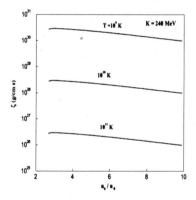

Figure 3: Hyperon bulk viscosity coefficient is exhibited as a function of normalised baryon density for different temperatures.

We compare the bulk viscosity coefficient due to antikaon condensate with bulk viscosity due to the nonleptonic weak interaction $n + p \rightleftharpoons p + \Lambda$ involving hyperons. The coefficient of bulk viscosity due to nonleptonic process involving hyperons is shown in Fig 3 as a function of normalised baryon density for different temperatures [38]. Comparing the two figures 2 and 3, we can infer that the bulk viscosity coefficient in antikaon condensed matter is suppressed by several orders of magnitude in comparison with nonsuperfluid hyperon bulk viscosity. It must also be noted that the antikaon bulk viscosity is independent of temperature, while hyperon bulk viscosity increases with decrease in temperature. Hence, hyperon bulk viscosity can act as an effective damping mechanism at low temperatures.

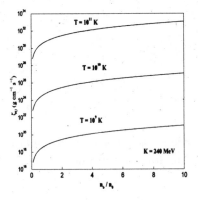

Figure 4: Density dependence of modified Urca bulk viscosity is shown for a range of temperatures.

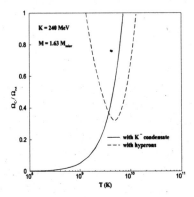

Figure 5: Critical angular velocities for 1.63 M$_\odot$ neutron star are plotted as a function of temperature.

We can obtain critical angular velocities as a function of temperature by solving $\frac{1}{\tau_r} = 0$ in (33) for a rotating neutron star of mass 1.63 M_{solar}. We calculate the bulk viscosity coefficient due to modified Urca process due to nucleons using (35). The modified Urca bulk viscosity is plotted as a function of normalised baryon density for a range of temperatures in Fig 5. It is evident from the figure that bulk viscosity due to modified Urca process involving nucleons increases with increase in temperature. Hence r-mode instability is effectively suppressed by modified Urca bulk viscosity at high temperatures. The ratio of critical angular velocities to the rotational velocity of the neutron star are plotted as a function of temperature in Fig 4 for nonleptonic processes involving antikaons as well as hyperons. In this figure, we observe that there exists a window of instability for matter with hyperons. We can interpret that hyperon bulk viscosity damps the r-mode instability at low temperatures, whereas it is effectively suppressed at high temperatures by the bulk viscosity due to the modified Urca process involving nucleons. For K^- condensed matter, we can infer that r-mode instability is damped by nucleonic modified Urca process at high temperatures. From the figure, it is evident that antikaon bulk viscosity is not an effective mechanism to damp the instability at low temperatures. The argument is further justified by the fact that the critical velocity curves obtained for both parameter sets K= 240 MeV and K = 300 MeV are dictated purely by the bulk viscosity due to nucleonic modified Urca process. The bulk viscosity due to antikaon condensation is not sufficient to suppress the r-mode within the temperature range considered here. Hence the instability window is wider for K^- condensed matter, and neutron stars with K^- condensate in its interior could be possible sources of gravitational waves.

5   Summary

The role of K^- condensation on bulk viscosity and r-mode instability has been investigated in this paper. We have estimated the bulk viscosity coefficient and the corresponding damping time scale due to the nonleptonic process $n \rightleftharpoons p + K^-$ and compared them with those associated with the nonleptonic process $n + p \rightleftharpoons p + \Lambda$ involving hyperons. We have considered a first order phase transition from the nuclear to the antikaon condensed phase, and the equation of state has been constructed using relativistic mean field theoretical models. We find that the bulk viscosity coefficient in K^- condensed is suppressed by several orders of magnitude in comparison with the nonsuperfluid hyperon bulk viscosity [3, 16, 38]. We also infer that the antikaon bulk viscosity is unable to damp the r-mode instability unlike hyperon bulk viscosity. If the instability is not suppressed by the viscosity, the star is forced to lose angular momentum as it cools down via gravitational radiation. So, r-modes in neutron stars with antikaon condensate in their core are possible candidates of gravitational waves.

References

[1] P.B. Jones, Phys. Rev. Lett. **86**, 1384 (2001).

[2] P.B. Jones, Phys. Rev. D **64**, 084003 (2001).

[3] L. Lindblom and B.J. Owen, Phys. Rev. D **65**, 063006 (2002).

[4] E.N.E. van Dalen and A.E.L. Dieperink, Phys. Rev. C **69**, 025802 (2004).

[5] A. Drago, A. Lavagno and G. Pagliara, Phys. Rev. D **71**, 103004 (2005).

[6] J. Madsen, Phys. Rev. D **46**, 3290 (1992).

[7] J. Madsen, Phys. Rev. Lett. **85**, 10 (2000).

[8] J. Boguta and A.R. Bodmer, Nucl. Phys. **A292**, 413 (1977).

[9] B.D. Serot and J.D. Walecka, Adv. in Nucl. Phys. **16**, 1 (1986).

[10] N.K. Glendenning and J. Schaffner-Bielich, Phys. Rev. C **60**, 025803 (1999).

[11] N.K. Glendenning, Phys. Rev. D **46**, 1274 (1992).

[12] L.D. Landau and E.M. Lifshitz, Fluid Mechanics, 2nd ed. (Butterworth-Heinemann, Oxford, 1999).

[13] N. Andersson, Class. Quant. Grav. **20**, R105 (2003).

[14] R.E. Marshak, Riazuddin and C.P. Ryan, Theory of weak interactions in particle physics (Wiley-Interscience, New York, 1969).

[15] J. Schaffner-Bielich, R. Mattiello and H. Sorge, Phys. Rev. Lett. **84**, 4305 (2000).

[16] M. Nayyar and B.J. Owen, Phys. Rev. D **73**, 084001 (2006).

[17] L. Lindblom, G. Mendell and B.J. Owen, Phys. Rev. D **60**, 064006 (1999).

[18] L. Lindblom, B.J. Owen and S. M. Morsink, Phys. Rev. Lett. **80**, 4843 (1998).

[19] R.F. Sawyer, Phys. Rev. D **39**, 3804 (1989).

[20] N.K. Glendenning and S.A. Moszkowski, Phys. Rev. Lett. **67**, 2414 (1991).

[21] S. Banik and D. Bandyopadhyay, Phys. Rev. C **64**, 055805 (2001).

[22] E. Friedman, A. Gal and C.J. Batty, Nucl. Phys. **A579**, 518 (1994); C.J. Batty, E. Friedman and A. Gal, Phys. Rep. **287**, 385 (1997).

[23] E. Friedman, A. Gal, J. Mareš and A. Cieplý, Phys. Rev. C **60**, 024314 (1999).

[24] V. Koch, Phys. Lett. B **337**, 7 (1994).

[25] T. Waas and W. Weise, Nucl. Phys. **A625**, 287 (1997).

[26] G.Q. Li, C.-H. Lee and G.E. Brown, Phys. Rev. Lett. **79**, 5214 (1997); Nucl. Phys. **A625**, 372 (1997).

[27] S. Pal, C.M. Ko, Z. Lin and B. Zhang, Phy. Rev. C **62**, 061903(R) (2000).

[28] J. Schaffner and I.N. Mishustin, Phys. Rev. C **53**, 1416 (1996).

[29] J. Schaffner and I.N. Mishustin, Phys. Rev. C **53**, 1416 (1996).

[30] C. B. Dover and A. Gal, Prog. Part. Nucl. Phys. **12**, 171 (1984).

[31] J. Schaffner, C.B. Dover, A. Gal, D. J. Millener, C. Greiner and H. Stöcker, Ann. Phys. (N.Y.) **235**, 35 (1994).

[32] R. E. Chrien and C. B. Dover, Annu. Rev. Nucl. Part. Sci. **39**, 113 (1989).

[33] T. Fukuda et al., Phys. Rev. C **58**, 1306 (1998).

[34] P. Khaustov et al., Phys. Rev. C **61**, 054603 (2000).

[35] E. Friedman, A. Gal and C. J. Batty, Nucl. Phys. **A579**, 518 (1994); C. J. Batty, E. Friedman and A. Gal, Phys. Rep. **287**, 385 (1997).

[36] S. Bart et al., Phys. Rev. Lett. **83**, 5238 (1999).

[37] J. Schaffner, C.B. Dover, A. Gal, C. Greiner and H. Stöcker, Phys. Rev. Lett. **71**, 1328 (1993).

[38] D. Chatterjee and D. Bandyopadhyay, Phys. Rev. D **74**, 023003 (2006); D. Chatterjee and D. Bandyopadhyay, astro-ph/0607005.

[39] D. Chatterjee and D. Bandyopadhyay, astro-ph/0702259.

Physics and Astrophysics of Hadrons and Hardronic Matter
Editor: A. B. Santra

Neutron Star Properties with Accurately Calibrated Field Theoretical Model

Shashi K. Dhiman[a]*, B. K. Agrawal[b], Raj Kumar[a]

a) *Department of Physics, H.P. University, Shimla - 171005, India.*
b) *Saha Institute of Nuclear Physics, Kolkata - 700064, India.*
* email: skd_phy_hpu@yahoo.com

Abstract

We study the properties of neutron star by using accurately calibrated parameterizations of the field theoretical model. We find that the radius $R_{1.4}$ of neutron star with canonical mass should be at least 12.8 km, provided, only those parameterizations are considered for which M_{max} is larger than $1.6 M_\odot$ which is the highest measured mass with 95% confidence limit for PSR J0751+1807.

1 Introduction

The knowledge of neutron star properties is necessary to probe the high density behaviour of the equation of state (EOS) for the baryonic matter in β- equilibrium. The EOS for the densities higher than $\rho_0 = 0.16 fm^{-3}$ can be well constrained if radii for the neutron stars over a wide range of masses are appropriately known. Even the accurate information on the maximum neutron star mass M_{max} and radius $R_{1.4}$ for the neutron star with canonical mass $(1.4 M_\odot)$ would narrow down the choices for the plausible EOSs to just a few. Till date, the neutron stars with masses only around $1.4 M_\odot$ are accurately measured[1, 2]. Recent measurement of mass of the pulsar PSR J0751+1807 imposes lower bounds on the maximum mass of the neutron star to be $1.6 M_\odot$ and $1.9 M_\odot$ with 95% and 68% confidence limits, respectively[3, 4]. The increase in the lower bounds of the neutron star maximum mass could eliminate the family of EOSs in which exotica appear and substantial softening begins around 2 to 4 ρ_0 leading to appreciable reduction of the maximum mass. The available data on the neutron star radius have large uncertainties[?, 6, 7]. The main source of the uncertainties in the measurements of the neutron star radii are (i) the unknown chemical composition of the atmosphere, (ii) high magnetic field ($\sim 10^{12}$ G) and (iii) inaccuracies in the star's distance.

Theoretically, the mass-radius relationship and compositions of the neutron stars are studied using various models which can be broadly grouped into (i) potential models[8], (ii) non-relativistic mean-field models[9, 10], (iii) field theoretical based relativistic mean-field models (FTRMF)[11, 12] and (iv) Dirac-Brueckner-Hartree-Fock model[13]. The values of M_{max} and $R_{1.4}$ obtained for the FTRMF models lie in the range of $1.2 - 2.8 M_\odot$ and $10 - 15$ km, respectively. Such a large variations in the M_{max} and $R_{1.4}$ are mainly due to the fact that the density dependence of the symmetry energy coefficient and behaviour of EOS for dense matter are at variance for the different FTRMF models. Further the values of nuclear matter incompressibility coefficient and effective nucleon mass also vary over a wide range for these models even though they are reasonably constrained by the experimental data.

In the present work, we appropriately calibrate the parameters of the FTRMF model and study the variations in the properties of neutron stars resulting from the differences in the behaviour of the density dependence of symmetry energy coefficient and EOS of dense baryonic matter which are not yet well constrained. We mainly study the properties of neutron star with canonical mass and the maximum mass of neutron star. In Sec. 2 we outline very briefly the Lagrangian density and corresponding energy density for the FTRMF model. In Sec. 3 we discuss our various parameterizations for, different combinations of neutron thickness (Δr) in ^{208}Pb nucleus, self interaction coupling constant (ζ) of ω meson, and hyperon-meson couplings, obtained by fitting the FTRMF results to exactly the same set of data for the total binding energies, charge rms radii for several closed shell normal and exotic nuclei. In this section we also discuss about the quality of the fits to the finite nuclei for these parameterizations along with the results for the neutron star properties. Finally our main conclusions are presented in Sec. 4.

2   Field Theoretical Model

In the FTRMF model, the Lagrangian density describes the interactions among baryons through the exchange of mesons and photons. The baryons considered in the present work are nucleons (n and p) and hyperons (Λ, Σ and Ξ). The mesons considered are σ, ω and ρ. In addition we also consider two strange mesons σ^*, and ϕ to describe the hyperon-hyperon interaction [14].

$$\mathcal{L} = \mathcal{L}_{BM} + \mathcal{L}_\sigma + \mathcal{L}_\omega + \mathcal{L}_\rho + \mathcal{L}_{\sigma\omega\rho} + \mathcal{L}_{em} + \mathcal{L}_{e\mu} + L_{YY}. \tag{1}$$

Where the baryonic and mesonic Lagrangian $\mathcal{L}_{BM}$ can be written,

$$\mathcal{L}_{BM} = \sum_B \overline{\Psi}_B [i\gamma^\mu \partial_\mu - (M_B - g_{\sigma B}\sigma) - (g_{\omega B}\gamma^\mu \omega_\mu + \frac{1}{2}g_{\rho B}\gamma^\mu \tau_B.\rho_\mu)]\Psi_B. \tag{2}$$

Here, the sum on B is taken over the baryon octet that consists of $p, n, \Lambda, \Sigma^+, \Sigma^0, \Sigma^-$, Ξ^0 and Ξ^-. For the calculation of finite nuclei properties only neutron and proton has been considered. τ_B are the isospin matrices. The Lagrangian describing self interactions of σ, ω, and ρ mesons can be written as,

$$\mathcal{L}_\sigma = \frac{1}{2}(\partial_\mu \sigma \partial^\mu \sigma - m_\sigma^2 \sigma^2) - \frac{\overline{\kappa}}{3!}g_{\sigma N}^3 \sigma^3 - \frac{\overline{\lambda}}{4!}g_{\sigma N}^4 \sigma^4, \tag{3}$$

$$\mathcal{L}_\omega = -\frac{1}{4}\omega_{\mu\nu}\omega^{\mu\nu} + \frac{1}{2}m_\omega^2\omega_\mu\omega^\mu + \frac{1}{4!}\zeta g_{\omega N}^4(\omega_\mu\omega^\mu)^2, \tag{4}$$

$$\mathcal{L}_\rho = -\frac{1}{4}\rho_{\mu\nu}\rho^{\mu\nu} + \frac{1}{2}m_\rho^2\rho_\mu\rho^\mu + \frac{1}{4!}\xi g_{\rho N}^4(\rho_\mu\rho^\mu)^2. \tag{5}$$

The $\omega^{\mu\nu}$, $\rho^{\mu\nu}$ are field tensors corresponding to the ω and ρ mesons, and can be defined as $\omega^{\mu\nu} = \partial^\mu\omega^\nu - \partial^\nu\omega^\mu$ and $\rho^{\mu\nu} = \partial^\mu\rho^\nu - \partial^\nu\rho^\mu$. The mixed interactions of σ, ω, and ρ mesons $\mathcal{L}_{\sigma\omega\rho}$ can be written as,

$$\mathcal{L}_{\sigma\omega\rho} = g_{\sigma N}g_{\omega N}^2\sigma\omega_\mu\omega^\mu\left(\overline{\alpha_1} + \frac{1}{2}\overline{\alpha_1}'\sigma\right) + g_{\sigma N}g_{\rho N}^2\sigma\rho_\mu\rho^\mu\left(\overline{\alpha_2} + \frac{1}{2}\overline{\alpha_2}'\sigma\right) \tag{6}$$

$$+ \frac{1}{2}\overline{\alpha_3}'g_{\omega N}^2 g_{\rho N}^2\omega_\mu\omega^\mu\rho_\mu\rho^\mu \tag{7}$$

The $\mathcal{L}_{em}$ is Lagrangian for electromagnetic interactions and can be expressed as,

$$\mathcal{L}_{em} = -\frac{1}{4}F_{\mu\nu}F^{\mu\nu} - \sum_B e\overline{\Psi}_B\gamma_\mu\frac{1+\tau_{3B}}{2}A_\mu\Psi_B, \tag{8}$$

where, $F^{\mu\nu} = \partial^\mu A^\nu - \partial^\nu A^\mu$.

The hyperon-hyperon interaction has been included by introducing two additional meson fields (σ^* and ϕ), and the corresponding Lagrangian L_{YY} ($Y = \Lambda, \Sigma$, and Ξ) can be written as,

$$\mathcal{L}_{YY} = \frac{1}{2}\left(\partial_\nu\sigma^*\partial^\nu\sigma^* - m_{\sigma^*}^2\sigma^{*2}\right) - \frac{1}{4}S_{\mu\nu}S^{\mu\nu} + \frac{1}{2}m_\phi^2\phi_\mu\phi^\mu \tag{9}$$

$$+ \sum_B \overline{\Psi}_B\left(g_{\sigma^*B}\sigma^* - g_{\phi B}\gamma^\mu\phi_\mu\right)\Psi_B. \tag{10}$$

The leptonic contributions to the total Lagrangian density can be written as,

$$\mathcal{L}_{e\mu} = \sum_{\ell=e,\mu}\overline{\Psi}_\ell\left(i\gamma^\mu\partial_\mu - M_\ell\right)\Psi_\ell. \tag{11}$$

The equation of motion for baryons, mesons and photons can be derived from the Lagrangian density defined in Eq. (1). The energy density of the uniform matter in the FTRMF models is given by

$$\mathcal{E} = \sum_{j=B,l}\frac{1}{\pi^2}\int_0^{k_j} k^2\sqrt{k^2 + M_j^{*2}}dk + \sum_B g_{\omega B}\omega\rho_B + \sum_B g_{\rho B}\tau_{3B}\rho \tag{12}$$

$$+ \frac{1}{2}m_\sigma^2\sigma^2 + \frac{\overline{\kappa}}{6}g_{\sigma N}^3\sigma^3 + \frac{\overline{\lambda}}{24}g_{\sigma N}^4\sigma^4 - \frac{\zeta}{24}g_{\omega N}^4\omega^4 - \frac{\xi}{24}g_{\rho N}^4\rho^4 \tag{13}$$

$$- \frac{1}{2}m_\omega^2\omega^2 - \frac{1}{2}m_\rho^2\rho^2 - \overline{\alpha_1}g_{\sigma N}g_{\omega N}^2\sigma\omega^2 - \frac{1}{2}\overline{\alpha_1}'g_{\sigma N}^2g_{\omega N}^2\sigma^2\omega^2 \tag{14}$$

253

$$-\overline{\alpha_2}g_{\sigma N}g_{\rho N}^2\sigma\rho^2 - \frac{1}{2}\overline{\alpha_2}'g_{\sigma N}^2g_{\rho N}^2\sigma^2\rho^2 - \frac{1}{2}\overline{\alpha_3}'g_{\omega N}^2g_{\rho N}^2\omega^2\rho^2 \qquad (15)$$

$$+\frac{1}{2}m_{\sigma^*}^2\sigma^{*2} + \sum_B g_{\phi B}\phi\rho_B - \frac{1}{2}m_\phi^2\phi^2. \qquad (16)$$

The pressure of the uniform matter is given by

$$P \quad = \sum_{j=B,l}\frac{1}{3\pi^2}\int_0^{k_j}\frac{k^4 dk}{\sqrt{k^2 + M_j^{*2}}} - \frac{1}{2}m_\sigma^2\sigma^2 - \frac{\overline{\kappa}}{6}g_{\sigma N}^3\sigma^3 - \frac{\overline{\lambda}}{24}g_{\sigma N}^4\sigma^4 \qquad (17)$$

$$+\frac{\zeta}{24}g_{\omega N}^4\omega^4 + \frac{\xi}{24}g_{\rho N}^4\rho^4 + \frac{1}{2}m_\omega^2\omega^2 + \frac{1}{2}m_\rho^2\rho^2 + \overline{\alpha_1}g_{\sigma N}g_{\omega N}^2\sigma\omega^2 \qquad (18)$$

$$+\frac{1}{2}\overline{\alpha_1}'g_{\sigma N}^2g_{\omega N}^2\sigma^2\omega^2 + \overline{\alpha_2}g_{\sigma N}g_{\rho N}^2\sigma\rho^2 + \frac{1}{2}\overline{\alpha_2}'g_{\sigma N}^2g_{\rho N}^2\sigma^2\rho^2 \qquad (19)$$

$$+\frac{1}{2}\overline{\alpha_3}'g_{\omega N}^2g_{\rho N}^2\omega^2\rho^2 - \frac{1}{2}m_{\sigma^*}^2\sigma^{*2} + \frac{1}{2}m_\phi^2\phi^2. \qquad (20)$$

3 Results and discussions

We generate several parameter sets for the FTRMF model corresponding to different combinations of the ω meson self-coupling ζ and neutron-skin thickness Δr in ^{208}Pb nucleus. Firstly, the hyperon-meson couplings g_{iY} ($i = \sigma, \omega, \rho, \sigma^*$ and ϕ) in Eqs. (2,9) are set to zero. Then, the remaining coupling parameters appearing in Eqs. (2 - 6) are determined by fitting the FTRMF results to the experimental data for the total binding energies and charge radii for many closed shell normal and exotic nuclei. In addition, we also fit the value of neutron-skin thickness for ^{208}Pb nucleus to constrain the linear density dependence of symmetry energy coefficient. We generate parameter sets for seven different values of neutron-skin thickness taken to be 0.16, 0.18..., 0.28 fm for the ^{208}Pb nucleus. The best fit parameters are obtained by minimizing the χ^2 function as discussed in ref. [15] by using simulated annealing method. In Table 1, we list the values of parameters for selected sets generated in the present work.

The coupling strength of the hyperon-hyperon interactions are calculated by hidden strange meson exchange of a scalar, σ^* and a vector, ϕ meson. In addition, the baryon-baryon interactions are mediated by scalar, σ meson, vector, ω meson, and isovector, ρ meson, exchange. The coupling constants of hyperon to the ρ, σ^*, and ϕ mesons are fixed by using the SU(6) symmetry. The neutron star properties are quite sensitive to the values of σ meson, $g_{\sigma Y}$ and ω meson, $g_{\omega Y}$ couplings. The coupling constant $g_{\sigma Y}$ is constrained by the hypernuclear potential in nuclear matter of $U_\Lambda^{(N)} = -28$MeV, $U_\Sigma^{(N)} = +30$MeV, and $U_\Xi^{(N)} = -18$ MeV.

The various properties associated with the nuclear matter are obtained using FTRMF parameter sets. We find that $B/A = 16.11 \pm 0.04$ MeV, $K = 230.24 \pm 9.80$ MeV, $M^*/M = 0.605 \pm 0.004$ and $\rho_{sat} = 0.148 \pm 0.003$ fm^{-3}. The values of the

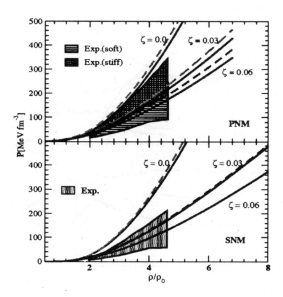

Figure 1: The EOS for pure neutron matter (upper panel) and symmetric nuclear matter (lower panel). The solid and dashed curves corresponds to $\Delta r = 0.16$ and 0.28 fm, respectively. The shaded regions represent the experimental data taken from Ref. [19].

symmetry energy coefficient J and its linear density dependence, $L = 3\rho \frac{dJ}{d\rho}|_{\rho_{sat}}$ are also calculated at saturation density. The values of L lie in the range of 80 ± 20 MeV for Δr varying in between 0.16 to 0.28 fm which is in reasonable agreement with the recent predictions. The relative errors in the total binding energy and rms charge radius for the nuclei included in the fits are more or less the same as we have obtained in our earlier work [15]. The rms errors on the total binding energies is 1.5 - 1.8 MeV which is comparable with one obtained using NL3 parameterizations as most commonly used. The rms error of charge radii for the nuclei considered in the fit lie within the 0.025 − 0.040 fm. The parameters of FTRMF model are so calibrated that the quality of fit to finite nuclei, the properties of nuclear matter at saturation density and hyperon-nucleon potentials are almost the same for each of the parameterizations. Thus, these parameterizations provide the right starting point to study the variations in the properties of neutron star resulting from the uncertainties in the EOS of dense matter and density dependence of symmetry energy coefficient.

The properties of static neutron stars are determined by integrating the Tolman-Oppenheimer-Volkoff (TOV) equations. To solve the TOV equations we use the EOS for the matter consisting of nucleons, hyperons and leptons. The composition of dense matter at fixed total baryon density is obtained by satisfying the charge neutrality condition and chemical equilibrium conditions. For densities higher than

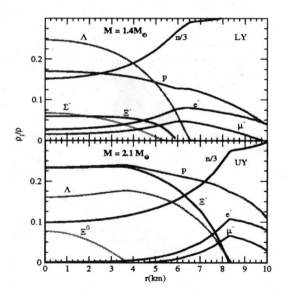

Figure 2: Particles fractions as a function of radial coordinate obtained at maximum mass for LY (upper panel) and UY (lower panel) parameterizations. The curves labeled by "n/3" should be multiplied by three to get the actual neutron fractions

$0.5\rho_0$, the baryonic part of EOS is evaluated within the FTRMF model. The Fermi gas approximation is used for electrons and muons. At densities lower than $0.5\rho_0$ down to $0.4 \times 10^{-10}\rho_0$ we use the EOS of Baym-Pethick-Sutherland (BPS)[16]. In Fig. 1 we plot the EOS for the pure neutron matter and symmetric nuclear matter as a function of baryon density for the selected combinations of ζ and Δr. We see that the EOS for $\zeta = 0.0$ is the most stiffest, and as ζ increases the EOS becomes softer. The softening of EOS with ζ is more pronounced at higher densities.

We now consider various neutron star properties resulting from two different parameter sets referred hereafter as LY and UY. These parameter sets are obtained using different combinations of Δr, ζ and $X_{\omega Y}$ ($X_{\omega Y} = \frac{g\omega Y}{g\omega N}$ for Λ, Σ and $X_{\omega Y} = 2\frac{g\omega Y}{g\omega N}$ for Ξ). The parameters of LY set are obtained with $\Delta r = 0.16 fm$, $\zeta = 0.06$ and $X_{\omega Y} = 0.5$. Whereas, UY parametrization is obtained with $\Delta r = 0.28 fm$, $\zeta = 0.0$ and $X_{\omega Y} = 0.8$ and yield stiffest EOS. For the comparison, we also present our results for the L0 and U0 parameter sets similar to LY and UY parameterizations, but, without hyperons.

In Fig. 2 we plot the particle fractions as a function of radial coordinate. These fractions are calculated for the neutron stars with $M_{max} = 1.4 M_\odot$ and $2.1 M_\odot$ corresponding to the LY (upper panel) and UY (lower panel) parameterizations, respectively. The neutron fractions in Fig. 2 are plotted after dividing them by a factor of three. We see that the compositions of the neutron stars shown in the upper and lower panels are not the same. For the case of LY parameterizations, Ξ^-

Table 1: New coupling strength parameters for the Lagrangian of the FTRMF model as given in Eq. 1. The seven different parameter sets correspond to the different values of the neutron skin-thickness Δr for the ^{208}Pb nucleus used in the fit. The value of ω meson self-couping ζ is equal to 0.00, 0.03 and 0.06 for all these parameterizations. The values of Δr are in fm, the parameters $\bar{\kappa}$, $\bar{\alpha}_1$, and $\bar{\alpha}_2$ are in fm^{-1} and m_σ are in MeV. The masses for other mesons are taken to be $m_\omega = 782.5$MeV, $m_\rho = 763$ MeV, $m_\sigma^* = 975$ MeV and $m_\phi = 1020$ MeV. For the masses of nucleons and hyperons we use $M_N = 939 MeV$, $M_\Lambda = 1116 MeV$, $M_\Sigma = 1193 MeV$ and $M_\Xi = 1313 MeV$. The values of $\bar{\kappa}$, $\bar{\lambda}$, $\bar{\alpha}_1'$, $\bar{\alpha}_2$, $\bar{\alpha}_2'$, and $\bar{\alpha}_3'$ are multiplied with 10^2.

ζ	0.00			0.03			0.06		
Δr	0.16	0.22	0.28	0.16	0.22	0.28	0.16	0.22	0.28
$g_{\sigma N}$	10.51369	10.50339	10.32009	10.6286	10.7942	10.600110	11.05170	11.01908	11.03151
$g_{\omega N}$	13.48789	13.80084	13.45113	13.65991	14.12534	14.03101	14.59582	14.77458	15.01572
$g_{\rho N}$	14.98497	12.12975	10.09608	14.99076	12.19156	10.00441	14.98725	11.94837	10.00666
$\bar{\kappa}$	2.62556	3.39711	2.82791	1.38118	1.61820	1.78793	0.66576	0.78002	0.80797
$\bar{\lambda}$	-0.73495	-1.15784	-1.13890	0.58536	1.06102	0.74676	2.46427	2.47238	2.41320
$\bar{\alpha}_1$	0.22672	0.44021	0.23357	0.00366	0.10650	0.16088	0.00601	0.01469	0.02073
$\bar{\alpha}_1'$	0.07325	0.00987	0.04733	0.02717	0.06526	0.01669	0.00203	0.01559	0.01109
$\bar{\alpha}_2$	3.05925	2.56759	0.60739	2.89393	2.77747	0.47146	2.86236	2.02292	0.55325
$\bar{\alpha}_2'$	1.55587	0.51396	0.33057	1.59659	0.22126	0.52816	1.55176	0.90169	0.16326
$\bar{\alpha}_3'$	1.50060	1.04562	0.30434	1.52088	0.45581	0.32358	1.55307	0.96305	0.72768
m_σ	502.23217	491.48257	491.86209	506.50582	497.20745	494.93882	503.43838	497.27203	490.68907

257

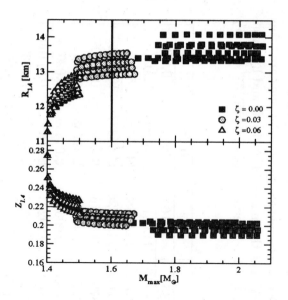

Figure 3: Variations of radius $(R_{1.4})$ and redshift $(Z_{1.4})$ for neutron star with the canonical mass as a function of maximum neutron star mass obtained using parameterizations for all combinations of Δr, ζ and $X_{\omega Y}$. The vertical line at $M_{max} = 1.6 M_\odot$ in the upper panel corresponds to the mass of the PSR J0751+1807 measured with 95% confidence limit.

and Σ^- hyperons appear more or less simultaneously. For the UY case, Ξ^0 hyperons appear instead of Σ^- hyperons. It is noteworthy that, for the case with UY parametrization, the hyperons are the dominant particles at the interior $(r < 4\text{km})$ of the neutron star leading to complete deleptonization. We see from Fig. 2 that the proton fractions for both the cases are greater than the critical value ($\sim 15\%$) for the Direct Urca process to occur [18].

In Fig. 3 we plot the variations of radius and the redshift for the neutron star with the canonical mass $(1.4 M_\odot)$ as a function of maximum mass. The vertical line at $M_{max} = 1.6 M_\odot$ corresponds to the mass of the PSR J0751+1807 measured with 95% confidence limit. We see that M_{max} varies between $1.4 - 2.1 M_\odot$ and $R_{1.4}$ varies between $11.3 - 14.1$ km. Our results clearly indicate strong correlations of M_{max} with $R_{1.4}$ and $Z_{1.4}$. For a given value of M_{max}, the spread in the values of $R_{1.4}$ is 0.7 ± 0.1 km. Only for the $M_{max} \sim 1.4 M_\odot$ we find that spread in the values of $R_{1.4}$ is ~ 0.3 km. To understand it better, we list in Table 2 the values of M_{max}, R_{max}, and $R_{1.4}$ obtained with the parameter sets for selected combinations of Δr, ζ and $X_{\omega Y}$. It is clear from the table that for smaller ζ the values of $R_{1.4}$ vary with Δr and are independent of $X_{\omega Y}$. Of course, this is due to the fact that for smaller ζ, central density for neutron star with mass $1.4 M_\odot$ is lower or almost equal to the threshold density for hyperons. But, as ζ increases, the central density becomes larger than

258

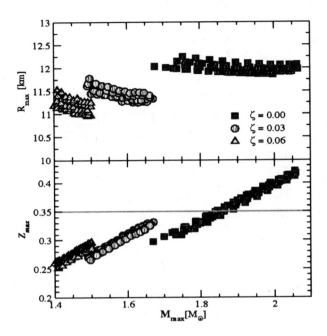

Figure 4: Variations of radius (R_{max}) and redshift (Z_{max}) for neutron star with the maximum mass as a function of maximum neutron star mass obtained using parameterizations for all combinations of Δr, ζ and $X_{\omega Y}$. The horizontal line in the lower panel corresponds to the measured value of the redshift, $Z = 0.35$, for the neutron star EXO 0748-676 [6].

the threshold densities for various hyperons, thus, $R_{1.4}$ depends on Δr as well as $X_{\omega Y}$. In Fig. 4 we plot the variations of R_{max} and Z_{max} versus the maximum neutron star mass. We see that the correlations in the values of M_{max} and R_{max} are stronger than the ones observed in the case of M_{max} and $R_{1.4}$. The spread in the values of R_{max} is only $0.2 \pm 0.1km$ for a fixed value of M_{max}. The values of R_{max} do not depend strongly on the choice of Δr as can be seen from Table 2. The horizontal line in the lower panel corresponds to the measured value of the redshift, $Z = 0.35$, for the neutron star EXO 0748-676 [6]. For $Z = 0.35$, we find that the M_{max} is $\sim 1.8 M_{\odot}$. It must be pointed out that the central densities in our case are in the range of $4 - 5\rho_0$ for neutron star with mass $\sim 1.8 M_{\odot}$.

4   Conclusions

The parameters of the FTRMF model are properly calibrated to obtain different parameterizations. These model parameters are so calibrated that most of the nuclear matter properties at saturation density, the quality of fit to the finite nuclei and hyperon-nucleon potentials are very much the same for each combinations of

Table 2: Values of the maximum neutron star mass and radius for neutron stars with canonical and maximum masses obtained with the parameter sets for some selected combinations of Δr, ζ and $X_{\omega Y}$. The values of Δr are given in units of fm.

$X_{\omega Y}$	Δr	$\zeta = 0.00$			$\zeta = 0.06$		
		M_{max}	$R_{1.4}$	R_{max}	M_{max}	$R_{1.4}$	R_{max}
0.50	0.16	1.8	13.4	12.0	1.4	11.3	11.3
	0.28	1.8	14.1	12.2	1.4	11.6	11.6
0.65	0.16	1.9	13.4	12.0	1.4	12.0	11.2
	0.28	1.9	14.1	12.1	1.4	12.6	11.4
0.80	0.16	2.1	13.4	12.0	1.5	12.3	11.0
	0.28	2.1	14.1	12.1	1.5	13.0	11.3

Δr, ζ and $X_{\omega Y}$,

The maximum variations in the properties of the neutron stars are obtained by performing the calculations for the two selected parameter sets named LY and UY. The parameter sets LY(UY) are obtained using $\Delta r = 0.16 fm (0.28 fm)$, $\zeta = 0.06(0.0)$ and $X_{\omega Y} = 0.5(0.8)$. The LY parameter set yields the most softest EOS, whereas, the EOS obtained using the UY parameter set is the most stiffest one among all different combinations of Δr, ζ and $X_{\omega Y}$ considered. We find that the maximum neutron star mass can vary between $1.4 - 2.1 M_{\odot}$. The radius for neutron star with the canonical mass can vary in the range of $11.3 - 14.1$ km. For the comparison, we also present our results for the L0 and U0 parameter sets analogous to LY and UY parameterizations, but, with no hyperons. In this case we find that maximum neutron star mass and the radius for neutron star with the canonical mass lie in the range of $1.7 - 2.4 M_{\odot}$ and $12.5 - 14.1$ km, respectively. For a fixed value of maximum mass, the radius for neutron star with the canonical mass can be predicted within 0.7 ± 0.1 km. The radius for canonical neutron star mass is found to be at least 12.8 km provided, one considers only those parameterizations for which the maximum mass is larger than $1.6 M_{\odot}$ which is the highest mass measured for PSR J0751+1807 with 95% confidence limit[4].

References

[1] S. E. Thorsett and D. Chakrabarty, Astrophys. J **512**, 288 (1999).

[2] A. Lyne, et al., Science **303**, 1153 (2004).

[3] J. M. Lattimer and M. Prakash, asto-ph/06

[4] D. Nice et al., Astrophys. J. **634**, 1242 (2005).

[5] R. Rutledge, et al., Astrophys. J. **578**, 405 (2002); **577**, 346 (2002).

[6] F. Ozel, Nature 441 1115 (2006).

[7] C. O. Heinke, et al., Astrophys. J. **644**, 798 (2006).

[8] A. Akmal, V. R. pandharipande and D.G. Ravenhall, Phys. Rev. C**58**, 1804 (1998).

[9] F. Douchin and P. Haensel, Astrophys and Arstono. **380**, 151 (2001).

[10] B. K. Agrawal, S. K. Dhiman, and R. Kumar, Phys. Rev. C **73**, 034319 (2006).

[11] N. K. Glendenning and J. Schaffner-Bielich, Phys. Rev. C**60**, 025803 (1999).

[12] S.K. Dhiman, B.K. Agrawal, and R. Kumar, Phys. Rev. C (submitted).

[13] L. Engvik, E. Osnes, M. Hjorth-Jensen, G. Bao, and E.Ostgaard, Astrophys. J., **469**, 794 (1996).

[14] B. D. Serot and J. D. Walecka, Int. J. Mod. Phys. E6, 515, (1997).

[15] R. Kumar, B. K. Agrawal, and S. K. Dhiman, Phys. Rev. C **74**, 034323 (2006).

[16] G. Baym, C. Pethick, and P. Sutherland, Astrophys. J. **170**, 299, (1971).

[17] J. M. Lattimer and M. Prakash, Astrophys. J. **550**, 426 (2001).

[18] J. M. Lattimer, C. J. Pethick, M. Prakash, and P. Haensel, Phys. Rev. Lett. **66**, 2701 (1991).

[19] P. Danielewacz, R. Lacey, and G. lynch, Science **298**, 1592 (2002).

Physics and Astrophysics of Hadrons and Hardronic Matter
Editor: A. B. Santra

Effect of Hyperons on Nuclear Equation of State and Neutron Star structure

T. K. Jha[a]*, P. K. Raina[a], P. K. Panda[b], S. K. Patra[b]

a) *Indian Institute of Technology, Kharagpur, India - 721302*
b) *Institute of Physics, Bhubaneswar, India - 751005*
* email: tkjha@phy.iitkgp.ernet.in

Abstract

We study the effect of hyperons on nuclear equation of state (EOS) for dense matter in the core of the compact star and calculate the global properties of compact stars in an effective model in the mean field approach. With varying incompressibility and effective nucleon mass, we analyse the resulting EOS with hyperons and and the gross properties of the compact star sequences. The maximum mass of the compact star lies in the range $1.21 - 1.96$ $M_\odot$ for five different EOS obtained in the model.

1   Introduction

Neutron stars are the most compact objects found in the universe, which represents the ned point of the life cycle of stellar evolution. Thus we need to study dense matter properties and behaviour which not only deals with astrophysical problems such as the evolution of neutron stars, the supernovae mechanism but also reviews the implications from heavy-ion collisions.

Neutron stars are not purely composed of neutrons, but for the β-stability conditions, neutron star is much closer to neutron matter than the symmetric nuclear matter [1]. However, with increasing densities, the fermi energy of the occupied baryon states reaches eigenenergies of other species such as $\Lambda^0(1116)$, $\Sigma^{-,0,+}(1193)$ and $\Xi^{-,0}(1318)$ and the possibility of these hyperonic states are speculated in the dense core of neutron stars ([2]-[4]). Theoretically also, it has been found that the inclusion of hyperons in neutron star cores lowers the energy and pressure of the system resulting in the lowering of the maximum mass of neutron stars, in the range of observational limits.

Presently we apply an effective hadronic model to study the equation of state (EOS) for neutron star matter in the mean-field framework. Along with non-linear terms, which ensure reasonable saturation properties of nuclear matter, the model embodies dynamical generation of the vector meson mass that ensures a reasonable incompressibility. Therefore, one of the motivation for the present study is to check the applicability of the model to the study of high density matter. Secondly, the parameter sets of the model are in accordance with recently obtained heavy-ion data [5] which allows us to correlate the implications of heavy-ion collision data and the underlying dense matter equation of state.

we start with a brief description of the hadronic model that we implement in our calculations and then then look at the gross properties of the neutron stars in comparison with the predictions from other field theoretical models and experimental observations. Some constraints on the mass and radius of the star will also be discussed.

2 The equation of state

We start with an effective Lagrangian generalized to include all the baryonic octets interacting through mesons [6]:

$$
\mathcal{L} = \bar{\psi}_B \left[(i\gamma_\mu \partial^\mu - g_{\omega B} \gamma_\mu \omega^\mu - \frac{1}{2} g_{\rho B} \vec{\rho}_\mu \cdot \vec{\tau} \gamma^\mu) - g_{\sigma B} \left(\sigma + i\gamma_5 \vec{\tau} \cdot \vec{\pi} \right) \right] \psi_B
$$
$$
+ \frac{1}{2} \left(\partial_\mu \vec{\pi} \cdot \partial^\mu \vec{\pi} + \partial_\mu \sigma \partial^\mu \sigma \right) - \frac{\lambda}{4} \left(x^2 - x_0^2 \right)^2 - \frac{\lambda B}{6} \left(x^2 - x_0^2 \right)^3 - \frac{\lambda C}{8} \left(x^2 - x_0^2 \right)^4
$$
$$
- \frac{1}{4} F_{\mu\nu} F_{\mu\nu} + \frac{1}{2} g_{\omega B}^2 x^2 \omega_\mu \omega^\mu - \frac{1}{4} \vec{R}_{\mu\nu} \cdot \vec{R}^{\mu\nu} + \frac{1}{2} m_\rho^2 \vec{\rho}_\mu \cdot \vec{\rho}^\mu . \tag{1}
$$

Here $F_{\mu\nu} \equiv \partial_\mu \omega_\nu - \partial_\nu \omega_\mu$ and $x^2 = \vec{\pi}^2 + \sigma^2$, ψ_B is the baryon spinor, $\vec{\pi}$ is the pseudoscalar-isovector pion field, σ is the scalar field. The subscript $B = n, p, \Lambda, \Sigma$ and Ξ, denotes for baryons. In this model for hadronic matter, the baryons interact via the exchange of the σ, ω and ρ-meson. The Lagrangian includes a dynamically generated mass of the isoscalar vector field, ω_μ, that couples to the conserved baryonic current $j_\mu = \bar{\psi}_B \gamma_\mu \psi_B$. Here we shall be concerned only with the normal non-pion condensed state of matter, so we take $\vec{\pi} = 0$ and also the pion mass $m_\pi = 0$. The interaction of the scalar and the pseudoscalar mesons with the vector boson generate the mass through the spontaneous breaking of the chiral symmetry. Then the masses of the baryons, scalar and vector mesons, which are generated through x_0, are respectively given by

$$
m_B = g_{\sigma B} x_0, \quad m_\sigma = \sqrt{2\lambda} x_0, \quad m_\omega = g_{\omega B} x_0 . \tag{2}
$$

In the above, x_0 is the vacuum expectation value of the σ field, $\lambda = (m_\sigma^2 - m_\pi^2)/(2f_\pi^2)$, with m_π, the pion mass and f_π the pion decay constant, and $g_{\omega B}$ and $g_{\sigma B}$ are the coupling constants for the vector and scalar fields, respectively. In the mean-field treatment we ignore the explicit role of π mesons.

The equation of motion for the scalar field is given by,

$$
\sum_B \left[(1 - Y^2) - \frac{B}{c_{\omega B}} (1 - Y^2)^2 + \frac{C}{c_{\omega B}^2} (1 - Y^2)^3 + \frac{2c_{\sigma B} c_{\omega B} \rho_B^2}{m_B^2 Y^4} - \frac{2c_{\sigma B} \rho_{SB}}{m_B Y} \right] = 0 ,
$$
$$
\tag{3}
$$

where the effective mass of the baryonic species is $m_B^* \equiv Y m_B$ and $c_{\sigma B} \equiv g_{\sigma B}^2 / m_\sigma^2$

and $c_{\omega B} \equiv g_{\omega B}^2/m_\omega^2$ are the usual scalar and vector coupling constants respectively. The equation of motion for the ω field is then calculated as

$$\omega_0 = \sum_B \frac{\rho_B}{g_{\omega B}x^2} , \tag{4}$$

The quantity k_B is the Fermi momentum for the baryon and γ is the spin degeneracy. Similarly, the equation of motion for the ρ−meson is obtained as:

$$\rho_{03} = \sum_B \frac{g_{\rho B}}{m_\rho^2} I_{3B}\rho_B , \tag{5}$$

where I_{3B} is the 3rd-component of the isospin of the baryon species. For a particular baryon, the scalar density (ρ_{SB}) and the baryon density (ρ_B) are,

$$\rho_{SB} = \frac{\gamma}{(2\pi)^3} \int_o^{k_B} \frac{m_B^* d^3k}{\sqrt{k^2 + m_B^{*2}}}, \tag{6}$$

$$\rho_B = \frac{\gamma}{(2\pi)^3} \int_o^{k_B} d^3k, \tag{7}$$

Then the total energy density ε and pressure P for a given baryon density is:

$$\begin{aligned}
\varepsilon &= \frac{2}{\pi^2} \int_0^{k_B} k^2 dk \sqrt{k^2 + m_B^{*2}} + \frac{m_B^2(1 - Y^2)^2}{8c_{\sigma B}} - \frac{m_B^2 B}{12c_{\omega B}c_{\sigma B}}(1 - Y^2)^3 \\
&+ \frac{m_B^2 C}{16c_{\omega B}^2 c_{\sigma B}}(1 - Y^2)^4 + \frac{1}{2Y^2}c_{\omega B}\rho_B^2 + \frac{1}{2}m_\rho^2\rho_{03}^2 \\
&+ \frac{1}{\pi^2} \sum_{\lambda=e,\mu^-} \int_0^{k_\lambda} k^2 dk \sqrt{k^2 + m_\lambda^2} ,
\end{aligned} \tag{8}$$

$$\begin{aligned}
P &= \frac{2}{3\pi^2} \int_0^{k_B} \frac{k^4 dk}{\sqrt{k^2 + m_B^{*2}}} - \frac{m_B^2(1 - Y^2)^2}{8c_{\sigma B}} + \frac{m_B^2 B}{12c_{\omega B}c_{\sigma B}}(1 - Y^2)^3 \\
&- \frac{m_B^2 C}{16c_{\omega B}^2 c_{\sigma B}}(1 - Y^2)^4 + \frac{1}{2Y^2}c_{\omega B}\rho_B^2 + \frac{1}{2}m_\rho^2\rho_{03}^2 \\
&+ \frac{1}{3\pi^2} \sum_{\lambda=e,\mu^-} \int_0^{k_\lambda} \frac{k^4 dk}{\sqrt{k^2 + m_\lambda^2}}
\end{aligned} \tag{9}$$

Computing the energy density and the pressure for the neutron star sequences, we calculate the properties of neutron stars.

Table 1: Parameter sets for the model.

set	$c_{\sigma N}$ (fm^2)	$c_{\omega N}$ (fm^2)	B (fm^2)	C (fm^4)	K (MeV)	$m_N^\star/m_N$
I	8.86	1.99	-12.24	-31.59	210	0.85
II	6.79	1.99	-4.32	0.165	300	0.85
III	5.36	1.99	1.13	22.01	380	0.85
IV	8.5	2.71	-9.26	-40.73	300	0.80
V	2.33	1.04	9.59	46.99	300	0.90

The equations for the structure of a relativistic spherical and static star composed of a perfect fluid were derived from Einstein's equations by Oppenheimer and Volkoff [7] which are,

$$\frac{dp}{dr} = -\frac{G}{r} \frac{[\varepsilon + p]\,[M + 4\pi r^3 p]}{(r - 2GM)}, \tag{10}$$

$$\frac{dM}{dr} = 4\pi r^2 \varepsilon, \tag{11}$$

with G as the gravitational constant and $M(r)$ as the enclosed gravitational mass. We have used $c = 1$. Given an EOS, these equations can be integrated from the origin as an initial value problem for a given choice of central energy density, (ε_c). The value of r $(= R)$, where the pressure vanishes defines the surface of the star.

3   Results and discussion

The parameter sets (Table I) satisfies the nuclear saturation properties, E_B, energy per nucleon, -16 MeV at saturation density $0.153\ fm^{-3}$, effective nucleon Landau mass $0.8 - 0.9\ m_N$, incompressibility, and asymmetry energy coefficient value (≈ 32 MeV for $c_{\rho N} = 4.66$ fm^2).

The EOS at high density is very sensitive to the underlying hyperon couplings and is in turn reflected in the structural properties of the compact stars [3, 9]. In our present work, we take $x_\sigma = g_{\sigma H}/g_{\sigma N} = 0.7$, $x_\omega = g_{\omega H}/g_{\omega N} = 0.783$ and $x_\omega = x_\rho$, to calculate the EOS for the neutron star matter. Here, binding of Λ^0 in nuclear matter: $(B/A)_\Lambda = x_\omega g_\omega \omega_0 + m_\Lambda^\star - m_\Lambda \approx -30$ MeV.

Figure 1(Left) displays the equation of state for the five parameter sets. From the figure, it is to be noted that parameter set I, II, and III follows similar trend upto ten times normal nuclear matter density, whereas sets IV and V represents the stiff and soft character of the EOS respectively, which can be attributed to their different effective mass values. For all the cases, we find the dip in the curve at $\varepsilon \approx$ 1.5-2 fm^{-4}, which is the signature of the appearance of first members of the

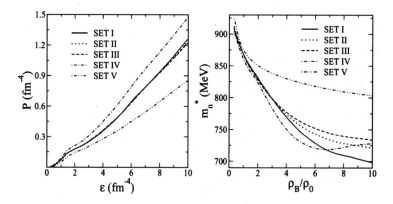

Figure 1: (Left): Equation of state (P vs ε) of neutron star matter with hyperons for the parameter sets listed in Table 1. (Right): Effective nucleon mass as a function of baryon density in the neutron star matter upto $10\rho_0$.

hyperons family namely Λ^0 and Σ^- states. Similar feature is noticeable in most of the other relativistic field theoretical models [11].

The nucleon effective mass ($m_N^\star \equiv Y m_N$) as a function of baryon density upto ten times normal nuclear matter density is displayed in figure 1(Right). Set I, II, III and IV follows the same trend till 6-7ρ_0 but then the mass increases slowly in case of Set IV where the strong repulsive component is responsible for the feature. In case of set V the decrease is much slower after a steep decrease till $2\rho_0$. This gradual decrease is as a result of strong scalar component in set V.

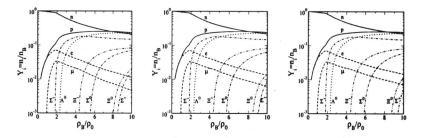

Figure 2: Relative particle population for neutron star matter with hyperons as a function of baryon density for parameter set I(Left), II(Middle), III(Right).

Fig. 2 displays the relative particle composition of neutron star matter for parameter sets I, II and III, which varry in incompressibility. From the figure it is noticeable that the difference in incompressibility does not affect the dense matter composition very much. The relative particle composition of neutron star matter

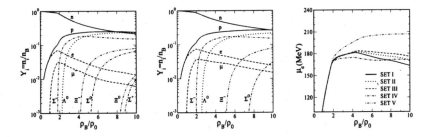

Figure 3: Relative particle population for neutron star matter with hyperons as a function of baryon density for parameter set IV (Left) and V (Middle). (Right): Electron chemical potential plotted as a function of baryon density.

for parameter sets IV and V is displayed in Fig. 3. In all the cases, the hyperons start appearing at around $2\rho_0$, where Σ^- appears first, closely followed by Λ^0. At higher densities other baryon thresholds are attained and they also start appearing. It can be seen that the difference in effective mass is very much pronounced and is reflected in the respective particle composition of the neutron star matter. Set V don't predict the presence of Σ^+ state even upto $10\rho_0$. At higher density hyperons forms a sizable population in neutron star matter. The electron chemical potentials for the all sets are displayed in Figure 3(Right). Leptons primarily gets used up in maintaining the charge neutrality of the neutron star matter which is the cause for the rapid deleptonisation in dense matter composition. Since parameter set V does not predict any charged hyperon species after $5\rho_0$, the electron chemical potential remains constant thereafter.

Figure 4(Left) shows the maximum baryonic mass M_b ($M_\odot$) obtained as a function of star mass for the five parameter sets. The difference between the two is defined as the gravitational binding of the star. Gravitational mass of the neutron star as a function of central density of the star is plotted in Figure 4(Right). Beyond the maximum mass, gravity overcomes and results in the collapse of the star. Set I, II and III predicts almost same central density for the star at maximum mass (denoted by filled circles in the plot). Recent observations of neutron star masses ([12]-[16]) predicts massive stars. Our results agrees remarkably with these observed masses except for set V.

Figure 5(Left) displays the maximum mass of the neutron star as a function of the star radius. In order to calculate the radius, we included the results of Baym, Pethick and Sutherland [17] EOS at low baryonic densities. These results are quite comparable to other works in the relativistic regime, like the non-linear walecka model (NLWM) and the quark-meson coupling model [18] predictions are 1.90 $M_\odot$ and 1.98 $M_\odot$ respectively.

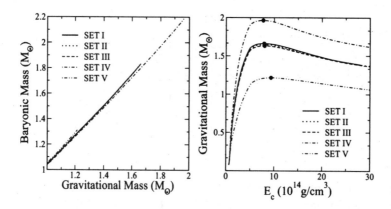

Figure 4: (Left): Baryonic mass ($M_\odot$) of the star as a function of Maximum mass ($M_\odot$) for the five sets. (Right): Maximum mass of the neutron star sequences as a function of central density of the star (in $10^{14} gcm^{-3}$).

Table 2: Properties of Neutron star as predicted by the model

SET	$M(M_\odot)$	$E_c(10^{14}gcm^{-3})$	R (Km)	$M_b(M_\odot)$	Z
I	1.66	7.90	16.78	1.83	0.19
II	1.65	7.99	16.70	1.81	0.19
III	1.63	7.99	16.62	1.79	0.19
IV	1.96	7.72	17.44	2.18	0.22
V	1.21	9.34	15.03	1.31	0.15

Recently a constraint to M-R plane was reported [19] based on the observation of two absorption features in the source spectrum of the IE 1207.4-5209 neutron star, which limits M/R=(0.069 - 0.115) $M_\odot$/km. The region enclosed in figure 5(Left) by two solid lines denotes the area enclosed in accordance with the observed range. All the parameter set of the present model satisfies the criterian very well. Another important aspect of compact stars are the observed gravitational redshift which is given by

$$Z = \frac{1}{\sqrt{1 - 2GM/Rc^2}}. \tag{12}$$

The gravitational redshift interpreted by the M/R ratio for the isolated neutron star comes lies in the range Z=0.12-0.23, which is plotted in figure 5(Right). For

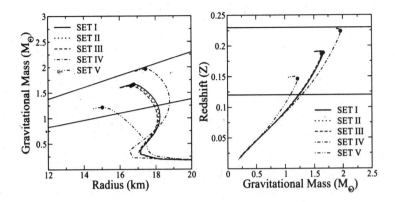

Figure 5: (Left): Maximum mass of the neutron star (in solar mass) as a function of radius (in Km) for the five Sets. The two solid lines corresponds to $M/R = 0.069$ and $M/R = 0.115$. (Right): Gravitational Redshift (Z) as a function of Maximum mass of the neutron star for the five parameter sets. (The solid points represent the values at maximum mass). The area enclosed between solid horizontal lines represents the redshift values $Z = (0.12 - 0.23)$ [19].

all the parameter sets, the redshift obtained at maximum mass lies in the range $(0.15 - 0.22)$, which corresponds to $R/M = (8.8 - 14.2)$ km/$M_\odot$. Our calculations predicts R/M in the range $(8.90 - 12.40)$ km/$M_\odot$, which is consistent with the observed value.

The overall results of our calculation are presented in Table 2.

4 Summary and outlook

We studied the equation of state of high density matter in an effective model and calculated the gross properties for neutron stars like mass, radius, central density and redshift. We analysed five set of parameters with incompressibility values K=210, 300 and 380 MeV and effective masses $m^\star$ =0.80, 0.85 and 0.90 m_n. It was found that the difference in nuclear incompressibility is not much reflected in either equation of state or neutron star properties, but nucleon effective masses were quite decisive. The maximum mass obtained for the the five EOS lies in the range 1.21-1.96$M_\odot$ [6].

The results were also found to be in good agreement with recently imposed constrains on neutron star properties in the M–R plane, and the redshift interpreted therein. In case of radii, the values are still unknown, however some estimates are expected in a few years, which would further constrain the EOS of neutron star in the M-R plane based on observed neutron star mass, the redshift measurements and also constraints from URCA cooling mechanism. In future, we intend to study the effect of rotation to neutron star structure and also the phase transition aspects in the model.

References

[1] S.L. Shapiro and S.A. Teukolski, *Black holes, white dwarfs, and Neutron stars* (Wiley, New York, 1983).

[2] N.K. Glendening, Phys. Lett. **B114**, 392 (1982); N. K. Glendening, Astrophys. J. **293**, 470 (1985); N. K. Glendening, Z. Phys. **A 326**, 57 (1987).

[3] M. Prakash, I. Bombaci, M. Prakash, P.J. Ellis, J.M. Lattimer and R. Knorren, Phys. Rep. **280**, 1 (1997).

[4] J. Schaffner-Beilich and I.N. Mishustin, Phys. Rev. **C 53**, 1416 (1996).

[5] P. Danielewicz, R. Lacey, W.G. Lynch, Science **298**, 1592 (2002).

[6] T. K. Jha, P. K. Raina, P. K. Panda and S. K. Patra, Phys. Rev. **C 74**, 055803 (2006).

[7] J.R. Oppenheimer and G.M. Volkoff, Phys. Rev **55**, 374 (1939); R.C. Tolman, Phys. Rev **55**, 364 (1939).

[8] P. Möller, W.D. Myers, W.J. Swiatecki and J. Treiner, At. Data Nucl. Data Tables **39**, 225 (1988).

[9] N.K. Glendening Phys. Rev. **C 64**, 025801 (2001).

[10] S.A. Moszkowski, Phys. Rev. **D 9**, 1613 (1974).

[11] A. Mishra, P.K. Panda and W. Greiner, J. Phys. **G 28**, 67 (2002).

[12] D.J. Nice, E.M. Spalver, I.H. Stairs, O. Loehmer, A. Jessner, M. Kramer and J.M. Cordes, Astrophys. J. **634** 1242 (2005).

[13] D. Barret, J.F. Olive and M.C. Miller, *astro-ph/0605486*.

[14] O. Barziv, L. Kaper, M.H. van Kerkwijk, J.H. Telting and J. van Paradijs, Astron. & Astrophys. **377** 925 (2001).

[15] J. Casares, P.A. Charles and E. Kuulkers, Astro. J. **493** L39 (1998).

[16] J.A. Orosz and E. Kuulkers, Mon. Not. R. Astron. Soc. **305** 132 (1999).

[17] G. Baym, C. Pethick and P. Sutherland, Astrophys. J. **170**, 299 (1971).

[18] P.K. Panda, D.P. Menezes, C. Providñcia, Phys.Rev. **C69**, 025207 (2004); P.K. Panda, D.P. Menezes, C. Providñcia, Phys.Rev. **C69**, 058801 (2004); D.P. Menezes, P.K. Panda and C. Providência, Phys. Rev. **C 72**, 035802 (2005).

[19] D. Sanyal, G.G. Pavlov, V.E. Zavlin and M.A. Teter, Astrophys. J. **574** L61 (2002).

Physics and Astrophysics of Hadrons and Hardronic Matter
Editor: A. B. Santra

Two Step Conversion of Neutron Star to Strange Star

Abhijit Bhattacharyya[a]*, Sanjay K. Ghosh[b,c], Partha S. Joarder[c]
Ritam Mallick[b], Sibaji Raha[b,c]

a) *Department of Physics, University of Calcutta,*
92, A.P.C Road; Kolkata - 700009; INDIA
b) *Centre for Astroparticle Physics and Space Science, Bose Institute*
93/1, A.P.C Road, Kolkata - 700009; INDIA
c) *Department of Physics, Bose Institute*
93/1, A. P. C. Road; Kolkata - 700 009, INDIA
* email: abphy@caluniv.ac.in

Abstract

Neutron star is envisaged to be a laboratory for studying the phase transition process in strongly interacting matter. Such transition may be viewed as a two step process in which the hadronic matter first gets converted to a two-flavour quark matter that, in turn, converts to a strange quark matter in the second step of the process. Here a preliminary study of such phase transition in neutron stars has been presented.

1 Introduction

The strange quark matter (SQM) is conjectured to be the true ground state of strongly interacting matter [1] at high density and/or temperature. Normal nuclear matter at high enough density and/or temperature, would be unstable against conversion to two-flavor quark matter. The two-flavor quark matter would be metastable and would eventually decay to Strange Quark Matter (SQM) in a weak interaction time scale. Such a two step conversion process may take place in the interior of a neutron star where the densities can be as high as $(8\text{-}10)\rho_0$ with ρ_0 being the nuclear matter density at saturation [2].

There are several ways in which conversion may be triggered at the center of the star. Alcock *et al* [3] proposed that as the star comes in contact with a seed of external strange quark nugget the seed grows by 'eating up' baryons. Glendenning [4] suggested that a sudden spin down of the star may increase the density at its core thereby triggering the conversion process spontaneously. Olinto [5] viewed the conversion process to proceed via weak interactions as a propagating slow-combustion (i.e. a deflagration) front and derived the velocity of such a front. Olesen and Madsen [6] and Heiselberg *et al* [7] estimated the speed of such conversion front to range between 10 m/s to 100 km/s. Collins and Perry [8] assumed that the hadronic matter gets converted first to a two-flavor quark matter that eventually decays to a three-flavor strange matter through weak interactions. Lugones *et al* [9] argued that the hadron to SQM conversion process may rather proceed as a detonation than as a deflagration even in the case of strangeness production occurring through seeding mechanisms [3]. Horvath and Benvenuto [10] inferred that a convective instability may increase the velocity of the deflagration front, so that a transition from slow combustion to detonation may occur which may as well be responsible for the type II supernova explosions [11]. Recently, Tokareva *et al* [12] argued that the mode of conversion would vary with temperature of SQM and with the value of bag constant in the Bag model Equation of State (EOS).

Here we model the conversion of nuclear matter to SQM in a neutron star as occurring through a two step process. Deconfinement of nuclear matter to a two (up and down) flavor quark matter takes place in the first step in strong interaction time scale. The second step concerns with the generation of strange quarks from the excess of down quarks via a weak process. The paper is organized as follows. In section 2, we discuss the conversion to two-flavor quark matter. Conversion to three-flavor SQM is discussed in section 3. In section 4, we summaries the results.

2 Conversion to two-flavor matter

In order to describe the conversion of nuclear matter to two flavour quark matter we use non-linear Walecka model [13] for the hadronic matter and Bag model for the quark matter. Here we consider the conversion of nuclear matter consisting of only nucleons (*i.e.* without hyperons) to a two-flavor quark matter. The final composition of the quark matter is determined from the nuclear matter EOS by enforcing the baryon number conservation during the conversion process. That is, for every neutron two down and one up quarks and for every proton two up and one down quarks are produced, electron number being same in the two phases. While describing the state of matter for the quark phase we consider a range of values for the bag constant.

We heuristically assume the existence of a combustive phase transition front in this paper. Using the macroscopic conservation conditions, we examine the range of densities for which such a combustion front exists. We next study the outward propagation of this front through the model star by using the hydrodynamic (*i.e.* Euler) equation of motion and the equation of continuity for the energy density flux [14]. In this study, we consider a non-rotating, spherically symmetric neutron star. The geometry of the problem effectively reduces to a one dimensional geometry for which radial distance from the center of the model star is the only independent spatial variable of interest.

Let us now consider the physical situation where a combustion front has been generated in the core of the neutron star. This front propagates outwards through the neutron star with a certain hydrodynamic velocity, leaving behind a u–d–e matter. In the following, we denote all the physical quantities in the hadronic sector by subscript 1 and those in the quark sector by subscript 2.

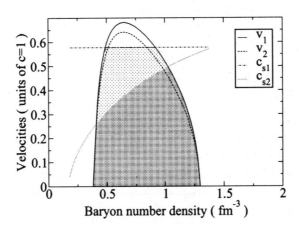

Figure 1: Variation of different velocities with baryon number density for $B^{1/4} = 160$ MeV and strange quark mass $m_s = 200$ MeV. The dark-shaded region correspond to deflagration, light-shaded region correspond to detonation and the unshaded region correspond to supersonic conversion processes.

Condition for the existence of a combustion front is given by [2]

$$\epsilon_2(p, X) < \epsilon_1(p, X), \tag{1}$$

where p is the pressure and $X = (\epsilon + p)/n_B^2$, n_B being the baryon density.

The velocities of the matter in the two phases are written as [14]:

$$v_1^2 = \frac{(p_2 - p_1)(\epsilon_2 + p_1)}{(\epsilon_2 - \epsilon_1)(\epsilon_1 + p_2)}, \qquad v_2^2 = \frac{(p_2 - p_1)(\epsilon_1 + p_2)}{(\epsilon_2 - \epsilon_1)(\epsilon_2 + p_1)}. \tag{2}$$

It is possible to classify the various conversion mechanisms by comparing the velocities of the respective phases with the corresponding velocities of sound, denoted by c_{si}, in these phases. These conditions may be obtained in ref.[2]. For the conversion to be physically possible, velocities should satisfy an additional condition, namely, $0 \leq v_i^2 \leq 1$. We here find that the velocity condition, along with the eq.(1), puts severe constraint on the allowed equations of state.

To examine the nature of the hydrodynamical front, arising from the neutron to two-flavor quark matter conversion, we plot, in fig.1, the quantities v_1, v_2, c_{s1} and c_{s2} as functions of the baryon number density (n_B). We find that the energy condition and velocity condition ($v_i^2 > 0$) both are satisfied only for a small window of $\approx \pm 5.0$ MeV around the bag pressure $B^{1/4} = 160$ MeV. The constraint imposed by the above conditions results in the possibility of deflagration, detonation or supersonic front as shown in the figure 1.

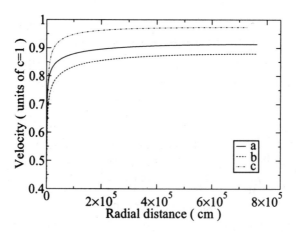

Figure 2: Variation of velocity of the conversion front with radius of the star for three different values of the central densities, namely, (a) $4.5\rho_0$, (b) $3\rho_0$ and (c) $7\rho_0$, respectively. Here ρ_0 is the nuclear density.

To examine the evolution of a combustion front, we move to a reference frame in which the nuclear matter is at rest. The speed of the combustion front in such a frame is given by $v_f = -v_1$ with v_1 being the velocity of the nuclear matter in the rest frame of the front. The relevant equations are the equation of continuity and the Euler's equation. These equations are solved for a static star configuration. The solution gives us the variation of the velocity with the position as a function of time of arrival of the front, along the radius of the star. Using this velocity profile, we can calculate the time required to convert the whole star using the relation $v = dr/d\tau$.

Figure 2, shows the variation of the velocity with radius of the star for three values of central densities. The respective initial velocities corresponding to such central densities are taken to be 0.66, 0.65 and 0.47. The figure shows that the velocity of the front, for all the central densities, shoots up near the center and then saturates at a certain velocity for higher radius. The numerically obtained saturation velocity varies from 0.92 for central baryon density $3\rho_0$ to 0.98 for $7\rho_0$. A comparison with fig.1 shows that for the densities $3\rho_0$ and $4.5\rho_0$, the conversion starts as weak detonation and stays in the same mode throughout the star. On the other hand, for $7\rho_0$, initial detonation front changes over to weak detonation and the velocity of front becomes almost 1 as it reaches the outer crust. The corresponding time taken by the combustion front to propagate inside the star is plotted against the radius in figure 3. The time taken by the front to travel the full length of the star is of the order of few milliseconds. According to the present model, the initial neutron star thus becomes a two-flavor quark star in about 10^{-3} sec.

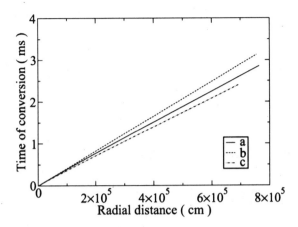

Figure 3: Variation of the time of arrival of the conversion front at a certain radial distance inside the star as a function of that radial distance from the center of the star for three different central densities. The notations are same as in figure 2

3   The conversion to three-flavor SQM

In this section we discuss the conversion of two-flavor quark matter to three-flavor SQM in a compact star. Similar to the discussion above, we assume the existence of a conversion front at the core of the star that propagates radially outward leaving behind the SQM as the combustion product.

The conversion to SQM starts at the center ($r = 0$) of the two-flavor star. As the combustion front moves radially towards the surface of the star excess d quarks get converted to s quark through the non leptonic weak process. We now define a quantity,

$$a(r) = [n_d(r) - n_s(r)]/2n_B \tag{3}$$

such that, $a(r = 0) = a_0$ at the core of the star. The quantity a_0 is the number density of the strange quarks, at the center, for which the SQM is stable and its value lies between 0 and 1. Ideally at the center of the star a_0 should be zero for strange quark mass $m_s = 0$. Since s quark has a mass $m_s \sim 150 - 200$ MeV, at the center of the star a_0 would be a small finite number, depending on the EOS. The s quark density fraction, however, decreases along with the decrease of the baryon density towards the surface of the star, so that, $a(r \rightarrow R) \rightarrow 1$ with R being the radius of the star. At any point along the radius, say $r = r_1$, initial $a(r_1)$, before the arrival of the front, is decided by the initial two-flavor quark matter EOS. The final $a(r_1)$, after the conversion, is obtained from the equilibrium SQM EOS at the density corresponding to r_1. The conversion to SQM occurs via decay of down quark to strange quark and the diffusion of the strange quark across the front [5].

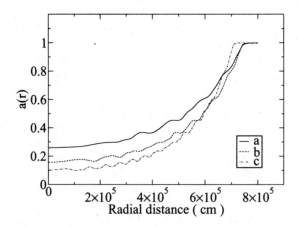

Figure 4: Variation of $a(r)$ (as in the text) with radius of star, for different central densities, where, (a) corresponds to case for which the central density is $3\rho_0$, (b) for $4.5\rho_0$ and (c) for $7\rho_0$.

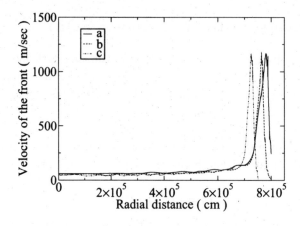

Figure 5: Variation of velocity of the two to three-flavor quark conversion front with radius of the star for different central densities, where, the notations are same as figure 2.

The corresponding rate of change of $a(r)$ with time is governed by following two equations:

$$\frac{da}{dt} = -R(a), \qquad \frac{da}{dt} = D\frac{d^2a}{dr^2}, \tag{4}$$

where $R(a)$ is the rate of conversion and D is the diffusion constant. Assuming one dimensional steady state solution these equations give:

$$Da'' - va' - R(a) = 0, \tag{5}$$

276

where v is the velocity and $a' = \frac{da}{dr}$. The subscripts q and s denote the two–flavor quark matter phase and SQM phase, respectively. The baryon flux conservation condition yields the initial boundary condition at any point r along the radius of the star:

$$a'(r) = -\frac{v}{D}(a_i(r) - a_f(r)), \tag{6}$$

where $a_i(r)$ and $a_f(r)$ are the values for the $a(r)$ before and after the combustion, respectively.

Our calculation proceeds as follows. First we get the star characteristics for a fixed central baryon density ρ_c. For a given ρ_c, number densities of u, d and s quarks, in both the sectors, are known at any point. That means $a_i(r)$ and $a_f(r)$ is fixed. Then one can obtain the diffusion constant and hence the radial velocity of the front [2].

The variation of $a(r)$ with the radius of the star is given in figure 4. The plot shows that $a(r)$ increases radially outward, which corresponds to the fact that as density decreases radially, the number of excess down quark which is being converted to strange quark by weak interaction also decreases. Hence, it takes less time to reach a stable configuration and hence the front moves faster, as shown in the figs.5 and 6. In fig.5, we have plotted the variation of velocity along the radius of the star. The velocity shows an increase as it reaches sufficiently low density and then drops to zero near the surface as $d \rightarrow s$ conversion rate becomes zero. Fig.6 shows the variation of time taken to reach a stable configuration at different radial position of the star. The total time needed for the conversion of the star, for different central densities, is of the order of 100 seconds, as can be seen from the figure.

4 Summary and discussion

We have studied the conversion of a neutron star to strange star. This conversion takes place in two stages. In the first stage a detonation wave is developed in the hadronic matter (containing neutrons, protons and electrons). We have described this hadronic matter with a relativistic model. For such an equation of state the density profile of the star is obtained by solving Tolman-Openheimer-Volkoff equations. The corresponding quark matter equation of state is obtained by using the bag model. However, this quark matter equation of state is not equilibrated and contains two flavors. Matter velocities in the two media, as measured in the rest frame of the front, have been obtained using conservation conditions. These velocities have been compared with the sound velocity in both phases.

For a particular density inside the star, flow velocities of the matter on the two sides of the front is now fixed. Starting from a point, infinitesimally close to the center, hydrodynamic equations are solved radially outward. The solution of the hydrodynamic equations gives the velocity profiles for different central densities. The velocity of the front shoots up very near to the core and then saturates at a value close to 1. The mode of combustion is found to be weak detonation for lower

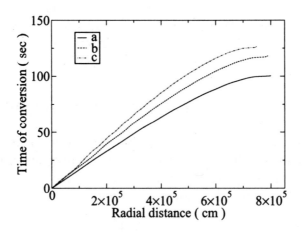

Figure 6: Variation of time taken for the two to three-flavor quark conversion front with radius of the star, for different central densities, where, the notations are same as figure 2.

central densities. For higher central densities, the initial detonation becomes weak detonation as the the front moves radially outward inside the star. This result is different from that of ref.[15], where the conversion process always correspond to a strong deflagration. The time required for the conversion of the neutron star to a two-flavor quark star is found to be of the order of few milliseconds. After this front passes through, leaving behind a two-flavor matter a second front is generated. This second front converts the two-flavor matter via weak interaction processes. The velocity of the front varies along the radius of the star. As the front moves out from the core to the crust, its velocity increases, implying faster conversion. The time for the second conversion to take place comes out to be ~ 100 seconds. This is comparable to the time scale obtained in ref.[5].

The comparison of the time of conversion from neutron star to two-flavor quark star and the weak interaction time scale suggests that at some time during the passage of the first combustion front, the burning of two-flavor matter to strange matter should start. This means that at some point of time, there should be two fronts moving inside the star. So there will be some interaction between the two fronts. Furthermore we have studied the burning of the nuclear matter to two-flavor quark matter using special relativistic hydrodynamic equations. The actual calculation should involve general relativity, taking into account the curvature of the front for the spherical star. We propose to explore all these detailed features in our subsequent papers.

Acknowledgements : R.M. would like to thank CSIR, Govt. of India for financial support. S.K.G., S.R. and P.S.J. thank DST, Govt. of India for support under the IRHPA scheme.

References

[1] E. Witten, Phys. Rev. D **30**, 272 (1984).

[2] A. Bhattacharyya, S. K. Ghosh, P. Joarder, R. Mallick and S. Raha, Phys. Rev. C **74** 065804 (2006).

[3] C. Alcock, E. Farhi and A. Olinto, Astrophys. J. **310**, 261 (1986).

[4] N. K. Glendenning, Nucl. Phys. B (Proc. Suppl.) **24**, 110 (1991); Phys. Rev. D **46**, 1274 (1992).

[5] A. Olinto, Phys. Lett. B **192**, 71 (1987); Nucl. Phys. B (Proc. Suppl.) **24**, 103 (1991).

[6] M. L. Olesen, J. Madsen, Nucl. Phys. B (Proc. Suppl.) **24**, 170 (1991).

[7] H. Heiselberg, G. Baym and C. J. Pethick, Nucl. Phys. B (Proc. Suppl.) **24**, 144 (1991).

[8] J. Collins and M. Perry, Phys. Rev. Lett. **34**, 1353 (1975).

[9] G. Lugones, O. G. Benvenuto and H. Vucetich, Phys. Rev. D **50**, 6100 (1994).

[10] J. E. Horvath and O. G. Benvenuto, Phys. Lett. B **213**, 516 (1988).

[11] O. G. Benvenuto, J. E. Horvath, and H. Vucetich, Int. J. Mod. Phys. A **4**, 257 (1989); O. G. Benvenuto and J. E. Horvath, Phys. Rev. Lett. **63**, 716 (1989).

[12] I. Tokareva, A. Nusser, V. Gurovich and V. Folomeev, Int. J. Mod. Phys. D **14**, 33 (2005).

[13] J. Ellis, J. I. Kapusta and K. A. Olive, Nucl. Phys. B **348**, 345 (1991).

[14] L. D. Landau and E. M. Lifshitz, *Fluid Mechanics*, Pergamon Press, New York (1987).

[15] A. Drago, A. Lavagno and G. Pagliara, Phys. Rev. D **69**, 057505 (2004).

Physics and Astrophysics of Hadrons and Hardronic Matter
Editor: A. B. Santra

Model Study of Speed of Sound and Susceptibilities in Quark-Gluon Plasma

Sanjay K. Ghosh[a]*, Tamal K. Mukherjee[b]

[a] Centre for Astroparticle Physics and Space Science
Bose Institute; 93/1, A.P.C Road, Kolkata - 700009; INDIA
[b] Department of Physics, Bose Institute
93/1, A.P.C Road, Kolkata - 700 009, INDIA
* email: sanjay@bosemain.boseinst.ac.in

Abstract

The speed of sound and susceptibilities have been studied in Density dependent quark mass model and PNJL model. The results are compared with the recent Lattice studies.

1   Introduction

The theory of strong interaction, Quantum chromodynamics (QCD) predicts two phase transitions above some critical value of density and/or temperature. These two phase transitions, namely the confinement-deconfinement transition and the chiral phase transition, are defined within two extreme quark mass limit. For infinite quark mass limit quarks in the hadron becomes deconfined giving rise to the confined to deconfined phase transition. On the other hand in the limit of vanishing quark mass we can have chiral phase transition which is basically chiral symmetry restoring transition. So, for the realistic value of quark masses one expects the two transitions to occur beyond a critical value of temperature and/or chemical potential. Whether these two transitions occur simultaneously or one preceeds the other remain an open question. While certain study [1, 2] predicts that the chiral transition occur after the deconfinement transition, the Lattice result on the other hand, finds that they occur simultaneously and in fact they are related in such a way that one drives the other.

Order of the phase transition is another debatable issue yet to be understood. In fact the order of the transition depends on the value of current quark masses, the number of colours and the number of flavours involved. Current understanding indicates that for low temperature and high baryonic chemical potential there is a first of order phase transition which ends at some critical end point (CEP) and beyond which for higher temperature and low baryonic chemical potential there is a rapid crossover. The exact location of the critical end point and the structure of the phase boundary line are currently being explored in the Lattice Gauge Theory (LGT) as well as in the effective models. But definite result are yet not been found.

Here we are going to discuss about the behaviour of sound velocity, quark number susceptibility (QNS), and some other thermodynamic variables, like specific heat and conformal measure, as the quark gluon plasma moves from the non-perturbative to the perturbative domain of QCD. These thermodynamic variables are related to various fluctuations, which can be measured in heavy-ion collisions. The study of fluctuations is important as the difference in fluctuations of various conserved quantities in the hadronic and deconfined phases is supposed to signal the phase transition between these two phases [3].

The speed of sound supplies the information about the functional relationship between the pressure and the energy density, known as the equation of state (EOS). The knowledge of the speed of sound can be used to determine the flow properties in heavy-ion collisions [4–7] as well as in the analysis of the rapidity distribution of the secondary particles [8]. QNS on the other hand, are related to charge fluctuations and susceptibility in general, are related to various fluctuations via the fluctuation-dissipation theorem. Another important aspect of the study of QNS is that they can be used as an independent check on the theoretical models which try to explain the Lattice results [9]. Specific heat (at constant volume) C_V is important as it is related to the event-by-event temperature fluctuations [10] and mean transverse momentum fluctuations [11].

Two types of models, the density dependent quark mass (DDQM) model [12, 13] and the polyakov loop + Nambu-Jona-Lasinio (PNJL) model [14] are used in this study. The DDQM model is basically one parameter model and the dependence of the computed quantities on the model parameter are found to be small. It is basically a quasiparticle model and the performance of the model is improved by incorporating the thermodynamic consistency as suggested by Birò et al. [15]. The other model, the PNJL model, is a polyakov loop embedded NJL model. Such a synthesis enabled us to study the chiral physics as well as the confinement-deconfinement physics within a single framework. While the polyakov loop acts as the order parameter for the deconfinement transition, the chiral condensate serve as the order parameter for the chiral transition.

The paper is organized as follows. In section 2, we discuss our study in the framework of thermodynamically consistent DDQM model. Section 3, discusses about the PNJL model and the calculation of quark number susceptibility along with other thermodynamic variables. Finally in section 4 we summarize our results.

2 DDQM model

We have studied the behaviour of QGP incorporating the thermodynamic consistency. The essence of DDQM model is to explain confinement from a phenomenological point of view. Confinement is achieved through a parameterization of the effective quark mass as a function of density. In the low density limit, the mass of the 'free' quark is made infinitely heavy (confinement) and all the interactions are taken care of by the mass.

Following [13], the quark mass at finite temperature is defined as,

$$m(T, n_q) = \frac{B}{\Sigma_i(n_i^+ + n_i^-)}, \tag{1}$$

The quark (antiquark) number density of the i-th flavour $n_i{}^+$ $(n_i{}^-)$ is given by,

$$n_i^{\pm} = \frac{3}{\pi^2} \int dp p^2 \{\exp[T^{-1}(\epsilon_i \mp \mu_i)] + 1\}^{-1}, \tag{2}$$

With,

$$\epsilon_i(p, n_q, T) = [p^2 + m^2(T, n_q) + \Delta_i(p)]^{1/2}. \tag{3}$$

Here ϵ_i and μ_i are the single particle energy and the chemical potential of the i-th flavour quark. Baryon number density, temperature and momentum of the quasiparticle are represented by n_q, T and p respectively. The term $\Delta_i(p)$ present in the expression of the single particle energy takes care of the divergent term in the mass at finite temperature. As mass takes care of all interactions at zero density limit, $\Delta_i(0) = 0$, in this limit.

The expression for $\Delta_i(p)$, is given by [13]:

$$d\Delta_i(p_{f,i}) \equiv \left\{ \frac{6B\pi^2 c_i}{p_{f,i}{}^4} \times \frac{[p_{f,i}{}^2 - (B\pi^2 c_i/p_{f,i}{}^3)^2]}{[p_{f,i}{}^2 + (B\pi^2 c_i/p_{f,i}{}^3)^2]^{1/2}} + \frac{6B^2\pi^4 c_i{}^2}{p_{f,i}{}^7} \right\} dp_{f,i}, \tag{4}$$

With $c_i = n_i/n_q$.

After defining the mass relation and the number density, all relevant thermodynamic quantities pertaining to the quark sector are calculated. Expression for the energy density and pressure in the gluonic sector are taken from ref.[13]. The equation of state for the QGP is then obtained by summing the contributions from the quark and the gluonic sector, i.e,

$$\mathcal{E}_{QGP} = \mathcal{E}_q + \mathcal{E}_g, P_{QGP} = P_q + P_g. \tag{5}$$

Expressions for the individual quantities are given below:

$$\mathcal{E}_q = \sum_{i=u,d} \frac{3}{\pi^2} \int_0^{\infty} dp[p^2 \epsilon_i(p, n_q, T)(\{\exp[T^{-1}(\epsilon_i - \mu_i)] + 1\}^{-1}$$

$$+ \{\exp[T^{-1}(\epsilon_i + \mu_i)] + 1\}^{-1})], \tag{6}$$

$$P_q = \frac{1}{3} \sum_{i=u,d} \frac{3}{\pi^2} \int_0^\infty p \frac{\partial \epsilon}{\partial p} d^3 p (\{\exp[T^{-1}(\epsilon_i - \mu_i)] + 1\}^{-1}$$

$$+ \{\exp[T^{-1}(\epsilon_i + \mu_i)] + 1\}^{-1}), \qquad (7)$$

$$\mathcal{E}_g = \frac{8}{15} \pi^2 T^4 (1 - \frac{15\alpha_c}{4\pi}), \qquad (8)$$
$$P_g = \frac{8}{45} \pi^2 T^4 (1 - \frac{15\alpha_c}{4\pi}).$$

Here, α_c is the effective gluon-gluon coupling. The form of α_c is taken from [13].

At low density and temperatures, energy density and the pressure becomes negative, implying confinement of gluons (gluon condensate). As long as gluons remain in the condensate, they do not contribute to the thermodynamic quantities (except for an additive constant) and we have, $\mathcal{E}_g = P_g = 0$ for $\alpha_c \geq \frac{4\pi}{15}$ [13]. Above condition implies a second order phase transition and thus we expect a finite discontinuity in the plot of velocity of sound with respect to temperature, marking the built in phase transition within the EOS.

After defining the EOS our next step is to incorporate the thermodynamic consistency into our model. To do this, we have to introduce a background field Φ. The thermodynamic consistency conditions in terms of the background field Φ [15] reads,

$$\frac{\partial \Phi}{\partial T} + \sum_{j=u,d} d_j \int \frac{d^3 k}{(2\pi)^3} \frac{\partial \epsilon_{kj}}{\partial T} \nu_{kj} = 0, \qquad (9)$$

$$\frac{\partial \Phi}{\partial n_i} + \sum_{j=u,d} d_j \int \frac{d^3 k}{(2\pi)^3} \frac{\partial \epsilon_{kj}}{\partial n_i} \nu_{kj} = 0, \qquad (10)$$

where, the subscript 'kj' stands for the j-th particle with momentum k, n_i is the i-th flavour quark number density, d_j is the degeneracy factor and ν_{kj} is the average quasiparticle occupation number, given by:

$$\nu_{kj} = (\{\exp[T^{-1}(\epsilon_j - \mu_j)] + 1\}^{-1} + \{\exp[T^{-1}(\epsilon_j + \mu_j)] + 1\}^{-1})]. \qquad (11)$$

In the present context above two consistency conditions translates into the following two equations:

$$\frac{\partial \Phi}{\partial T} + \sum_{j=u,d} n_j^{(s)} \frac{\partial m_j}{\partial T} = 0, \qquad (12)$$

$$\frac{\partial \Phi}{\partial n_i} + \sum_{j=u,d} n_j^{(s)} \frac{\partial m_j}{\partial n_i} = 0. \qquad (13)$$

Where $n_j^{(s)}$ is the scalar quasiparticle density of j-th flavour quark,

$$n_j^{(s)} = d_j \int \frac{d^3k}{(2\pi)^3} \nu_{kj} \frac{m_j}{\epsilon_j}. \tag{14}$$

In our present analysis we have considered a two flavour QGP system with equal masses of u and d quarks. Under this, our system contains only one type of quasiparticle (except for the degeneracy factor 2 for two flavours) and the expression for the background field Φ reduces to

$$\Phi = -\int \sum_{j=u,d} n_j^{(s)} dn_j \tag{15}$$

Carrying out the integration and identifying the integration constant with $\Phi = B$ at $T = 0$, $n_B = 0$, background field Φ can be obtained. At zero temperature, Φ naturally depends on the baryon density (n_B) alone. With this background field, we define thermodynamically consistent energy density and pressure as,

$$\mathcal{E}_{QGP} = \mathcal{E}_q + \mathcal{E}_g + \Phi(n_B, T) \tag{16}$$
$$P_{QGP} = P_q + P_g - \Phi(n_B, T) \tag{17}$$

at zero temperature, the total energy density and the pressure become B and zero respectively. Here 'B' is just a phenomenological parameter but not the bag pressure. The confinement is achieved in the DDQM model through infinitely heavy mass in the zero density limit and hence there is no bag pressure present in this model (which if present, would amount to double counting [12]). In fig.1 we have plotted variation of Φ with temperature at three different baryon number densities for $B^{1/4} = 145 MeV$. As expected, we see the effect of Φ is greater at lower densities. This is because as we go to the higher density regime we are more closer to the perturbative region of the QCD theory. From the graph, we see Φ changes rapidly with temperature near the phase transition region and then saturates at higher temperature. Hence, thermodynamic quantities are expected to be greatly modified from the contribution of Φ near the transition temperature. But at higher temperatures it merely gives an additive contribution. It is to be noted here that transition temperature changes with the change in density and the respective transition temperatures are given in caption of the figure. Here we have taken nuclear matter saturation density to be 0.17 fm^{-3}.

Next in figure 2 we have shown the plot for the scaled interaction measure $(\mathcal{E} - 3P)/T^4$ with T/T_c. Two curves are for with and without background field contribution. We see, with the inclusion of Φ two curves differ drastically near the transition temperature but at higher temperature they are almost equal. We know the background field mimics the nonperturbative effects and therefore with the in-

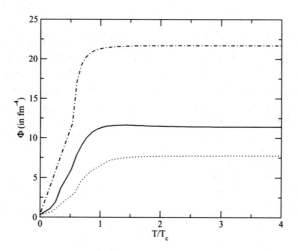

Figure 1: Variation of Φ with temperature for $B^{1/4}$=145 MeV. Dash-dot, solid and dotted curves are for densities n_B=0, n_B=3n_0 and n_B=4n_0 respectively, n_0 being the nuclear matter saturation density. Critical temperature for n_B=0 is 330 MeV, n_B=3n_0 is 290

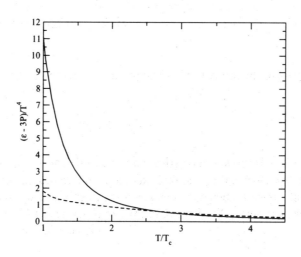

Figure 2: Variation of $(\mathcal{E}\text{-}3P)/T^4$ with temperature for $B^{1/4}$=145 MeV, n_B=3n_0 and $T_c = 290 MeV$. Solid and dash curves are for with and without Φ contributions respectively.

clusion of thermodynamic consistency nonperturbative effects are enhanced near the phase transition region.

Figures 3 and 4 are the variation of sound velocity as a function of T/T_c. Here

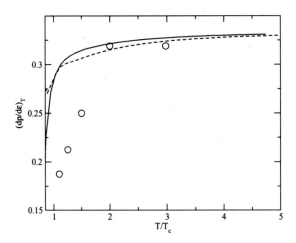

Figure 3: Variation of $(\frac{\partial P}{\partial \mathcal{E}})_T$ with temperature for $B^{1/4}$=145 MeV, n_B=3n_0 and $T_c = 290 MeV$, calculated from the model. Lattice data taken from [16]. Solid and the broken lines are for with and without ϕ contributions respectively and the circles are from Lattice

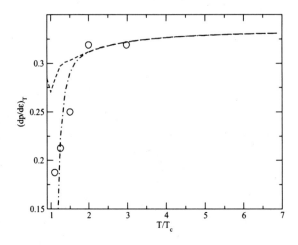

Figure 4: Variation of $(\frac{\partial P}{\partial \mathcal{E}})_T$ with temperature for $B^{1/4}$=145 MeV, n_B=3n_0 and $T_c = 200 MeV$. Lattice data taken from [16]. Dash-dot and the broken lines are for with and without ϕ contributions respectively and the circles are from Lattice calculation.

we have plotted $(\frac{\partial P}{\partial \mathcal{E}})_T$ instead of $(\frac{\partial P}{\partial \mathcal{E}})_S$ (which is the true sound velocity). The reason being that they do not differ much [13] and the first one is easier to compute. The two curves are for with and without Φ contribution and the circles are the

Lattice data for pure glue theory. The discontinuity in the plot is due to the built in second order phase transition in the EOS. Inclusion of Φ leads to the modification of sound velocity near the phase transition region. As seen from the curve sound velocity rises steeply at low temperature and then gradually saturates at higher temperature slightly above the the ideal gas value. It is to be noted here, the transition temperature of the figure 3 and 4 are different. One is calculated from our model (fig.3) $T_c = 290 MeV$ and in fig.4 it is taken to be $T_c = 200 MeV$. The Lattice result is computed only for gluonic contribution. Our result can be viewed as a prediction to be confirmed by the future Lattice calculations. For detailed discussion see [17].

3 PNJL model

In this section we are going to consider the PNJL model (Polyakov loop + Nambu-Jona-Lasinio model) to calculate the quark number susceptibility (QNS) and some other thermodynamic quantities of importance. PNJL model is basically a synthesis of NJL model with the polyakov loop. The advantage of PNJL model is that one can study the chiral and confinement-deconfinement physics together within a single theoretical framework. While the NJL model gives the chiral physics, the polyakov loop simulates the confinement-deconfinement physics. But like any other phenomenological model, the PNJL model has also some limitations. The gluons within this model are contained in a static temporal background. So, this model will be applicable upto an upper temperature limit where the transverse gluons do not contribute significantly. For this reason we present our result upto $2.5T_c$, where T_c is the critical temperature of the phase transition as transverse degrees of freedom will be important for $T > 2.5T_c$ [18].

To compute the quark number susceptibility and other thermodynamic variables we start with the thermodynamic potential pre unit volume given by,

$$
\begin{aligned}
\Omega =\ & \mathcal{U}\left(\Phi, \bar{\Phi}, T\right) + \frac{\sigma^2}{2G} \\
& - 2N_f T \int \frac{d^3 p}{(2\pi)^3} \left\{ \ln\left[1 + 3\left(\Phi + \bar{\Phi} e^{-(E_p - \mu_0)/T}\right) e^{-(E_p - \mu_0)/T} + e^{-3(E_p - \mu_0)/T}\right] \right. \\
& + \left. \ln\left[1 + 3\left(\bar{\Phi} + \Phi e^{-(E_p + \mu_0)/T}\right) e^{-(E_p + \mu_0)/T} + e^{-3(E_p + \mu_0)/T}\right] \right\} \\
& - 6N_f \int \frac{d^3 p}{(2\pi)^3} E_p \theta\left(\Lambda^2 - \vec{p}^{\,2}\right).
\end{aligned}
\tag{18}
$$

Where, $\mathcal{U}(\Phi, \bar{\Phi}, T)$ is the effective potential, the functional form of this potential being,

$$\frac{\mathcal{U}\left(\Phi, \bar{\Phi}, T\right)}{T^4} = -\frac{b_2\left(T\right)}{2}\bar{\Phi}\Phi - \frac{b_3}{6}\left(\Phi^3 + \bar{\Phi}^3\right) + \frac{b_4}{4}\left(\bar{\Phi}\Phi\right)^2 \quad , \tag{19}$$

with

$$b_2\left(T\right) = a_0 + a_1\left(\frac{T_0}{T}\right) + a_2\left(\frac{T_0}{T}\right)^2 + a_3\left(\frac{T_0}{T}\right)^3 \tag{20}$$

Here, Φ and $\bar{\Phi}$ are the traced polyakov loop and its conjugate respectively. T_0 is the transition temperature for pure gauge theory and the coefficients a_i and b_i are fitted from Lattice data of pure gauge theory. For details of the parameterization please refer to [19]. The value for T_0 is important as the QCD transition temperature depends on the chosen value of T_0. Lattice data for T_0 is 270 MeV [20–22], but some authors choose a value 190 MeV [19] to match the QCD transition temperature with the Lattice data. Here we work with the value $T_0 = 270MeV$. For such a value we find 5 MeV difference between chiral and deconfinement transition; whereas for $T_0 = 190MeV$, the difference is about 25 MeV. With $T_0 = 270MeV$ we find transition temperature at $T_c = 227MeV$. For details please see [23]. It is to be noted here at this point that with the coupling to NJL model the transition does not remain first order but it is rather crossover transition.

Among the other notations, σ is the auxiliary field, $<\sigma> = G < \bar{\psi}\psi >$ is the chiral condensate and G is the effective coupling strength of chirally symmetric scalar-pseudoscalar four point interaction of quark fields. N_f is the number of flavours, which in our case is 2. $E_p = \sqrt{\vec{p}^{\,2} + m^2}$ is the single particle energy and $m = m_0 - \langle\sigma\rangle$, with $m_0 = m_u = m_d$ being the value of current quark mass. Λ is the 3-momentum cutoff in the NJL model and $\mu_0 = 0$ is the quark number chemical potential.

After obtaining the thermodynamic potential we calculate the pressure as:

$$P(T, \mu_0) = -\Omega(T, \mu_0) \tag{21}$$

Next we minimize the potential with respect to the fields σ, Φ and $\bar{\Phi}$ using the following set of equations,

$$\frac{\partial\Omega}{\partial\sigma} = 0 \, , \frac{\partial\Omega}{\partial\Phi} = 0 \, , \frac{\partial\Omega}{\partial\bar{\Phi}} = 0 \, . \tag{22}$$

The fields so obtained from above equations are then put back into Ω to obtain pressure from (21). All the necessary quantities can now be calculated from the Taylor expansion of the pressure.

We expand the pressure as,

$$\frac{P(T,\mu_0)}{T^4} = \sum_{n=0}^{\infty} c_n(T) \left(\frac{\mu_0}{T}\right)^n \quad , \tag{23}$$

where,

$$c_n(T) = \frac{1}{n!} \frac{\partial^n \left(P(T,\mu_0)/T^4\right)}{\partial \left(\frac{\mu_0}{T}\right)^n}\bigg|_{\mu_0=0} \tag{24}$$

The expansion is done around $\mu_0 = 0$ and the odd terms in the expansion vanish due to CP symmetry. The quark number density is defined as the first derivative of the thermodynamic potential with respect to $\mu_0 = 0$ and the second derivative with respect to $\mu_0 = 0$ gives the QNS. So, from the above expansion $c_2(T)$ gives the QNS and the higher coefficients give the higher order derivatives of the QNS. Details of the computation can be found in [23].

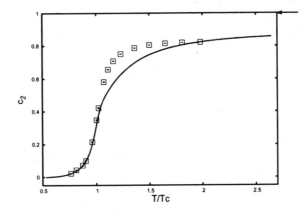

Figure 5: The QNS as a function of T/T_c. Symbols are Lattice data [24]. Arrow on the right indicates ideal gas value.

Figure 5 and 6 show the variation of c_2 and c_4 as a function of T/T_c respectively. As mentioned above c_2 is the QNS and c_4 can ne treated as the susceptibility of c_2. The variation of c_2 with T/T_c shows an order parameter like behaviour, remains zero at low temperature and gets saturated at high temperature. At saturation, it reaches 85% of the ideal gas value and the result is consistent with the Lattice data. The curve for c_4 shows a peak at the transition temperature and structure around T_c agrees well with the Lattice data [24, 25]. But, it differs significantly above T_c. The Lattice data [24] converges to the SB limit and our result shows only a weak convergence towards the SB limit. Our overestimation may be due to the mean field

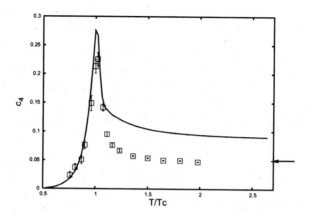

Figure 6: The c_4 as a function of T/T_c. The solid line is from PNJL model, Symbols are Lattice data [24]. Arrow on the right indicates ideal gas value.

approximation. But it is to be noted here that there are questions/confusions regarding the convergence of c_4 to SB limit at higher temperature and further detailed study is needed to say something conclusively in this regard.

Specific heat at constant volume is defined as,

$$C_V = \left.\frac{\partial \epsilon}{\partial T}\right|_V = -\left.T\frac{\partial^2 \Omega}{\partial T^2}\right|_V \tag{25}$$

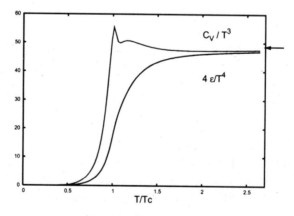

Figure 7: C_V/T^3 and $4\epsilon/T^4$ as function of T/T_c. The arrow on the right shows the ideal gas value.

290

In figure 7 we have plotted C_V along with $4\epsilon/T^4$ as a function of T/T_c. The reason for plotting them together is that specific heat is expected to coincide to $4\epsilon/T^4$ for a conformal gas. We see they almost coincide but ref.[16] shows a better agreement. Around T_c, C_V develops a peak and above T_c it saturates to just below the ideal gas value. It is worth mentioning here that for a continuous phase transition one expects a divergence in C_V which will translate into highly enhanced transverse momentum fluctuations or highly suppressed temperature fluctuations if the system passes close to CEP

Next we consider the velocity of sound v_s^2 and the conformal measure $\mathcal{C}$, defined respectively as,

$$v_s^2 = \left.\frac{\partial P}{\partial \epsilon}\right|_S = \left.\frac{\partial P}{\partial T}\right|_V \bigg/ \left.\frac{\partial \epsilon}{\partial T}\right|_V = \left.\frac{\partial \Omega}{\partial T}\right|_V \bigg/ T\left.\frac{\partial^2 \Omega}{\partial T^2}\right|_V \tag{26}$$

$$\mathcal{C} = \frac{\Delta}{\epsilon} = \frac{\epsilon - 3P}{\epsilon} \simeq 1 - 3v_s^2 \quad . \tag{27}$$

From the above relation we find that minima of the sound velocity (as expected near

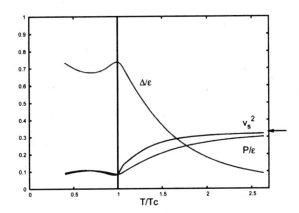

Figure 8: Squared velocity of sound v_s^2 and conformal measure $\mathcal{C} = \Delta/\epsilon$ as function of T/T_c. The arrow on the right shows the ideal gas value for v_s^2. For comparison with v_s^2 we also plot the ratio P/ϵ.

a phase transition or crossover) means a corresponding maxima for the conformal measure. Figure 8 shows that they really behaved in such a correlated manner. At T_c we have minima for the sound velocity and maxima for the conformal measure $\mathcal{C}$. Above T_c, $\mathcal{C}$ decreases sharply with temperature, whereas sound velocity v_s^2 increases with temperature and approaches towards the ideal gas value.

4 Discussions and Summary

The DDQM model and the PNJL model are used to calculate various thermodynamic quantities of relevance. The results are also compared with the recent Lattice data. We find that with the incorporation of thermodynamic consistency into the DDQM model, the non-perturbative effects get enhanced. Thermodynamic quantities are greatly modified by this non-perturbative effects near the phase transition region. Quark number susceptibility, and other thermodynamic quantities calculated within the PNJL model agree well with the recent Lattice data. Our results are presented upto $2.5T_c$ as the model are expected to work well within this temperature range. The behaviour of c_2 and c_4 are in agreement with the Lattice data. But, above T_c there is a significant difference between our result and the Lattice data. This may be due to the mean field approximation we are using here. Also it is to be noted here that Lattice data presented here are for large current quark mass compared to 5.5 MeV in our case. The other quantities like, specific heat, sound velocity and the conformal measure are also in close agreement with the Lattice data. From the result it seems, high temperature physics is dominated by the gluonic degrees of freedom.

References

[1] T. Banks and A. Casher, Nucl. Phys. B **169** 103 (1980).

[2] B. A. Campbell, J. Ellis and K. A. Olive, Nucl. Phys. B **345** 57 (1990).

[3] M. Asakawa, U. Heinz and B. Müller, *Phys. Rev. Lett.* **85** 2072 (2000); S. Jeon and V. Koch, *Phys. Rev. Lett.* **85** 2076 (2000).

[4] J. Ollitrault, *Phys. Rev.* D **46** 229 (1992).

[5] H. Sorge, *Phys. Rev. Lett.* **82** 2048 (1999).

[6] P. F. Kolb *et al.* , *Phys. Lett.* B **459** 667 (1999); P. F. Kolb *et al.* , *Nucl. Phys.* A **661** 349 (1999).

[7] D. Teaney *et al.* , *Phys. Rev. Lett.* **86** 4783 (2001); D. Teaney *et al.* , nucl-th/0110037.

[8] B. Mohanty and J. Alam, *Phys. Rev.* C **68** 064903 (2003).

[9] R.V.Gavai, S.Gupta, Eur. Phys. J. C **43** 31 (2005).

[10] L. Stodolsky, *Phys. Rev. Lett.* **75** 1044 (1995).

[11] R. Korus *et al.* , *Phys. Rev.* C **64** 054908 (2001).

[12] G.N.Fowler, S.Raha, R.M.Weiner, Z. Phys. C **9** 271 (1981).

[13] M.Plümer, S.Raha, R.M.Weiner, Phys. Lett. **139B** 198 (1984).

[14] P. N. Meisinger and M. C. Ogilvie, *Phys. Lett.* B **379** 163 (1996); *Nucl. Phys.* B (Proc. Suppl.) **47** 519 (1996).

[15] T.S.Birò, A.A.Shanenko, V.D. Toneev, Phys. Atom. Nucl. **66** 982 (2003).

[16] R.V.Gavai, S.Gupta, S.Mukherjee hep-lat/0506015.

[17] S. K. Ghosh, T. K. Mukherjee and S. Raha, Mod. Phys. Lett. A **21** 2067 2006.

[18] P. N. Meisinger, M. C. Ogilvie and T. R. Miller, *Phys. Lett.* B **585** 149 (2004).

[19] C. Ratti, M.A. Thaler and W. Weise, *Phys. Rev.* D **73** 014019 (2006).

[20] G. Boyd *et al.* , *Nucl. Phys.* B **469** 419 (1996);
      Y. Iwasaki *et al.* *Phys. Rev.* D **56** 151 (1997).

[21] B. Beinlich *et al.* , *Eur. Phys.* C **6** 133 (1999);
      M. Okamoto *et al.* , *Phys. Rev.* D **60** 094510 (1999).

[22] P. de Forcrand *et al.* , *Nucl. Phys.* B **577** 263 (2000);
      Y. Namekawa *et al.* , *Phys. Rev.* D **64** 074507 (2001).

[23] S. K. Ghosh, T. K. Mukherjee, M. G. Mustafa and R. Ray, Phys. Rev. D **73** 114007 2006.

[24] C. R. Allton *et al.* , *Phys. Rev.* D **71** 054508 (2005).

[25] R. V. Gavai and S. Gupta, *Phys. Rev.* D **72** 054006 (2005).

Physics and Astrophysics of Hadrons and Hardronic Matter
Editor: A. B. Santra

Collapse/Flattening of Nucleonic Bags in Ultra-Strong Magnetic Field

Somenath Chakrabarty

Department of Physics, Visva-Bharati
Santiniketan 731 235, India
email: somenath@vbphysics.net.in

Abstract

It is shown explicitly using MIT bag model that in presence of ultra-strong magnetic fields, a nucleon either flattens or collapses in the direction transverse to the external magnetic field in the classical or quantum mechanical picture respectively. Which gives rise to some kind of mechanical instability. Alternatively, it is argued that the bag model of confinement may not be applicable in this strange situation.

1 Introduction

One of the oldest subject of physics- "the effect of strong magnetic field on dense matter" has gotten a new life after the observational discoveries of a few strongly magnetized exotic stellar objects- known as magnetars [1–5]. These uncommon objects are believed to be strongly magnetized young neutron stars and their strong magnetic fields are supposed to be the possible sources of X-rays from anomalous X-ray pulsars (AXP) and low energy γ-radiation form the soft gamma-ray repeaters (SGR). It is believed that such objects may also act as the central engine for gamma ray bursts (GRB). The measured value of magnetic field strength at the surface of these objects are $\sim 10^{14} - 10^{15}$G. Then it can very easily be shown by scalar virial theorem that the magnetic field strength at the core region may go up to 10^{18}G. These objects are also assumed to be too young compared to the decay/expulsion time scale of magnetic fields from the core region. Now in presence of such intense magnetic fields, most of the physical properties of dense stellar matter, e.g., equation of states, quark-hadron phase transitions etc., must change significantly [6–8]. Not only that, some of the physical processes [9, 10], in particular, weak and electromagnetic decays and reactions, neutrino opacities etc., at the core region of compact neutron stars will also be affected in presence of ultra-strong magnetic fields. The transport properties (e.g, shear and bulk viscosities, thermal and electrical conductivities) of dense neutron star matter also change both qualitatively and quantitatively in presence of strong magnetic field [11, 12]. Furthermore, these intense magnetic fields could cause structural deformation of the exotic objects. In the classical general relativistic theory, it is shown by using Maxwell stress tensor that such exotic objects get flattened [13–15] for the macroscopic field B_m, whereas in the quantum mechanical scenario they collapse in the direction transverse to the magnetic field [16, 17]. In the case of ultra-strong magnetic field, the structure of these objects could become either disk like (in classical picture) or cigar like (in quantum mechanical scenario) from their usual spherical shapes. In the extreme case, they may be converted to black disks or black strings. Therefore, in some sense these strange stellar objects become mechanically unstable in presence of ultra-strong magnetic field. Long ago Chandrasekhar and Fermi in their studies on the stability of magnetized white dwarfs explained the possibility of such strange behavior [18]. Those conclusions are also valid for strongly magnetized neutron stars, where the white dwarf parameters have to be replaced by typical neutron star parameters; the upper limit of magnetic field strength for a stable neutron star of typical character is found to be 10^{18}G. In a recent work we have shown that if the magnetic field is extremely high to populate only the zeroth Landau level (with fully polarized spin states) of electrons, then stable neutron star/proto-neutron star matter can not exist in the β-equilibrium condition [19, 20]. It was also shown by Bander and Rubinstein in the context of stability of neutron and protons in a strong magnetic field that in presence of extremely strong magnetic field, protons becomes unstable by gaining effective mass, whereas neutrons, loosing effective mass and becomes stable [21]. In their calculations a delicate interplay between the anomalous magnetic moments of neutron and proton makes the neutron stable and proton becomes unstable; decays into neutron via e^+ and neutrino emission.

In this article following the recent work of Martínez et al [16, 17] and Kohri et al [22], we shall show that even the nucleonic (proton or neutron) bags can not be stable in presence of ultra-strong magnetic field- they either collapse or elongated in the transverse direction of ultra-strong external magnetic field. We have shown that either the nucleons are mechanically unstable or the bag model calculations can not be well suited for the conditions referred to above. In this work we have therefore studied the mechanical stability of a neutron/proton placed in an ultra-strong magnetic field. On the other hand in ref. [21], Bander and Rubinstein have studied the stability of these objects from the effective mass point of view and showed that neutrons are much more stable energetically than protons in this situation. The paper is organized in the following manner: in section 2, we have reviewed very briefly the MIT bag Lagrangian approach of color confinement. In section 3, we have studied the collapse of nucleons in the transverse direction following the ideas of Martínez et al [16, 17]. In section 4, following the model proposed by Kohri et al [22] in the context of anisotropic e^+e^- pressure, we have shown that nucleons get flattened in the transverse direction. The conclusions and discussions are presented in the last section.

2 Color Confinement- a Brief Overview

To study the mechanical stability of neutron/proton bags in presence of ultra-strong magnetic fields, in the flat space time coordinate, we have considered the MIT bag model of quark confinement [23–25]. We have taken into account both the gluonic interaction of quarks and the bag pressure B to confine quarks within the bag. Before we go into the detailed discussion on the mechanical instability problem of nucleonic bags in presence of intense magnetic fields, we give a brief overview of bag model Lagrangian approach to obtain the pressure balance at the surface of the nucleons. The usual form of bag Lagrangian density is given by

$$\mathcal{L}_{\text{MIT}} = [i\{\bar{\psi}\gamma^\mu\partial_\mu\psi - (\partial_\mu\bar{\psi})\gamma^\mu\psi\} \ + \ g\bar{\psi}\frac{\lambda_a}{2}\gamma^\mu V_\mu^a\psi - \bar{\psi}m\psi$$
$$- \frac{1}{4}F_{\mu\nu}^a F^{\mu\nu a} - B]\theta_v(x) - \frac{1}{2}\bar{\psi}\psi\Delta_s \quad (1)$$

where g is the strong coupling constant, λ_a's are the $SU(3)$ generators, with $a = 1, 2, ...8$, the gluonic color index, V_μ^a is the gluonic field four vector, $F_{\mu\nu}^a$ is the corresponding field tensor, m is the current mass of quarks, B is the bag constant, $\theta_v = 1$ inside the bag and $= 0$ outside the bag, $\partial\theta_v/\partial x^\mu = n_\mu\Delta_s$, Δ_s is the surface delta-function and n_μ is the space-like unit vector normal to the surface, The sum over flavors and color quantum numbers carried by quarks have not been shown explicitly. To obtain the pressure balance at the bag surface, we consider the energy momentum tensor of the bag, given by

$$T^{\mu\nu} \ = \ -g^{\mu\nu}\mathcal{L} + \left(\frac{\partial\mathcal{L}}{\partial(\partial_\mu\psi)}\partial^\nu\psi + \partial^\nu\bar{\psi}\frac{\partial\mathcal{L}}{\partial(\partial_\mu\bar{\psi})}\right)$$
$$= \ -g^{\mu\nu}\mathcal{L} + \frac{i}{2}\left(\bar{\psi}\gamma^\mu\partial^\nu\psi - (\partial^\nu\bar{\psi})\gamma^\mu\psi\right)\theta_v \quad (2)$$

and using the energy momentum conservation, given by $\partial_\mu T^{\mu\nu} = 0$, we have

$$B\Delta_s n^\nu + \frac{i}{2}\left(\bar{\psi}\gamma^\mu\partial^\nu\psi - (\partial^\nu\bar{\psi})\gamma^\mu\psi\right)n_\mu\Delta_s = 0 \quad (3)$$

and

$$\partial_\mu\left(\bar{\psi}\psi\Delta_s\right) = 0 \quad (4)$$

Now considering the surface boundary condition, given by (obtained from standard Euler-Lagrange equation)

$$in_\mu\gamma^\mu\psi = \psi \quad (5)$$

we obtain on the bag surface

$$Bn^\mu = \frac{1}{2}\frac{\partial}{\partial x_\mu}(\bar{\psi}\psi) \quad (6)$$

This equation is nothing but the pressure balance equation. Since $n^\mu n_\mu = -1$, we have on the bag boundary

$$B = -\frac{1}{2} n_\mu \partial^\mu (\bar{\psi}\psi) \tag{7}$$

In the case of spherical bag, $n^\mu \equiv (0, \hat{r})$ and this pressure balance equation reduces to

$$B = -\frac{1}{2} \frac{\partial}{\partial r} (\bar{\psi}\psi) \tag{8}$$

Which means that outward pressure of the quarks is exactly balanced by the inward vacuum pressure B on the surface of the bag.

3 Collapse of Nucleonic Bags

Now we shall consider the nucleonic bag (either neutron or proton) as an interacting thermodynamic system in equilibrium. The constituents are valance quarks, sea quarks and gluons. Then the total kinetic pressure of the system is given by

$$P_{in} = P_{in}^{(v)} + P_{in}^{(s)} + P_{in}^{(g)} \tag{9}$$

where v, s and g represent the valance quarks, sea quarks and gluonic contributions respectively. As discussed before, this internal kinetic pressure has to be balanced by the external bag pressure to maintain the stability of the system. Then we can write down the effective thermodynamic potential per unit volume of the system as

$$-\Omega = P_{in} - B \tag{10}$$

and it should be zero. Then following Martínez et al [16, 17], we have in presence of ultra-strong magnetic field of strength B_m, the thermodynamic potential per unit volume (we have chosen the gauge $A^\mu \equiv (0, -yB_m/2, xB_m/2, 0)$, so that B_m is a constant magnetic field along Z-axis)

$$\begin{aligned} T_\mu^\nu &= \left(T \frac{\partial \Omega}{\partial T} + \sum_r \mu_r \frac{\partial \Omega}{\partial \mu_r} \right) g_\mu^4 g_4^\nu \\ &+ 4 F_{\mu\lambda} F^{\nu\lambda} \frac{\partial \Omega}{\partial F^2} - g_\mu^\nu \Omega \end{aligned} \tag{11}$$

Hence the longitudinal component of pressure (along the direction of field) is given by

$$T_{zz} = P_{\parallel} = -\Omega = 0 \tag{12}$$

and the transverse part of total pressure

$$T_{xx} = T_{yy} = P_\perp = -\Omega - \mathcal{M} B_m = P_{\parallel} - \mathcal{M} B_m \tag{13}$$

where $\mathcal{M}$ is the effective magnetic dipole moment density of the bag. Since $\Omega = 0$, nucleons will therefore be inflated or collapsed in the transverse direction in presence of ultra-strong magnetic field depending on the overall sign of $\mathcal{M}$. The system will collapse if $\mathcal{M}$ is positive, else it will be inflated in the transverse direction. In order to have an order of magnitude estimate of extra in/out-ward pressure, we choose the contribution to $\mathcal{M}$ from valance quarks only (in fact the valance quarks only contribute in the evaluation of magnetic dipole moment of the nucleons). The magnetic dipole moment density of the ith. component ($i = u$ or d-quarks) is given by

$$\mathcal{M}_i = -\frac{\partial \Omega_i}{\partial B_m} \tag{14}$$

and the total value is given by

$$\mathcal{M} = \sum_{i=u,d} \mathcal{M}_i \tag{15}$$

where

$$\Omega_i = \frac{g_i q_i B_m}{4\pi^2} \sum_{\nu=0}^{\nu_{max}} \sum_{s=\pm1} \left[\mu_i(\mu_i^2 - M_{i,\nu,s}^2)^{1/2} - M_{i,\nu,s}^2 \ln\left(\frac{\mu_i + (\mu_i^2 - M_{i,\nu,s}^2)^{1/2}}{M_{i,\nu,s}} \right) \right] \tag{16}$$

is the thermodynamic potential density of the component i, g_i and q_i are respectively the degeneracy and charge of the ith. species, $M_{i,\nu,s}^2 = \{(p_\perp^2 + m_i^2)^{1/2} + sQ_iB_m\}^2$, m_i is the current quark mass ($= 5\text{MeV}$), $p_\perp = (2\nu q_i B_m)^{1/2}$ is the transverse component of momentum and Q_i is the anomalous magnetic dipole moment of the ith. quark species ($Q_u = 1.852\mu_N$ and $Q_d = -0.972\mu_N$, μ_N is the nuclear magneton). The maximum value of Landau quantum number is given by

$$\nu_{max}^{(i)} = \left[\frac{(\mu_i^2 - sQ_iB_m)^2 - m_i^2}{2q_iB_m} \right] \tag{17}$$

where [ ] indicates an integer less than the decimal number within the brackets. To obtain the chemical potentials for u and d quarks, we have made the following assumptions. The ith. quark species density within the nucleon is given by

$$n_i = \frac{g_i q_i B_m}{2\pi^2} \sum_{\nu=0}^{\nu_{max}} \sum_{s=\pm1} (\mu_i^2 - M_{i,\nu,s}^2)^{1/2} = \frac{\text{No.}(i)}{V} \tag{18}$$

where NO(i) is the number of ith. quarks species in the system. Therefore, NO(i) = NO(u) = 1 for neutrons and 2 for protons. Similarly, NO(i) = NO(d) = 2 for neutrons and 1 for protons and V is the nucleonic volume. We further assume that $r = 0.8\text{fm}$ as the radius of the nucleons. Solving numerically, we have obtained the

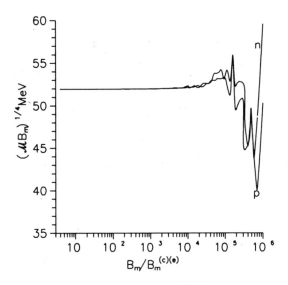

Figure 1: The variation of $\mathcal{M}B_m$ with $B_m/B_m^{(c)(e)}$ for neutron (indicated by the symbol n) and proton (indicated by the symbol p).

chemical potentials μ_i's for both u and d quarks and hence evaluated the magnetic dipole moment per unit volume for the system from eqns.(13)-(15).

In fig.(1) we have plotted $\mathcal{M}B_m$ for various values of B_m for both neutrons and protons. The product $\mathcal{M}B_m$ is always positive and oscillatory in the strong field regime ($\geq 10^{17}$G). The system will therefore collapse in the transverse direction and becomes ellipsoidal with cylindrical symmetry. The minor axes lengths b will therefore oscillate with the strength of magnetic field in particular, above 10^{17}G. Now in the study of mechanical stability of strongly magnetized neutron stars in quantum mechanical scenario, it has been shown that the system will either be inflated or collapsed if the magnetic dipole moment is negative or positive respectively. It has further been shown that neutron matter always behaves like a paramagnetic material with $\mathcal{M} > 0$, as a result, in the quantum mechanical picture a strongly magnetized neutron star always collapses in the transverse direction. Therefore we can infer that the conclusion drawn for such macroscopic objects like neutron stars is also valid in the microscopic level- e.g., neutrons or protons. We can then conclude that in a strong magnetic field, not only neutron stars, even their constituents, neutrons and protons become mechanically unstable. Alternatively, one could conclude that the bag model is perhaps not applicable in such strange situation, in that case the use of bag model for magnetized quark stars is also questionable. Therefore the investigations of this section show that both neutrons and protons become cigar like and in the extreme case they may reduce to what is called black string.

4 Flattening of Nucleons

In this section we shall evaluate the longitudinal and transverse parts of the kinetic pressures following ref.[22]. We choose the gauge $A^\mu \equiv (0, 0, xB_m, 0)$ so that $\vec{B}_m \equiv (0, 0, B_m)$. Then the solution of the Dirac equation is given by

$$\psi = \exp(-iE_n t) \begin{pmatrix} \phi \\ \chi \end{pmatrix} \tag{19}$$

where ϕ and χ are the upper and lower components. The upper component is given by

$$\phi = \exp(ip_y y + ip_z z) f_n \zeta_s, \tag{20}$$

where $n = 0, 1, 2, ..$ is the Landau quantum number, $s = \pm 1$, the spin quantum number, so that

$$\zeta_1 = \begin{pmatrix} 1 \\ 0 \end{pmatrix}, \tag{21}$$

$$\zeta_{-1} = \begin{pmatrix} 0 \\ 1 \end{pmatrix} \tag{22}$$

and

$$f_n(x, p_y) = \frac{1}{(2^n n! \pi^{1/2})^{1/2}} \exp\left(-\frac{\xi^2}{2}\right) H_n(\xi) \tag{23}$$

$\xi = (q_i B_m)^{1/2}(x - p_y/(q_i B_m))$ and $H_n(\xi)$ is the Hermite polynomial of order n. The lower component is given by

$$\chi = \frac{\vec{\sigma}.(\vec{p} - q_i \vec{A})}{E_n + m_i} \phi \tag{24}$$

The energy eigen value is given by $E_n = (p_z^2 + m_i^2 + q_i B_m(2n + 1 - s))^{1/2}$. Then we have from the first part of eqn.(2),

$$T^\mu_\nu = \text{diag}(E_n, -\hat{P}_x, -\hat{P}_y, -\hat{P}_z) \tag{25}$$

whereas all the off-diagonal terms are zero. Then it is very easy to show

$$\hat{P}_x = \hat{P}_y = \left(n + \frac{1}{2} - \frac{s}{2}\right)\frac{q_i B_m}{E_n}, \hat{P}_z = \frac{p_z^2}{E_n} \tag{26}$$

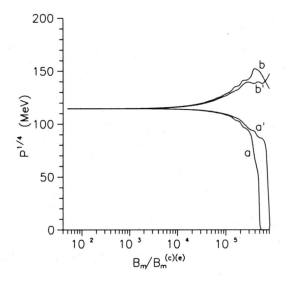

Figure 2: The variation of kinetic pressure ($P^{1/4}$ in MeV) with $B_m/B_m^{(c)(e)}$ for proton and neutron. Curve a is for longitudinal and b is for transverse components of kinetic pressures respectively for proton, curves a' and b' are the corresponding quantities for neutron.

These are called the dynamic pressure [22]. The ensemble average of these pressures are given by (at $T = 0$)

$$P_x = P_y = \frac{(q_i B_m)^2}{2\pi^2} \sum_{\nu=0}^{[\nu_{\max}]} \nu(2 - \delta_{0\nu}) \ln\left\{\frac{\mu_i + (\mu_i^2 - m_\nu^2)^{1/2}}{m_\nu}\right\} \tag{27}$$

which are the transverse part, where $2\nu = 2n + 1 - s$ and $m_\nu^2 = m_i^2 + 2q_i B_m \nu$. Similarly, we have the longitudinal component

$$P_z = \frac{q_i B_m}{4\pi^2} \sum_{\nu=0}^{[\nu_{\max}]} (2 - \delta_{0\nu}) \left[\mu_i(\mu_i^2 - m_\nu^2)^{1/2} - \ln\left\{\frac{\mu_i + (\mu_i^2 - m_\nu^2)^{1/2}}{m_\nu}\right\}\right] \tag{28}$$

Following the same numerical techniques as followed in previous section,

in fig.(2) we have plotted the longitudinal and transverse component of kinetic pressures with various magnetic field strengths for protons and neutrons respectively. The curves in this figure show that the longitudinal part of kinetic pressure is zero and / or very low for high magnetic field strength. Whereas the transverse part is high for high magnetic field. These two components saturate to some constant value for low or moderate magnetic field strength. Which indicates that the system reduces to pressure isotropic configuration at low magnetic field (as we generally see in conventional thermodynamic system). Therefore, according to this model, at very high magnetic field strength the system (neutron or proton) becomes oblet in shape and in the extreme case it reduces to a black disk.

5 Conclusions

In conclusion, we have studied the mechanical stability of neutrons and protons in a compact neutron star in presence of strong quantizing magnetic field. We have followed two entirely different approaches. In the so called quantum mechanical picture, in which the interaction of magnetic dipole moments of quark constituents with the external magnetic field has been considered, the shapes of both neutron and proton become prolate type from their usual spherical nature. The effect is more prominent at high field limit ($> 10^{16}$G). On the other hand, in the classical picture, both the systems acquire oblate shape. The effect is again prominent for high magnetic field. In the classical picture, it has been observed that such anisotropy of kinetic pressure is automatically removed at moderate ($\geq 10^{15-16}$G) values of magnetic field strength and both the systems become mechanically stable. However, in the quantum mechanical picture, there is always an extra in-word pressure in the transverse direction even for moderate values of magnetic field strength. This is because of non-zero finite values for $\mathcal{M}B_m$ of the systems, but the effect is not so significant. Therefore, the behavior of bulk objects like neutron stars and their constituents, e.g., neutrons and protons (which are of microscopic in sizes) are almost identical in an external strong magnetic field.

References

[1] R.C. Duncan and C. Thompson, Astrophys. J. Lett. **392**, L9 (1992); C. Thompson and R.C. Duncan, Astrophys. J. **408**, 194 (1993); C. Thompson and R.C. Duncan, MNRAS **275**, 255 (1995); C. Thompson and R.C. Duncan, Astrophys. J. **473**, 322 (1996).

[2] P.M. Woods et al., Astrophys. J. **519**, L139 (1999); C. Kouveliotou, et al., Nature **393**, 235 (1998); Astrophys. J. **510**, L115 (1999).

[3] S. Kulkarni and D. Frail, Nature **365**, 33 (1993).

[4] T. Murakami et al, Nature **368**, 127 (1994).

[5] K. Hurley, et al., Astrophys. Jour. **442**, L111 (1999); S. Mereghetti and L. Stella, Astrophys. Jour. **442**, L17 (1999); J. van Paradihs, R.E. Taam and E.P.J. van den Heuvel, Astron. Astrophys. **299**, L41 (1995); S. Mereghetti, astro-ph/99111252; see also A. Reisenegger, astro-ph/01003010.

[6] D. Bandopadhyaya, S. Chakrabarty, P. Dey and S. Pal, Phys. Rev. **D58**, 121301 (1998); T. Ghosh and S. Chakrabarty, Phys. Rev. **D63**, 0403006 (2001); T. Ghosh and S. Chakrabarty, Int. Jour. Mod. Phys. **D10**, 89 (2001).

[7] S. Chakrabarty, D. Bandopadhyay and S. Pal, Phys. Rev. Lett. **78**, 2898 (1997); D. Bandopadhyay, S. Chakrabarty and S. Pal, Phys. Rev. Lett. **79**, 2176 (1997).

[8] C.Y. Cardall, M. Prakash and J.M. Lattimer, Astrophys. J. **554**, 322 (2001); A. Broderic, M. Prakash and J.M. Lattimer, Astrophys. J. **537**, 351 (2000).

[9] E. Roulet, Astro-ph/9711206; L.B. Leinson and A. Pérez, Astro-ph/9711216; D.G. Yakovlev and A.D. Kaminkar, The Equation of States in Astrophysics, eds. G. Chabrier and E. Schatzman P.214, Cambridge Univ. Press, 1994; V.G. Bezchastnov and P. Haensel (astro-ph/9608090), Phys.Rev. **D54**, 3706 (1996).

[10] V.G. Bezchastrov and P. haensel, Astro-ph/9608090. Esteban Roulet; JHEP, **9801**, 013 (1998) (hep-ph/9711206).

[11] D.G. Yakovlev and D.A. Shalybkov, Sov. Astron. Lett. **16**, 86 (1990); D.G. Yakovlev and D.A. Shalybkov, Astrophys. Space Sc. **176**, 171 (1991); D.G. Yakovlev and D.A. Shalybkov, Astrophys. Space Sc. **176**, 191 (1991).

[12] Sutapa Ghosh, Sanchayita Ghosh, Kanupriya Goswami, Somenath Chakrabarty, Ashok Goyal, Int. Jour. Mod. Phys. **D11**, 843 (2002).

[13] K. Konno, T. Obata and Y. Kojima, Astron. Astrophys. **356**, 234 (2000).

[14] K. Konno, T. Obata and Y. Kojima, astro-ph/0001397.

[15] See also T. W. Baumgarte and S. L. Shapiro, astro-ph/0211339.

[16] M. Chaichian et al, Phys. Rev. Lett. **84**, 5261 (2000) (hep-ph/9911218).

[17] A. Pérez Martínez et al, hep-ph/0011399.; astro-ph/0303213.

[18] S. Chandrasekhar and E. Fermi, Astrophys. Jour. **118**, 116 (1953); see also D. Lai and S.L. Shapiro, Astrophys. Jour. **383**, 745 (1991).

[19] S. Ghosh, S. Mandal and S. Chakrabarty, astro-ph/0205445.

[20] S. Ghosh, S. Mandal and S. Chakrabarty, astro-ph/0207492.

[21] M. Bander and H.R. Rubinstein, Nucl. Phys. Proc. Suppl. **31**, 248 (1992).

[22] K. Kohri, S. Yamada and S. Nagataki, hep-ph/0106271.

[23] Models of the Nucleon, From Quarks to Soliton, R.K. Bhaduri, Addiison-Wesley Publ. Comp. Inc., New York, 1988.

[24] Quantum Chromodynamics, W. Greiner and A. Schäfer, Springer, 1989.

[25] see also T. DeGrand, R.L. Jaffe, K. Johnson and J. Kiskis, Phys. Rev. **D12**, 2060 (1975).

Physics and Astrophysics of Hadrons and Hardronic Matter
Editor: A. B. Santra

Probing the Response of Quark Gluon Plasma

Purnendu Chakraborty

Theory Division, Saha Institute of Nuclear Physics
1/AF Bidhannagar, Kolkata 700064, India.
email : purnendu.chakraborty@saha.ac.in

Abstract

Using linear response theory, we study the problem of passage of fast color charges through an equilibrium quark gluon plasma. In particular, we show how the potential of a (color) dipole is modified if it is moving relative to the background and discuss the consequences of possible new bound states and J/Ψ suppression in the quark gluon plasma. We also discuss the effect of the the spontaneous (color) electromagnetic fluctuations on the stopping power of heavy quark in QGP. We find that the electromagnetic fluctuations feeds the energy from the plasma to the heavy quark thereby reducing the total collisional energy loss of the heavy quark by $15 - 40\%$ for parameters relevant at RHIC energies.

A plasma is a statistical system of charged particles which move randomly, interact with themselves and respond to external disturbances. Therefore, it is capable of sustaining rich classes of physical phenomena. Screening of charges, damping of plasma modes, and plasma oscillations are important collective phenomena in plasma physics [1, 2]. A proper description of such phenomena may be obtained if we know how a plasma will respond macroscopically to a given external disturbance. The microscopic features of the particle interactions in the plasma are not completely lost in such a macroscopic description. But they can be implemented in a response function through the ways in which the mutually interacting particles adjust themselves to the external disturbance and the response function plays a crucial role in determining the properties of the plasma [1].

The quark gluon plasma is a special kind of plasma in which the electric charges are replaced by the color charges of quarks and gluons, mediating the strong interaction among them. Such a state of matter is expected to exist at extreme temperatures, above 150 MeV, or densities, above about 10 times normal nuclear matter density. These conditions could be achieved in the early universe for the first few microseconds after the Big Bang or in the interior of neutron stars. In accelerator experiments high-energy nucleus-nucleus collisions are used to search for the QGP. These collisions create a hot and dense fireball, which might consist of a QGP in an early stage (less than about $10\text{fm}/c$) [3]. Since the masses of the lightest quarks and of the actually massless gluons are much less than the temperature of the system, the QGP is an ultra-relativistic plasma. To achieve a theoretical understanding of the QGP, perturbative methods at finite temperature are adopted [4, 5]. Perturbative QCD should work at high temperatures far above the phase transition where the interaction between the quarks and gluons becomes weak due to asymptotic freedom.

In the high temperature quark gluon plasma, de-confinement and plasma screening are expected to modify the heavy quark-antiquark $Q\bar{Q}$ potential and may lead to dissociation of $Q\bar{Q}$ bound states. It was suggested that observation of this effect through a strong and sequential suppression of the J/Ψ peak in the dilepton invariant mass spectrum can be used to test the formation of the QGP in the heavy ion collision [6].

Generally, in the calculation of $Q\bar{Q}$ potential the pair is assumed to be rest relative to the medium. However, quarks and gluons coming from initial hard processes receive a transverse momentum which causes them to propagate through the QGP [7]. In addition, hydrodynamical models predict strong flow in the fireball [8]. Hence, it is of great interest to estimate the screening potential of a parton moving relatively to the QGP [9, 10].

High energy partons produced in initial partonic sub-processes in collisions between two heavy nuclei will lose their energy while propagating through the dense matter formed after such collisions, resulting in jet quenching. The amount of quenching depends upon the state of the fireball produced and the resulting quenching pattern may be used for identifying and investigating the plasma phase [11]. In order to quantitatively understand medium modifications of hard parton characteristics in the final state, the energy loss of partons in the QGP has to be determined.

There are two contributions to the energy loss of a parton in a QGP: one is caused by elastic collisions between the partons and the other is caused by radiative losses. It is generally believed that the radiative loss due to multiple gluon radiation (see [12, 13] for a review) dominates over the collisional one in the ultra-relativistic case. However, it has been shown recently in a number of publications that for realistic values of the parameters there is a wide range of parton energies in which the magnitude of the collisional loss is comparable to the radiative loss for heavy [14–17] as well as for light [14, 18] quark flavors.

Earlier estimates of collisional energy loss in the QGP [19, ?] were obtained by neglecting the microscopic fluctuations. On the other hand, in the case of non-relativistic QED plasma it is known for a long time that the electromagnetic fluctuations substantially modifies the stopping power of charges, particularly in the low velocity limit, by injecting the energy from the medium [1, 21, 22]. In order to use jet quenching as tomographic tool in the heavy ion collision, it is therefore obvious that this phenomenon deserves to be studied in this context also [23].

The article is organized is as follows : in the next section we will set up our notations and conventions. Screening potential of a moving dipole and the collisional energy loss of a heavy quark due to electromagnetic fluctuations in the QGP are discussed in section 2 and section 3 respectively and finally we conclude.

1 Response Functions of Quark Gluon Plasma

As we will be mainly interested in the (color) electromagnetic properties of the quark gluon plasma, it will be apt to select an external electric field as the disturbance. The response of the medium typically then appears in the form of an induced current density. If the system is stable against such a disturbance whose strength is weak, then the induced current may appropriately be expressed by that part of the response which is linear in the externally disturbing field. The *linear response* of a plasma to an external electromagnetic field has extensively been studied [1, 2] in plasma physics in which the external current is related to the total electric field by

$$\vec{j}^{a}_{\text{ind}}(\omega, \vec{\mathbf{k}}) = -\frac{i\omega}{4\pi} \left[\epsilon(\omega, \vec{\mathbf{k}}) - \mathbb{1} \right] \vec{\mathcal{E}}^{a}(\omega, \vec{\mathbf{k}}), \tag{1}$$

where $\epsilon(\omega, \vec{\mathbf{k}})$ is the (color independent) dielectric tensor, which is the sole agency describing the linear chromo-electromagnetic properties of the medium and $\mathbb{1}$ is the identity operator. Non-Abelian effects (beyond the color factors as in (1)) might be important at realistic temperatures. However, they cannot be treated by the method presented here.

For an isotropic medium, dielectric tensor can be written as a linear combination of two mutually independent components as

$$\epsilon_{ij}(\omega, k) = \mathcal{P}^{L}_{ij}\epsilon_{L}(\omega, k) + \mathcal{P}^{T}_{ij}\epsilon_{T}(\omega, k), \tag{2}$$

where, the projection operators $\mathcal{P}^{L/T}$ are given by,

$$\mathcal{P}^{L}_{ij} = \frac{k_i k_j}{k^2},$$

$$\mathcal{P}^{T}_{ij} = \delta_{ij} - \frac{k_i k_j}{k^2}, \tag{3}$$

and the longitudinal and the transverse dielectric functions $\epsilon_{L/T}$ are given by

$$\epsilon_{L}(\omega, k) = \frac{\epsilon_{ij} k_i k_j}{k^2},$$

$$\epsilon_{T}(\omega, k) = \frac{1}{2} \left[\text{Tr}\epsilon(\omega, \vec{\mathbf{k}}) - \epsilon_{L}(\omega, k) \right]. \tag{4}$$

The dielectric functions, both longitudinal and transverse, in (4) are related [4, 5, 24] to the self-energies of the gauge boson, i.e., gluon, in the medium as

$$\epsilon_{L}(\omega, k) = 1 - \frac{\Pi_{L}(\omega, k)}{K^2},$$

$$\epsilon_{T}(\omega, k) = 1 - \frac{\Pi_{T}(\omega, k)}{\omega^2}, \tag{5}$$

306

where $K = (\omega, k)$ with $k = |\vec{\mathbf{k}}|$. Π_L and Π_T are, respectively, the longitudinal and the transverse self energies of the gluon to be discussed below.

In covariant gauges the general form of the in-medium gluon polarization tensor can be written as [24]

$$\Pi_{\mu\nu}(\omega, \vec{\mathbf{k}}) = \mathcal{A}_{\mu\nu}\Pi_T(\omega, k) + \mathcal{B}_{\mu\nu}\Pi_L(\omega, k) \,, \tag{6}$$

where the transverse and longitudinal projection tensors [24] are

$$
\begin{aligned}
\mathcal{A}_{\mu\nu} &= \tilde{\eta}_{\mu\nu} - \frac{\tilde{K}_\mu \tilde{K}_\nu}{\tilde{K}^2} \,, \\
\mathcal{B}_{\mu\nu} &= -\frac{1}{K^2 \, k^2}\left[\left(k^2 u_\mu + \omega \tilde{K}_\mu\right)\left(k^2 u_\nu + \omega \tilde{K}_\nu\right)\right] \,,
\end{aligned}
\tag{7}
$$

in which u_μ is the four velocity of the fluid with $u^\mu u_\mu = 1$. In the rest frame of the medium $u^\mu = \delta_0^\mu$. The tensor $\tilde{\eta}_{\mu\nu}$ and the vector $\tilde{K}$ are orthogonal to u_μ and defined as

$$
\begin{aligned}
\tilde{\eta}_{\mu\nu} &\equiv \eta_{\mu\nu} - u_\mu u_\nu \,, \\
\tilde{K}_\mu &\equiv K_\mu - \omega u_\mu \,,
\end{aligned}
\tag{8}
$$

with the Minkowski metric tensor $\eta_{\mu\nu}$. Using the properties of the projection operators (7), one can obtain the scalar functions in (6) as

$$
\begin{aligned}
\Pi_L(\omega, k) &= -\frac{K^2}{k^2}\Pi_{00}(\omega, k) \,, \\
\Pi_T(\omega, k) &= \frac{1}{2}\left(\delta_{ij} - \frac{k_i k_j}{k^2}\right)\Pi_{ij}(K) = \frac{1}{2}\left[\Pi_\mu^\mu(\omega, k) - \Pi_L(\omega, k)\right] \,.
\end{aligned}
\tag{9}
$$

The in-medium gluon propagator follows from the Dyson-Schwinger equation

$$(\Delta_{\mu\nu}(K))^{-1} = (\Delta_{\mu\nu}^0(K))^{-1} + \Pi_{\mu\nu}(K) \,, \tag{10}$$

where $\Delta_{\mu\nu}^0(K)$ is the gluon propagator in vacuum. Combining (6), (7) and (10), the full in-medium gluon propagator in covariant gauge is obtained as

$$\Delta_{\mu\nu}(K) = -\frac{\mathcal{A}_{\mu\nu}}{K^2 - \Pi_T} - \frac{\mathcal{B}_{\mu\nu}}{K^2 - \Pi_L} + (\xi - 1)\frac{K_\mu K_\nu}{K^4} \,, \tag{11}$$

where ξ is the gauge parameter.

The one loop gluon self energy in the high temperature limit has been obtained as [24, 25]

$$\Pi_L(\omega, k) = -m_D^2 \frac{K^2}{k^2}\left[1 - \frac{\omega}{2k}\ln\frac{\omega + k}{\omega - k}\right] \,,$$

307

$$\Pi_T(\omega, k) = \frac{m_D^2}{2} \frac{\omega^2}{k^2} \left[1 - \left(1 - \frac{\omega^2}{k^2}\right) \frac{\omega}{2k} \ln \frac{\omega + k}{\omega - k}\right], \tag{12}$$

where the Debye mass is given by

$$m_D^2 = \Pi_L(\omega = 0, k) = \Pi_{00}(\omega = 0, k) = g^2 T^2 \left(1 + \frac{N_f}{6}\right), \tag{13}$$

in which N_f is the number of light quark flavors in the QGP. Though the gluon propagator is a gauge dependent quantity, the gluon self energies in (12) are gauge independent only for the leading term of the high temperature expansion [26]. The dielectric functions are therefore also gauge invariant and can be written combining (5) and (12) as

$$\epsilon_L(\omega, k) = 1 + \frac{m_D^2}{k^2} \left[1 - \frac{\omega}{2k} \left(\ln \left|\frac{\omega + k}{\omega - k}\right| - i\pi\Theta(k^2 - \omega^2)\right)\right],$$

$$\epsilon_T(\omega, k) = 1 - \frac{m_D^2}{2\omega^2} \left[\frac{\omega^2}{k^2} - \left(1 - \frac{\omega^2}{k^2}\right) \frac{\omega}{2k} \left(\ln \left|\frac{\omega + k}{\omega - k}\right| - i\pi\Theta(k^2 - \omega^2)\right)\right]. \tag{14}$$

2 Screening potential of a moving dipole in the QGP

We consider two color charges Q^a and Q^b separated by a distance r. The change in free energy of the system to bring the two widely separated color charges together [5, 27, 28] is

$$\Delta\mathcal{F} = \frac{1}{2} \int d^3\vec{x} \, J_{\text{ext}}^{\mu}(x) A_{\mu}(x)$$

$$= \frac{1}{2} \int \frac{d^3k}{(2\pi)^3} \int_{-\infty}^{\infty} \frac{d\omega}{(2\pi)} \int_{-\infty}^{\infty} \frac{d\omega'}{(2\pi)} e^{it(\omega+\omega')} J_{\text{ext}}^{\mu}\left(\omega, -\vec{k}\right) A_{\mu}\left(\omega', \vec{k}\right), \tag{15}$$

where $J_{\text{ext}}^{\mu}(x) = \left(\rho_{\text{ext}}, \vec{J}_{\text{ext}}\right)$ is the sum of the two external currents, $J_{\text{ext}}^{\mu}(x) = J_1^{\mu}(x) + J_2^{\mu}(x)$, and $A_{\mu}(\vec{x})$ is the associated gauge field. The entropy generation is neglected in (15).

We assume that the average value of A_{μ} vanishes in equilibrium $\langle A_{\mu} \rangle = 0$. The induced expectation value of the vector potential, A_{μ} follows from linear response theory [24] as

$$\langle A_{\mu} \rangle = 4\pi \Delta_{\mu\nu}(K) J_{\text{ext}}^{\nu}(K), \tag{16}$$

where, $\Delta_{\mu\nu}$ is the propagator of the gauge boson exchanged between the two currents which is given in (11). Combining (16) with (15) we can write

$$\Delta\mathcal{F} = 2\pi \int \frac{d^3k}{(2\pi)^3} \int_{-\infty}^{\infty} \frac{d\omega}{(2\pi)} \int_{-\infty}^{\infty} \frac{d\omega'}{(2\pi)} e^{it(\omega+\omega')} J_{\text{ext}}^{\mu}\left(\omega, -\vec{k}\right) \Delta_{\mu\nu}(\omega', \vec{k}) J_{\text{ext}}^{\nu}(\omega', \vec{k}). \tag{17}$$

Now for two comoving charges Q^a and Q^b the dipole current can be written as

$$\begin{aligned} J_\mu(t, \vec{x}) &= (1, \vec{v}) \left[Q^a \delta(\vec{x} - \vec{x}_1 - \vec{v}t) + Q^b \delta(\vec{x} - \vec{x}_2 - \vec{v}t) \right] \\ &\overset{\text{FT}}{=} 2\pi(1, \vec{v}) \delta(\omega - \vec{k} \cdot \vec{v}) \left[Q^a e^{-i\vec{k} \cdot \vec{x}_1} + Q^b e^{-i\vec{k} \cdot \vec{x}_2} \right], \end{aligned} \qquad (18)$$

Substituting (18) in (17) and using (11) we obtain, after some manipulation, the dipole potential as

$$\begin{aligned} \Delta \mathcal{F}^{ab}(r; \rho, z) &= \frac{Q^a Q^b}{2\pi^2} \int d^3 k \, \cos(\vec{k} \cdot \vec{r}) \left[\frac{v^2 - \frac{\omega^2}{k^2}}{K^2 - \Pi_T(\omega, k)} - \frac{1 - \frac{\omega^2}{k^2}}{K^2 - \Pi_L(\omega, k)} \right]_{\omega = \vec{k} \cdot \vec{v}}, \\ &= \frac{2 Q^a Q^b}{\pi} \int_0^\infty d\kappa \, \kappa J_0(\kappa \rho) \int_0^\infty dk_z \, \cos(k_z z) \, \times \\ &\qquad \left\{ \text{Re} \left[\frac{v^2 - \frac{\omega}{k^2}}{K^2 - \Pi_T(\omega, k)} - \frac{1 - \frac{\omega^2}{k^2}}{K^2 - \Pi_L(\omega, k)} \right]_{\omega = \vec{k} \cdot \vec{v}} \right\}. \end{aligned} \qquad (19)$$

Letting $v \to 0$ in (19), well known Yukawa potential for a static pair is recovered,

$$\Delta \mathcal{F}^{ab}(r) = \frac{Q^a Q^b}{2\pi^2} \int d^3 k \, \frac{\cos(\vec{k} \cdot \vec{r})}{\kappa^2 + k_z^2 + m_D^2} = Q^a Q^b \frac{e^{-m_D r}}{r}. \qquad (20)$$

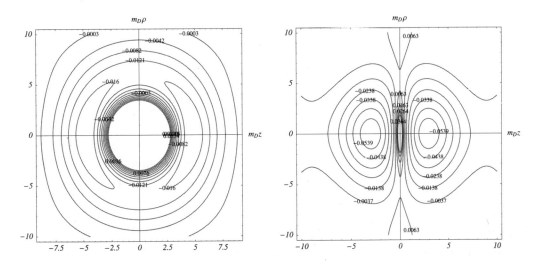

Figure 1: Spatial distribution of the scaled dipole potential for (Left panel) $v = 0.55c$ and (Right panel) $v = 0.99c$

The first term in (19) corresponds to the transverse (magnetic) interaction whereas the second term is due to the longitudinal (electric) interaction. The spatial distribution of the total potential for $v = 0.55c$ and $v = 0.99c$ are displayed in Fig. 1. The potential distributions show usual singularity at $r = 0$ ($z = 0$ and $\rho = 0$) and a completely symmetric behavior along with a pronounced negative minimum in the ($\rho - z$) plane, which are clearly reflected in both panels of Fig. 1. However, the detailed features in the specific direction can also be seen from both panels of Fig. 2 and Fig. 3 where potential distribution along the direction of motion and perpendicular to it have been delineated explicitly. The potential along the $|z|$-direction falls off like that of a static one, flips its sign, exhibits a negative minimum, and then oscillates around zero. On the other hand, the potential in the transverse direction falls off slowly compared to the static one, flips its sign, and exhibits a negative minimum at some values of ρ, again changes its sign to attain small positive maximum and then tends to zero at large ρ. It can be seen that the nature of the potential is largely dictated by the electric interaction rather than by the magnetic one.

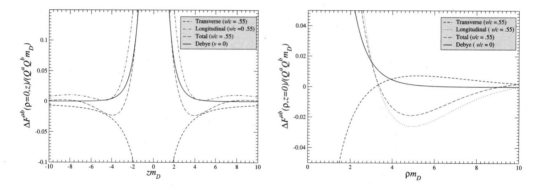

Figure 2: Scaled dipole (Left panel) along the the direction of motion (Right panel) normal to the direction of motion for $v = 0.55c$

With the increase of v, as seen from Fig. 3, the relative strength of the dipole potential grows strongly. The minimum of the dipole potential along the $|z|$ direction increases and it position shifts towards the center indicating a faster fall off as well as a larger anisotropy. This is due to the fact that the magnitude of the magnetic interaction becomes dominant. These features could have important consequences on various binary bound states in QGP as we will discuss.

In QCD the interaction between color charges in various channels is either attractive or repulsive. A quark and an antiquark correspond to the sum of irreducible color representations : $\mathbf{3} \otimes \bar{\mathbf{3}} = \bar{\mathbf{1}} \oplus \mathbf{8}$, where the interaction strength of the color singlet representation is $-16/3$ (attractive) whereas that of the color octet channel is $2/3$ (repulsive). Similarly, a two quark state corresponds to the sum of the irre-

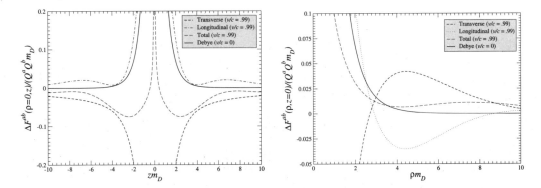

Figure 3: Scaled dipole (Left panel) along the the direction of motion (Right panel) normal to the direction of motion for $v = 0.99c$

ducible color representations: $\mathbf{3} \otimes \mathbf{3} = \bar{\mathbf{3}} \oplus \mathbf{6}$, where the antisymmetric color triplet is attractive ($-8/3$) giving rise to possible bound states [29]. The symmetric color sextet channel, on the other hand, is repulsive ($4/3$). Color bound states (*i.e.* diquarks) of partons at rest have been claimed by analyzing lattice data [29]. The situation is different when partons are in motion. The dipole potentials along the direction of propagation and also normal to it have both attractive and repulsive parts. This could lead to dissociation of bound states as well formation of colored bound states in the QGP.

Recent lattice results [30] using maximum entropy method indicate that charmonium states actually persist up to $2T_C$ and there are similar evidence for mesonic bound states made of light quarks as well [31]. Within our model such bound states as well other colored binary states in the QGP can experience quite a different potentials. One immediate consequence of this would be the modification of the transverse momentum, p_T, dependence of J/ψ spectra. As noticed by by Chu and Matsui [9], this will lead to more flattened p_T distribution of J/Ψ compared to the static case.

3 Energy gain of heavy quarks due to fluctuations in QGP

Here we will compute the energy loss (gain) of the heavy quarks due to electromagnetic fluctuations in the quark gluon plasma. The semiclassical approach has been shown to be equivalent to soft contribution to the energy loss [32]. Unless the partons are ultra-relativistic soft contribution gives the dominant contribution to the energy loss [17]. In the semiclassical approach, it is assumed that the energy lost by the particle per unit time is small compared to the energy of the particle itself so that the change in the velocity of the particle during the motion may be neglected, *i.e*, the particle moves in a straight line trajectory. The energy loss of a particle is determined by the work of the retarding forces acting on the particle in the plasma from the chromo-electric field generated by the particle itself while moving. So the energy loss of the particle per unit time is given by,

$$\frac{dE}{dt} = Q^a \vec{v} \cdot \vec{\mathcal{E}}^a|_{\vec{r}=\vec{v}t}, \tag{21}$$

where the field is taken at the location of the particle. In the Abelian approximation, the total chromo-electric field $\vec{\mathcal{E}}^a$ induced in the QGP can be related to the external current of the test charge by solving Maxwell's equations and the equation of continuity

$$\left[\epsilon_{ij}(\omega, k) - \frac{k^2}{\omega^2}\left(\delta_{ij} - \frac{k_i k_j}{k^2}\right)\right]\mathcal{E}_j^a(\omega, k) = \frac{4\pi}{i\omega}j_i^a(\omega, k), \tag{22}$$

with the color charge current $\vec{j}^a$.

Substituting (22) in (21) the collisional energy loss due to polarization of the medium follows [19],

$$\frac{dE}{dx} = -\frac{C_F \alpha_s}{2\pi^2 v} \int d^3 k \frac{\omega}{k^2}\left[\text{Im}\epsilon_L^{-1} + (v^2 k^2 - \omega^2)\,\text{Im}\,(\omega^2 \epsilon_T^{-1} - k^2)^{-1}\right], \tag{23}$$

where C_F is the quadratic Casimir invariant in fundamental representation. However, the previous formula for the energy loss in (21) does not take into account the field fluctuation in the plasma and the particle recoil in collisions. To accommodate these effects it is necessary to replace (21) with [21, 22],

$$\frac{dE}{dt} = \left\langle Q^a(t)\,\vec{v}(t) \cdot \vec{\mathcal{E}}_t^a(\vec{r}(t), t)\right\rangle, \tag{24}$$

where $\langle \cdots \rangle$ denotes the statistical averaging operation. It is to be noted that here two averaging procedures are performed: I) an ensemble average *w.r.t* the equilibrium density matrix II) a time average over random fluctuations in plasma. These two operations are commuting and only after both of them are performed the average quantity takes up a smooth value [33]. In the following, we will explictly denote the ensemble average by $\langle \cdots \rangle_\beta$ wherever required to avoid possible confusion.

The electric field $\vec{\mathcal{E}}_t$ in (24) consists of the induced field $\vec{\mathcal{E}}$ given by (22) and a spontaneously generated microscopic field $\vec{\tilde{\mathcal{E}}}$, the latter being a random function of position and time.

In the classical case, where we can use the concept of trajectory, the equations of motion of the particle have the form [34],

$$\frac{d\vec{p}}{dt} = Q^a(t)\left[\vec{\mathcal{E}}_t^a(\vec{r}(t),t) + \vec{v}\times\vec{B}_t^a(\vec{r}(t),t)\right],$$

$$\frac{dQ^a(t)}{dt} = -gf_{abc}Q^b(t)A_\mu^c v^\mu. \tag{25}$$

Integrating the equations of motion (25) we find,

$$\vec{v}(t) = \vec{v}_0 + \frac{1}{E_0}\int_0^t dt_1\, Q^a(t_1)\vec{\mathcal{F}}_t^a(\vec{r}(t_1),t_1),$$

$$\vec{r}(t) = r_0 + \frac{1}{E_0}\int_0^t dt_1\int_0^{t_1} dt_2\, Q^a(t_2)\vec{\mathcal{F}}_t^a(\vec{r}(t_2),t_2),$$

$$\tag{26}$$

$$Q^a(t) = \mathcal{P}\exp\left(-\int_{\vec{r}(t_1)}^{\vec{r}(t)} gf_{abc}A_\mu^b dx^\mu\right)Q^c(t_1)$$

$$= U_{ac}(\vec{r}(t),\vec{r}(t_1))\,Q^c(t_1), \quad . \tag{27}$$

where $\vec{\mathcal{F}}_t = \vec{\mathcal{E}}_t + \vec{v}\times\vec{B}_t$, E_0 is the initial parton energy and $\vec{B}_t$ has the same meaning as the $\vec{\mathcal{E}}_t$. The mean change in the energy of the particle per unit time is given by (24). The stochastic time dependence as embodied in (24) comes because of explicit fluctuations in time and motion of the particle from one field point to another. The dependence on the latter will be expanded about the mean rectilinear motion where the former will be left untouched. This is done by picking a time interval Δt sufficiently large with respect to time scale of random electromagnetic fluctuations in the plasma (τ_2) but small compared with the time during which the particle motion changes appreciably (τ_1),

$$\tau_1 \gg \Delta t \gg \tau_2. \tag{28}$$

Parametrically, $\tau_1 \sim (g^4 T\ln 1/g)^{-1}$ and $\tau_2 \sim (gT)^{-1}$ [35]. In the weak coupling limit, these scales are widely separated. Keeping only the leading order terms in the expansion we get,

$$\vec{v}(t) = \vec{v}_0 + \frac{1}{E_0}\int_0^t dt_1 Q^a(t_1)\vec{\mathcal{F}}_t^a(\vec{r}_0(t_1),t_1),$$

$$\vec{\mathcal{E}}_t^a(\vec{r}(t),t) = \vec{\mathcal{E}}_t^a(\vec{r}_0(t),t) + \frac{Q^b(t)}{E_0}\int_0^t dt_1\int_0^{t_1} dt_2\sum_j \mathcal{E}_{j,t}^b(\vec{r}_0(t_2),t_2)$$

$$\times\frac{\partial}{\partial r_{0j}}\vec{\mathcal{E}}_t^a(\vec{r}_0(t),t). \tag{29}$$

We substitute (27) and (29) in (24) and assume that the distribution of the color charges of the partons at a given instant t is random and independent of the color fields $\langle Q^a(t) Q^b(t) \rangle = C_F \dot{\alpha}_s \delta_{ab}$. Using consistently the Abelian approximation, $U_{ab} = U_{ab}|_{t_1=t} = \delta_{ab}$, and keeping in mind that $\langle \mathcal{E}_i^a \mathcal{B}_j^a \rangle_\beta = 0$ [36], we get

$$
\begin{aligned}
\frac{dE}{dt} = & \left\langle Q^a(t) \vec{v}_0 \cdot \vec{\mathcal{E}}_t^a(\vec{r}_0(t), t) \right\rangle + \frac{C_F \alpha_s}{E_0} \int_0^t dt_1 \left\langle \vec{\mathcal{E}}_t^a(\vec{r}_0(t_1), t_1) \cdot \vec{\mathcal{E}}_t^a(\vec{r}_0(t), t) \right\rangle_\beta \\
& + \frac{C_F \alpha_s}{E_0} \int_0^t dt_1 \int_0^{t_1} dt_2 \left\langle \sum_j \mathcal{E}_{t,j}^a(\vec{r}_0(t_2), t_2) \frac{\partial}{\partial r_{0j}} \vec{v}_0 \cdot \vec{\mathcal{E}}_t^a(\vec{r}_0(t), t) \right\rangle_\beta . \quad (30)
\end{aligned}
$$

Since the mean value of the fluctuating part of the field equals zero,

$$
\left\langle \vec{\mathcal{E}} \right\rangle_\beta = 0 ,
$$

$\left\langle \vec{\mathcal{E}}_t^a(\vec{r}(t), t) \right\rangle$ equals the chromo-electric field produced by the particle itself in the plasma. The first term in (30) therefore corresponds to the usual polarization loss of the parton calculated in (23) [19]. Provided there exists a hierarchy of scales (28), it can be shown that the polarization field does not contribute to leading order in the correlations functions appearing in the second and third terms in (30) [1, 37]. These terms correspond to the statistical change in the energy of the moving parton in the plasma due to the fluctuations of the chromo-electromagnetic fields as well as the velocity of the particle under the influence of this field. The temporal averaging in (30), by definition, are allowed to include many random fluctuations over the mean motion. However, the correlation function of these fluctuations are exponentially suppressed beyond their characteristic time scales. This allows us to formally extend the upper limits in time-integrations in (30) to ∞. Fourier transforming we obtain from (30) the energy loss of the parton due to fluctuations as,

$$
\left. \frac{dE}{dt} \right|_{\text{fl}} = \frac{C_F \alpha_s}{16\pi^3 E} \int d^3k \left[\frac{\partial}{\partial \omega} \left\langle \omega \vec{\mathcal{E}}_L^2 \right\rangle_\beta + \left\langle \vec{\mathcal{E}}_T^2 \right\rangle_\beta \right]_{\omega = \vec{k} \cdot \vec{v}} , \quad (31)
$$

where $\left\langle \vec{\mathcal{E}}_L^2 \right\rangle$ and $\left\langle \vec{\mathcal{E}}_T^2 \right\rangle$ denote longitudinal and transverse field fluctuations respectively and we have taken $E = E_0$ to be the initial energy of the parton.

Within linear response theory, the power spectrum of the chromo-electromagnetic fields follows from the fluctuation-dissipation theorem and is completely determined by the dielectric functions of the medium [21, 22],

$$
\left\langle \tilde{\mathcal{E}}_i^a \tilde{\mathcal{E}}_j^a \right\rangle_{\beta; \omega, k} = \frac{8\pi}{e^{\beta\omega} - 1} \left\{ \mathcal{P}_{ij}^L \frac{\operatorname{Im} \epsilon_L}{|\epsilon_L|^2} + \mathcal{P}_{ij}^T \frac{\operatorname{Im} \epsilon_T}{|\epsilon_T - \eta^2|^2} \right\} , \quad (32)
$$

where $\eta = k/\omega$. Combining (31) and (32) the part of the energy loss coming form the fluctuations can be summarized as,

$$\left. \frac{dE}{dt} \right|_{\text{fl}} = \frac{C_F \alpha_s}{8\pi^2 E v_0^3} \int_0^{k_{\max} v_0} d\omega \, \coth \frac{\beta\omega}{2} F(\omega,k)_{\omega, \frac{\omega}{v_0}}$$
$$+ \frac{C_F \alpha_s}{8\pi^2 E v_0} \int_0^{k_{\max}} dk \, k \int_0^{k v_0} d\omega \, \coth \frac{\beta\omega}{2} G(\omega,k) , \qquad (33)$$

where $F(\omega,k) = 8\pi\omega^2 \text{Im} \, \epsilon_l / |\epsilon_l|^2$ and $G(\omega,k) = 16\pi \text{Im} \, \epsilon_t / |\epsilon_t - k^2/\omega^2|^2$.

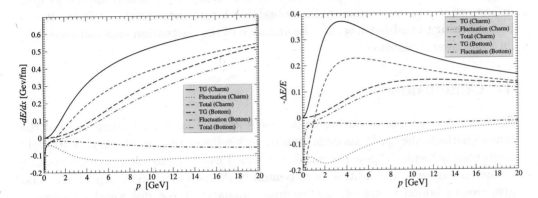

Figure 4: (Left) Various contributions to the energy loss of charm and bottom quarks in the QGP. (Right)Relative importance of the fluctuation loss compared to the collisional energy loss of We take the path length to be $L = 5$ fm.

It is to be noted here that since the spectral density of field fluctuations $\left\langle \vec{\mathcal{E}}_{L/T}^2 \right\rangle$ are positive for positive frequencies by definition, according to (33) the particle energy will grow due to interactions with the fluctuating fields. The contribution from field fluctuations to the heavy quark energy loss is shown in Fig. 4. Our choice of parameters are $N_f = 2$, $T = 250$ MeV, $\alpha_s = 0.3$ and we take the charm and bottom quark masses to be 1.25 and 4.2 GeV, respectively. For the upper cut-off $k_{\max}$ we take,

$$k_{\max} = \min \left\{ E, \frac{2q(E+p)}{\sqrt{m^2 + 2q(E+p)}} \right\} , \qquad (34)$$

where $q \sim T$ is the typical momentum of the thermal partons of the QGP.

It is evident that the effect of the fluctuations on the heavy quark energy loss is significant at low momenta and changes sign (*i.e.* there is net energy gain at low momentum). For momenta $4 - 20$ GeV the fluctuation effect reduces the collisional loss by $17 - 39\%$ for charm and $12 - 31\%$ for bottom.

Recently heavy quark probes at RHIC have posed new challenges to the theoretical understanding of the parton energy loss. As shown by Wicks et al [16] recent measurement of the non-photonic single electron data cannot be explained by the radiative loss alone. If the collisional energy loss is included, the agreement is better but not satisfactory. As seen it is obvious that the gain due to the field fluctuations will lead to reduction in collisional energy loss and hence more disagreement with single electron data. On the other hand, using the transport coefficients within pQCD energy loss calculations [38] the elliptic flow coefficient of single electrons v_2 is limited only to $2 - 3\%$ in semi-central Au-Au collisions, while the experimental values [39] reach up to 10% around transverse electron momenta of 2 GeV/c. It is suggested recently in Ref. [40] that not only the energy loss but also the energy gain in low momenta may be required for obtaining larger theoretical v_2 values. It will be interesting to find out whether the inclusion of the fluctuation gain can shed light on v_2 at lower momenta.

4 Conclusion

Screening potential of a moving color dipole and the collisional energy loss of the heavy quarks in the QGP has been studied within linear response theory. A strongly anisotropic screening dipole potential is obtained when the dipole is moving relative to the background. In addition, a minimum in the potential shows up which could give rise to bound states of, for example, diquarks if not destroyed by thermal fluctuations. The consequences, *e.g*, for the J/ψ yield should be investigated in more detail using hydrodynamical models or event generators for the space-time evolution of the fireball. The spontaneous electromagnetic fluctuations has been shown to somewhat reduce the total collisional e-loss of the heavy quark for hard momenta. However, at low momentum (typically in the hydrodynamic region) we find that there is net energy gain. It will be interesting to investigate the effect of fluctuations-loss on the elliptic flow parameter v_2 in detail using transport models.

References

[1] S. Ichimaru, *Basic Principles of Plasma Physics* (Benjamin, Reading, 1973).

[2] L.P. Pitaevskii and E.M. Lifshitz, *Physical Kinetics* (Pergamon Press, Oxford, 1981).

[3] B. Müller, *The Physics of the Quark-Gluon Plasma*, Lecturer Notes in Physics **225** (Springer-Verlag, Berlin, 1985).

[4] J.I. Kapusta, *Finite-Temperature Field Theory* (Cambridge University Press, Cambridge, 1989).

[5] M. LeBellac, *Thermal Field Theory* (Cambridge University Press, Cambridge, 2000).

[6] T. Matsui and H. Satz, Phys. Lett. B **178**, 416 (1986).

[7] M. Gyulassy, M. Plümer, M.H. Thoma, and X.N. Wang, Nucl. Phys. **A538**, 37c (1992).

[8] P. Huovinen, P.F. Kolb, U.W. Heinz, P.V. Ruuskanen, and S.A. Voloshin, Phys. Lett. B **503**, 58 (2001).

[9] M. C. Chu and T. Matsui, Phys. Rev. D **39**, 1892 (1989).

[10] P. Chakraborty, M. G. Mustafa and M. H. Thoma, Phys. Rev. D **74**, 094002 (2006).

[11] M. Gyulassy and M. Plümer, Phys. Lett. B **243**, 432 (1990).

[12] R. Baier, D. Schiff and B. G. Zakharov, Ann. Rev. Nucl. Part. Sci. **50**, 37 (2000).

[13] M. Gyulassy, I. Vitev, X. N. Wang and B. W. Zhang, in: Quark-Gluon Plasma 3, ed. R.C. Hwa (World Scientific, Singapore, 2003), p. 123.

[14] M. G. Mustafa and M. H. Thoma, Acta Phys. Hung. A **22**, 93 (2005).

[15] M. G. Mustafa, Phys. Rev. C **72**, 014905 (2005).

[16] S. Wicks, W. Horowitz, M. Djordjevic and M. Gyulassy, nucl-th/0512076.

[17] G. D. Moore and D. Teaney, Phys. Rev. C **71**, 064904 (2005).

[18] P. Roy, A. K. Dutt-Mazumder and Jane. Alam, Phys. Rev. C **73**, 044911 (2006).

[19] M. H. Thoma and M. Gyulassy, Nucl. Phys. B **351**, 491 (1991).

[20] S. Mrowczynski, Phys. Lett. B **269**, 383 (1991).

[21] A. G. Sitenko, *Electromagnetic Fluctuations in Plasma* (Academic Press, New York, 1967).

[22] A. I. Akhiezer et al, *Plasma Electrodynamics* (Pergamon Press, Oxford, 1975).

[23] P. Chakraborty, M. G. Mustafa and M. H. Thoma, arXiv:hep-ph/0611355.

[24] A. Weldon, Phys. Rev. **D26**, 1394 (1982).

[25] V. V. Klimov, Sov. Phys. JETP **55**, 199 (1982).

[26] U. Heinz, K. Kajantie and T. Toimela, Ann. Phys. (N.Y.) **176**, 218 (1987).

[27] J. P. Blaizot, J. Y. Ollitrault and E. Iancu, *Quark-Gluon Plasma-2*, Ed. R. C. Hwa (World Scientific, Singapore).

[28] A. Zee, *Quantum Field Theory in a nutshell* (Princeton University Press, 2003).

[29] E. V. Shuryak and I. Zahed, Phys. Rev. **D 70**, 054507 (2004).

[30] S. Datta, F. Karsch, P. Petreczky and I. Wetzorke, Phys. Rev. D **69**, 094507 (2004).

[31] F. Karsch, E. Laermann, P. Petreczky, S. Stickan and I. Wetzorke, Phys. Lett. B **530**,147 (2002);

F. Karsch, Nucl. Phys. A **715**, 701 (2003).

[32] E. Braaten and M. H. Thoma, Phys. Rev. D **44**, 1298 (1991); E. Braaten and M. H. Thoma, Phys. Rev. D **44**, 2625 (1991).

[33] G. Kalman and A. Ron, Ann. Phys. **16**, 118 (1961).

[34] U. Heinz, Ann. Phys. (N.Y.) **161**, 48 (1985); U. Heinz, Ann. Phys. (N.Y.) **168**, 148 (1986).

[35] P. Arnold, G. D. Moore and L. G. Yaffe, JHEP **0301**, 030 (2003).

[36] L. D. Landau and E. M. Lifshitz, *Statstical Physics*, (Pergamon, London, 1959).

[37] J. Hubbard, Proc. of Royal Soceity of London, **260**, 114 (1961).

[38] N. Armesto, M. Cacciari, A. Dainese, C. A. Salgado, and U. A. Wiedemann, Phys. Lett. B **637**, 362 (2006).

[39] S. S. Adler et al., [PHENIX Collaboration], Phys. Rev. C **72**, 024901 (2005).

[40] R. Rapp, H. van Hees, J. Phys. G **32**, S351 (2006).

Physics and Astrophysics of Hadrons and Hardronic Matter
Editor: A. B. Santra

Dark Matter – Possible Candidates and Direct Detection

Debasish Majumdar

Saha Institute of Nuclear Physics,
1/AF Bidhannagar, Kolkata 700 064, India
email: debasish.majumdar@saha.ac.in

Abstract

The cosmological observations coupled with theoretical calculations suggest the existence of enormous amount of unseen and unknown matter or dark matter in the universe. The evidence of their existence, the possible candidates and their possible direct detections are discussed.

1   Introduction

The observations by Wilkinson Microwave Anisotropy Probe or WMAP [1] for studying the fluctuations in cosmic microwave background radiation reveal that the universe contains 27% matter and the rest 73% is an unknown energy known as Dark Energy. Out of this 27%, only 4% accounts for the ordinary matter like leptons and baryons, stars and galaxies etc. The rest 23% is completely unknowm. Moreover, there are strong indirect evidence (gravitational) from various observations like velocity curves of spiral galaxies, gravitational lensing etc. in favour of the existence of enormous amount of invisible, nonluminous matter in the universe. The measurement of mass-luminousity ratio which can be used to determine the cosmological density parameter also estimates a very low value for luminous matter. This huge amount of unknown and "unseen" matter (which in fact constitutes more than 90% of the total matter content of the universe) is known as "Dark Matter".

Although the nature and identity of dark matter still remain a mystery, indirect evidence suggests that they are stable and probably heavy, non relativistic (Cold Dark Matter or CDM) and are weakly interacting. Therefore they are often known as Weakly Interacting Massive Particles or WIMPs.

In this article, the properties, types and the possible candidates of dark matter are discussed. The possibilities of their direct detection and theoretical detection rates are also given.

2 Cosmological Density Parameter

The space-time metric consistent with the homogeneity and isotropy of the universe – on large scales – can be given by the Robertson-Walker (RW) metric

$$ds^2 = -dt^2 + a^2(t)\left[\frac{dr^2}{1 - kr^2} + r^2(d\theta^2 + \sin^2\theta d\phi^2)\right] \tag{1}$$

Here $a(t)$ is a scale factor and k denotes the spatial curvature. Thus, $k = +1$ means the spatial section is positively curved, i.e. the space is locally isometric to 3-D spheres; $k = -1$ signifies that the space is locally hyperbolic (spatial section is negatively curved); and finally $k = 0$ signifies no spatial curvature, i.e. a flat geometry for the local space. Therefore for $k = 0$ the ordinary 3-dimensional geometry holds and the sum of three angles of a triangle in this space is indeed π.

The RW metric follows from the kinematic consequences. The dynamics, i.e. the time evolution of the scale factor $a(t)$ follows by applying Einstein's equation (with the cosmological constant Λ)

$$R_{\mu\nu} - \frac{1}{2}g_{\mu\nu}R = 8\pi G T_{\mu\nu} + \Lambda \tag{2}$$

to the RW metric. In the above, $R_{\mu\nu}$ are Ricci Tensors, R is the Ricci scalar, $g_{\mu\nu}$ is the spatial metric, $T_{\mu\nu} = (\rho + p)U_\mu U_\nu + pg_{\mu\nu}$ is the energy-momentum tensor contains the density ρ and pressure p. The Einstein's equation relates the geometry with the energy-momentum.

Applying Einstein's equation to cosmology, one gets the Friedmann's equation

$$\left(\frac{1}{a}\frac{da}{dt}\right)^2 = \frac{8\pi G}{3}\rho + \frac{\Lambda}{3} - \frac{k}{a^2} \tag{3}$$

Defining $\frac{1}{a}\frac{da}{dt} = H$, the expansion rate of the universe or formally Hubble constant, the above equation can be written as

$$\frac{k}{H^2a^2} = \frac{8\pi G}{3H^2}\rho + \frac{\Lambda}{3H^2} - 1 \tag{4}$$

Defining $\frac{3H^2}{8\pi G} = \rho_c$ – the critical density of the universe, the above equation takes the form

$$\begin{aligned}
\frac{k}{H^2a^2} &= \frac{\rho}{\rho_c} + \frac{\rho_\Lambda}{\rho_c} - 1 \\
&= \Omega_m + \Omega_\Lambda - 1
\end{aligned} \tag{5}$$

where $\Omega_m = \frac{\rho}{\rho_c}$ and $\Omega_\Lambda = \frac{\rho_\Lambda}{\rho_c}$ are the cosmological density parameters for matter and energy respectively. For a flat universe ($k = 0$) we have therefore

$$\Omega = \Omega_m + \Omega_\Lambda = 1 \tag{6}$$

The analysis of WMAP probe predicts curvature parameter $k = 0$ (the universe is spatially flat)[1] and therefore the matter density of the universe

$$\Omega_m = \Omega_{\text{visible}} + \Omega_{\text{DM}} = 0.27$$

out of which

$$\Omega_{\text{visible}} = 0.4 \quad \text{and} \quad \Omega_{\text{DM}} = 0.23$$

where 'DM' stands for the dark matter and the energy density (unknown dark energy) Ω_Λ is

$$\Omega_\Lambda = 0.73$$

3   Evidence of the existence of Dark Matter

The evidence of dark matter was first envisaged by the observation of motion of galaxies in cluster of galaxies like Virgo and Coma. A galaxy cluster is a gravitationally-bound group of galaxies[2]. Assuming the dynamical equilibrium of the cluster, it obeys the Virial theorem, $K + U/2 = 0$, where K is the kinetic term and U the potential. The kinetic term K was estimated by measuring the velocities of individual galaxies and is found to be much larger than the potential term U which was calculated by assuming that the mass of the cluster is the sum of the individual mass of the galaxies. This discrepancy indicates the existence of unseen and unknown mass in the cluster.

Stronger observational evidence exists by studying the rotational velocities of the stars inside a galaxy (rather than observing the galaxy itself inside a cluster). For a star in a spiral galaxy – which can be considered as a rotating disc with a central bulge where most of the galactic mass is concentrated – describing a circular orbit at a radial distance r from the centre of the galaxy, with a rotational velocity v_r, one has

$$\frac{mv_r^2}{r} = \frac{GM_r m}{r^2} \tag{7}$$

where m is the mass of the star and M_r is the mass inside the orbit of radius r. If the object is inside the central concentrated mass region, the mass M_r can be estimated as

$$M_r = \frac{4}{3}\pi r^3 \rho \tag{8}$$

---

[1]A stringent limit is however put for $\frac{k}{H^2 a^2} = \Omega_k = -0.003 \pm 0.010$ [2].

[2]A cluster can be rich with thousand(s) of galaxies or can be poor with $\sim$ 30 - 40 galaxies. The cluster, Local Group, to which our galaxy – Milky Way – belongs contains only about 30 galaxies.

where ρ is the average density of the central region. From Eq. (7) therefore we readily see

$$v_r \sim r \qquad (9)$$

Now, for a star outside the central bulge, one can approximate $M_r = M$ (a constant, neglecting the mass outside the central bulge) and in this case the nature of rotational velocity v_r becomes (from Eq. (7))

$$v_r \sim \frac{1}{r^{1/2}} \quad \text{(Keplerian Decline)} \qquad (10)$$

Hence normally, for rotational velocities $v_r(r)$, one would expect an initial rise with increase of radial distance r from the galactic centre (Eq. (7)) and then a Keplerian decline for radial distance r outside the central bulge.

Instead, the observation of rotation curves (variation of v_r with r) reveal the initial rise of v_r with r as expected but then $v_r(r)$ becomes a constant with the increase of r instead of suffering the $r^{-1/2}$ decline. Hence from Eq. (7), with v_r constant

$$M_r \sim r \qquad (11)$$

which suggests the existence of enormous unknown mass.

The evidence of dark matter is indicated from other observations like measurement of temperature and density of hot X-ray emitting gases from elliptical galaxies like M87.

The other evidence comes from the observance of the phenomenon of gravitational lensing. This occurs due to the bending of light in presence of gravitational potential. The mass of a cluster and hence Ω_m can be estimated by exploring the multiple lense effects of background galaxies produced by the cluster. These observations also point to $\Omega_m \sim 0.3$.

4   Types of Dark Matter

On the basis of the nature of the constituents, the dark matter can be divided into two types namely a) baryonic and b) non-baryonic. Different cosmic microwave anisotrpy (CMB) measurements predict a value of baryon density to be $\Omega_b \sim 0.04$ which is far less then the total dark matter density $\Omega_{DM} = 0.23$. This is indicative of the fact that the most of the dark matter in the universe is non-baryonic in nature.

Again, on the basis of their velocities, the dark matter can be broadly classified as a) Hot Dark Matter (HDM) and b) Cold Dark Matter (CDM). For HDM, the particle candidates are light and hence move with relativistic velocity while the CDM candidates are heavy and move with non-relativistic velocities. If a candidate falls in between the two categories they are sometimes referred to as Warm Dark Matter.

Neutrinos can be a possible candidate for Hot Dark Matter, but their relic density falls far short of the total dark matter density, 0.23, if the neutrinos are indeed light ($\sim$ eV). It is general wisdom that, most of the dark matter of the universe is Cold type (CDM) and non-baryaonic in nature.

5   Candidates for Dark Matter

The dark matter candidates still remain an enigma. But the fact that they constitute more than 90% of the matter content of the universe and their little or no interaction with any Standard Model particles indicate that they are made up of stable, neutral and very weakly (or almost non-) intearcting particles. Also most of their constituents are massive (heavy) to account for that large mass.

The known particles like baryons are proposed but as is discussed earlier baryons alone cannot explain the total dark matter of the universe. But some of the dark matter may need to be baryonic as $\Omega \lesssim 0.01$ in the galactic disk.

There are other candidates (baryonic) for dark matter like jupiter-like objects, dead massive stars etc. But they fail to account for the density $\Omega_{DM} = 0.23$.

Recently there has been experimental evidence of at least one form of dark matter namely Massive Astrophysical Compact Halo Objects or MACHOs in the halo of Milky Way galaxy. The light from a distant star, passing by a MACHO, bends due to the large gravitational field of the MACHO. The bending of light is a consequence of Einstein's General Theory of Relativity and as discussed above, is known as gravitational lensing. In the present case, since the lens is relatively small (compared to galaxy), multiple images are not observed. On the other hand, due to relative motion between the stars and MACHOs the lensing effect causes an increase in the brightness of that distant object. Using this phenomenon, known as gravitational microlensing, around 13-17 MACHOs have been detected in the Milky Way Halo.

A candidate for MACHOs has been proposed in Ref. [3]. It is suggested that MACHOs have evolved out of the strange quark nuggets (SQNs) formed during the first order phase transition of the early universe from quark phase to hadronic phase at a temperature around 100 MeV ($\sim 10^{-5}$ second after Big Bang). During this phase transition, hadronic matter starts to appear as individual bubbles in quark-gluon phase [4, 5]. With the progress of time more bubbles appear and they expand to form a network of such bubbles (percolation) in which the quark matter gets trapped. With further cooling of the universe, these trapped domains of quark matter shrink very rapidly without significant change of baryon number and eventually evolves to SQNs through weak interactions with almost nuclear density [6]. These objects are stable and calculation shows that to explain all the CDM, the baryon number of an SQN should be $\sim 10^{42-44}$ [7] assuming all SQNs to be of same size. These SQNs with masses $\sim 10^{44}$ GeV and size ~ 1 metre, would have very small kinetic energy compared to their mutual Gravitational potential.

Among the possible candidates of light non baryonic dark matter, come the relic neutrinos. But as briefly discussed earlier, the light neutrinos cannot account for the dark matter relic density obtained from, say, WMAP observation.

Another viable light dark matter candidate is axion. Axion is a pseudo-Goldstone boson and is introduced to solve the strong CP problem [8] (conservation of CP symmetry in Quantum chromodynamics or QCD). It arises as a consequence of a global $U(1)$ symmetry (Peccei-Quinn symmetry). The axions gets a small mass due to the breaking of this global $U(1)$ symmetry. Axions can also be produced in supernova. But the QCD consideration alongwith the production process of axions in supernova [9] (through nucleon-nucleon Bremsstrahlung), it is estimated that axion can be a dark matter candidate within a very limited window [10].

For the particle candidates of Cold Dark Matter or CDM that are non-baryonic in nature, there are various proposals. These candidates are not Standard Model (SM) particles and follow from the theories beyond SM like Supersymmetric theories or theories with extra dimensions. These particles if existed would have manifested themselves at higher energy scales during the very early phase of the universe. With the expansion of the universe, when the annihilation rate of these particles fall below the expansion rate of the universe, these particles get decoupled from the universe fluid and remain as they were. This phenomenon is known as "freeze out". After the freeze out takes place those particles float around as relics.

The popular and favourite candidate for non-baryonic CDM is proposed from theory of Supersymmetry or SUSY. Supersymmetry is the symmetry between fermions and bosons or rather more precisely the symmetry between the fermionic and bosonic degrees of freedom. This is introduced to address the so called "hierarchy problem" or "Weak scale instability problem". The hierarchy such as W-boson mass $m_W << M_p$ or the SM Higgs Boson mass $m_H << M_p$, where M_p is the Planck Mass ($1/\sqrt{G_N} \sim 10^{19}$ GeV) tends to be destroyed as a consequence of the higher order correction to the mass. The correction suffers a quadratic divergence. A fine tune of large orders of magnitude is required to restore the physical SM Higgs mass. This fine tuning in turn affects the masses of other SM fermions and gauge bosons and thus hierarchy. SUSY stabilises this hierarchy and peeps to the possibility of new physics beyond the electroweak energy scale of ~ 250 GeV.

In minimal supersymmetric standard model or MSSM (see e.g. [11]), each SM fermion has their bosonic SUSY partner and the gauge bosons have their fermionic SUSY partners. Thus in the MSSM framework, one generation in SM is to be represented by five left handed chiral superfields Q, U^c, D^c, L, E^c where the superfield Q contains quarks and their bosonic superpartner, squark $SU(2)$ doublets; U^c and D^c are the quark and squark singlets; L contains leptons and their bosonic superpartner slepton $SU(2)$ doublets and E^c contains lepton and slepton singlets. In the gauge sector however, in MSSM framework, in addition to the SM gauge bosons, we have eight gluinos, the fermionic superpartners of QCD gluons; three winos ($\tilde{W}$) the fermionic partner of $SU(2)$ gauge bosons and a bino ($\tilde{B}$), the fermionic partner of $U(1)_Y$ gauge boson. In the Higgs sector, one needs to introduce two Higgs superpartners $\tilde{H}_1$ and $\tilde{H}_2$ in order to break the $SU(2) \times U(1)_Y$. Without going into details, due to space constraints it is only mentioned that the two Higgsino doublet with hypercharge $Y = +1/2$ and $Y = -1/2$ make the model anomaly free (cancellation due to opposite hypercharge).

It is a general practice in MSSM (to ensure protection against rapid proton decay), to introduce a parity called R parity and it is assumed to be conserved. The R parity is defined as $R = (-1)^{3B+L+2S}$, where, B is the baryon number, L, the lepton number and S the spin. This ensures that the Lightest Supersymmetric Particle or LSP is stable and if it is neutral then can be a candidate for dark matter.

One such dark matter candidate is neutralino (χ) [12] which is the linear superposition of the fermionic superpartners of neutral SM gauge bosons and Higgs bosons and can be written as

$$\chi = \alpha \tilde{B} + \beta \tilde{W}^0 + \gamma \tilde{H}_1 + \delta \tilde{H}_2 \tag{12}$$

The coefficents can be obtained by diagonalizing the mass matrix (in the basis $\{ \tilde{B}, \tilde{W}^0, \tilde{H}_1, \tilde{H}_2 \}$)

$$\begin{pmatrix} M_2 & 0 & -M_Z \cos\beta \sin\theta_W & M_Z \sin\beta \sin\theta_W \\ 0 & M_1 & M_Z \cos\beta \cos\theta_W & -M_Z \sin\beta \cos\theta_W \\ -M_Z \cos\beta \sin\theta_W & M_Z \cos\beta \cos\theta_W & 0 & -\mu \\ M_Z \sin\beta \sin\theta_W & M_Z \sin\beta \cos\theta_W & -\mu & 0 \end{pmatrix} . \tag{13}$$

In the above, the parameters M_1 and M_2 are soft SUSY breaking terms, μ is the so called "μ term" in the superpotential (associated with two Higgs supermultiplates), $\tan\beta = \frac{v_2}{v_1}$, the ratio of the vev's of two Higgs.

The lightest neutralino eigenstate (LSP) of the mass matrix above (Eq. (13)) is considered to be a candidate for dark matter.

Another important proposal for dark matter candidates comes from the theories of extra higher dimensions. Although we live in a four dimensional world, there is apparently no reason to believe that extra dimensions do not exist. If dimensions > 4 do at all exist they must be so compactified that the effect due to them is not manifested in our 4-D world. The ideas and theories of extra dimensions have been proposed to look for new physics beyond standard model and to address the hierarchy problem mentioned earlier as also to explain the non SM particles like gravitons (unification of gravity and gauge interactions), cosmological constant problem etc.

The effect of compactification of one extra space dimension can be demonstrated by considering a Lagrangian density $\mathcal{L}$ for a massless 5 dimensional scalar field Φ, where one extra spatial dimension is inculded [13]. Thus (following [13])

$$\Phi \equiv \Phi(x_\mu, y), \qquad \mu = 0, 1, 2, 3 \, ; \ y \text{ is the extra spatial coordinate}$$

$$\mathcal{L} \;\; = \;\; -\frac{1}{2}\partial_A\Phi\partial^A\Phi \quad A = 0, 1, 2, 3, 4 \tag{14}$$

The extra 5th dimension is compactified over a circle of radius R so that at distance scales $>> R$, the radius of compactification, the effect of extra dimension is not manifested. It is to be noted that the field is periodic in $y \to y + 2\pi R$ ($\Phi(x, y) = \Phi(x, y + 2\pi R)$). Thus, expanding $\Phi(x, y)$ in y as

$$\Phi(x, y) = \sum_{n=-\infty}^{\infty} \phi_n(x)e^{iny/R} \tag{15}$$

(with $\phi_n^*(x) = \phi_{-n}(x)$) and substituting in the expression for $\mathcal{L}$ in Eq. (14) we have

$$\mathcal{L} = \frac{1}{2}\sum_{n,m=-\infty}^{\infty}\left(\partial_\mu\phi_n\partial^\mu\phi_m + \frac{nm}{R^2}\phi_n\phi_m\right)e^{i(n+m)y/R} \tag{16}$$

The action S is given by

$$S \;\; = \;\; \int d^4x \int_0^{2\pi R} dy \; \mathcal{L} \tag{17}$$

Replacing $\mathcal{L}$ (using Eq. (16)) and integrating out the 5th dimension to obtain the equivalent four dimensional result, the action S becomes

$$S = \int d^4x \left(-\frac{1}{2}\partial_\mu\psi_0\partial^\mu\psi_0\right) - \int d^4x \sum_{k=1}^{\infty}\left(\partial_\mu\psi_k\partial^\mu\psi_k^* + \frac{k^2}{R^2}\psi_k\psi_k^*\right) \tag{18}$$

where $\psi_n = \sqrt{2\pi R}\phi_n$. Thus, from Eq. (18) we see that for a massless scalar field in 5-dimension, compactification over a circle yields, in equivalent 4-dimensional theory, a zero mode (ψ_0) as real scalar field and an infinite number (tower) of massive complex scalar fields with tree level masses given by $m_k = k/R$. These modes are known as Kaluza-Klein modes (or Kaluza-Klein tower) and the integer k becomes a quantum number called Kaluza-Klein (KK) number which corresponds to the quantized momentum p_5 in the compactified dimension. The 5-D Lorentz invariance (local) of the tree level Lagrangian allows us to write the dispersion relation as

$$E^2 = \mathbf{p}^2 + p_5^2 = \mathbf{p}^2 + m_k^2 \tag{19}$$

where $\mathbf{p}$ is the usual 3-D momentum. The conservation of this KK number apparently seems to indicate that the Lightest Kaluza-Klein Particle or LKP is stable and can be a possible candidate for dark matter.

An LKP dark matter candidate is proposed by Cheng et al [14] in the model of universal extra dimension (UED) [15, 16]. According to UED model the extra dimension is accessible to all standard model fields. In other words all SM particles can propagate into the extra dimensional space. Therefore every SM particle has a KK tower. The proposed LKP candidate for dark matter in UED model is the first KK partner B^1, of the hypercharge gauge boson.

But in order to obtain chiral fermions in equivalent 4-D theory, the compactification over a circle (S^1) does not suffice. The simplest possibility for the purpose is to compactify the extra dimension over an orbifold S^1/Z_2 [17] where S^1 is the circle of compactification radius R and Z_2 is the reflection symmetry under which the 5th coordinate $y \rightarrow -y$. The fields can be even or odd under Z_2 symmetry. This orbifold can be looked as a line segment of length πR such that $0 \leq y \leq \pi R$ with the orbifold fixed points (boundary points) at $0, \pi R$ with two boundry conditions (Neumann and Dirichlet) for even and odd fields given by,

$$\partial_5 \phi = 0 \text{ For even fields}$$
$$\phi = 0 \text{ For odd fields} \tag{20}$$

A consistent assignment for chiral fermion ψ would be; (ψ_L even, ψ_R odd) or vice versa, for gauge field A; A_μ even ($\mu = 0, 1, 2, 3$), A_5 odd and the scalars can be either even or odd.

Now from Eq. (15) and using the orbifold compactification discussed above, the KK decomposition of Φ in even or odd fields looks as

$$\Phi_+(x, y) = \sqrt{\frac{1}{\pi R}} \phi_+^0 + \sqrt{\frac{2}{\pi R}} \sum_{n=1}^{\infty} \cos \frac{ny}{R} \phi_+^n(x)$$

$$\Phi_-(x, y) = \sqrt{\frac{2}{\pi R}} \sum_{n=1}^{\infty} \sin \frac{ny}{R} \phi_-^n(x) . \tag{21}$$

Thus, Φ_- (odd field) lacks a zero mode due to the effect of Z_2 symmetry and Eq. (21) satisfies the boundary conditions in Eq. (20). Thus we clearly see only left chiral or right chiral fermionic fields (by assigining $\psi_L(\psi_R)$ to even(odd) fields or vice versa) will have zero mode and chiral fermions can thus be identified in equivalent 4-D theory.

But this leads to problem as the boundary points $(0, \pi R)$ breaks the translational symmetry along the y direction. Thus under S^1/Z_2 orbifold compactification, the momentum p_5 is no more conserved and hence the KK number is also not conserved. This means that the stability of LKP is no more protected by the conservation of KK number.

However, it can be seen from Eq. (21) that, under a transformation πR in the y direction, the KK-modes remain invariant when the KK number n is even while the KK-modes with n odd change sign. Therefore, we readily have a quantity, $(-1)^{KK}$ which is a good symmetry and is conserved. This is called KK-parity. The conservation of this KK-parity ensures that the LKP is stable and therefore is a possible candidate for dark matter. In this context, the KK-parity serves the same purpose as the R-parity in supersymmetric models in terms of assuring stability to the dark matter candidate.

Note that the proposed dark matter candidate B^1 (as mentioned before) in universal extra dimension model is a bosonic neutral particle whereas the candidate (neutralino (χ)) in supersymmetric theory is a fermionic neutral particle. This dark matter candidate B^1 has been explored in several works (see e.g. [18–21]).

There are other possible dark matter candidates proposed from other models too. One such recently proposed candidate is lightest inert particle or LIP from the so called 'Inert Doublet' model [22]. This LIP dark matter has also been explored (see e.g. [23]).

6 Detection of Dark Matter

As the dark matter has no or very minimal interaction, it is extremely difficult to detect them. There are two types of detection processes, namely direct detection and indirect detection. In direct detection, the scattering of dark matter off the nucleus of the detecting material is utilised. As this cross-section is very small, the energy deposited by a dark matter candidate on the detector nucleus is also very small. In order to measure this small recoil energy ($\sim$ keV or less) of the nucleus, a very low threshold detector condition is required. In the indirect detection, the annihilation product of dark matter is detected. If the dark matter is entrapped by the solar gravitational field, they may annihilate with each other to produce a standard model particle such as neutrino. Such neutrino signal, if detected, is the signature of dark matter in the indirect process of their detection. In what follows, we will discuss the direct detection.

Differential detection rate of dark matter per unit detector mass can be written as

$$\frac{dR}{d|\mathbf{q}|^2} = N_T \Phi \frac{d\sigma}{d|\mathbf{q}|^2} \int f(v) dv \tag{22}$$

where N_T denotes the number of target nuclei per unit mass of the detector, Φ - the dark matter flux, v - the dark matter velocity in the reference frame of earth with $f(v)$ - its distribution. The integration is over all possible kinematic configurations in the scattering process. In the above, $|\mathbf{q}|$ is the momentum transferred to the nucleus in dark matter-nucleus scattering. Nuclear recoil energy E_R is

$$\begin{aligned} E_R &= |\mathbf{q}|^2/2m_{\mathrm{nuc}} \\ &= m_{\mathrm{red}}^2 v^2 (1 - \cos\theta)/m_{\mathrm{nuc}} \end{aligned} \tag{23}$$

$$m_{\text{red}} = \frac{m_\chi m_{\text{nuc}}}{m_\chi + m_{\text{nuc}}} \tag{24}$$

where θ is the scattering angle in dark matter-nucleus centre of momentum frame, m_{nuc} is the nuclear mass and m_χ is the mass of the dark matter.

Now expressing Φ in terms of local dark matter density ρ_χ, velocity v and mass m_χ and writing $|\mathbf{q}|^2$ in terms of nuclear recoil energy E_R with noting that $N_T = 1/m_{\text{nuc}}$, Eq. (22) takes the form

$$\frac{dR}{dE_R} = 2\frac{\rho_\chi}{m_\chi}\frac{d\sigma}{d|\mathbf{q}|^2}\int_{v_{min}}^{\infty} vf(v)dv,$$

$$v_{min} = \left[\frac{m_{\text{nuc}}E_R}{2m_{\text{red}}^2}\right]^{1/2} \tag{25}$$

Following Ref. [12] the dark matter-nucleus differential cross-section for the scalar interaction can be written as

$$\frac{d\sigma}{d|\mathbf{q}|^2} = \frac{\sigma_{\text{scalar}}}{4m_{\text{red}}^2 v^2}F^2(E_R) . \tag{26}$$

In the above σ_{scalar} is dark matter-nucleus scalar cross-section and $F(E_R)$ is nuclear form factor given by [24, 25]

$$F(E_R) = \left[\frac{3j_1(qR_1)}{qR_1}\right]\exp\left(\frac{q^2s^2}{2}\right) \tag{27}$$

$$R_1 = (r^2 - 5s^2)^{1/2}$$

$$r = 1.2A^{1/3}$$

where thickness parameter of the nuclear surface is given by $s \simeq 1$ fm, A is the mass number of the nucleus and $j_1(qR_1)$ is the spherical Bessel function of index 1.

The distribution $f(v_{\text{gal}})$ of dark matter velocity v_{gal} with respect to galactic rest frame, is considered to be of Maxwellian form. The velocity v (and $f(v)$) with respect to earth rest frame can then be obtained by making the transformation

$$\mathbf{v} = \mathbf{v}_{\text{gal}} - \mathbf{v}_\oplus \tag{28}$$

where $v_\oplus$ is the velocity of earth with respect to galactic rest frame and is given by

$$v_\oplus = v_\odot + v_{\text{orb}}\cos\gamma\cos\left(\frac{2\pi(t - t_0)}{T}\right) \tag{29}$$

In Eq. (29), $T = 1$ year, the time period of earth motion around the sun, $t_0 \equiv 2^{\text{nd}}$ June, v_{orb} is earth orbital speed and $\gamma \simeq 60^o$ is the angle subtended by earth orbital plane at galactic plane. The speed of solar system $v_\odot$ in the galactic rest frame is given by,

$$v_\odot \;=\; v_0 + v_{\text{pec}} \tag{30}$$

where v_0 is the circular velocity of the Local System at the position of Solar System and v_{pec} is speed of Solar System with respect to the Local System. The latter is also called peculiar velocity and its value is 12 km/sec. The physical range of v_0 is given by [26, 27] 170 km/sec $\le v_0 \le$ 270 km/sec (90 % C.L.). Eq. (29) gives rise to annual modulation of dark matter signal reported by DAMA/NaI experiment [28]. This phenomenon of annual modulation can be elaborated a little more. Due to the earth's motion around the sun, the directionality of the earth's motion changes over the year. This in turn induces an annual variation of the WIMP dark matter speed relative to the earth (maximum when the earth's rotational velocity adds up to the velocity of the Solar System and minimum when these velocities are in opposite directions). This phenomenon imparts an annual variation of dark matter detection rates at terrestrial detectors. Therefore investigation of annual variation of WIMP detection rate is a useful method to confirm the WIMP dark matter detection.

Defining a dimensionless quantity $T(E_R)$ as,

$$T(E_R) = \frac{\sqrt{\pi}}{2} v_0 \int_{v_{\text{min}}}^{\infty} \frac{f(v)}{v} dv \tag{31}$$

and noting that $T(E_R)$ can be expressed as [12]

$$T(E_R) = \frac{\sqrt{\pi}}{4v_\oplus} v_0 \left[\text{erf}\left(\frac{v_{\text{min}} + v_\oplus}{v_0} \right) - \text{erf}\left(\frac{v_{\text{min}} - v_\oplus}{v_0} \right) \right] \tag{32}$$

we obtain from Eqs. (25) and (26)

$$\frac{dR}{dE_R} = \frac{\sigma_{\text{scalar}}\rho_\chi}{4v_\oplus m_\chi m_{\text{red}}^2} F^2(E_R) \left[\text{erf}\left(\frac{v_{\text{min}} + v_\oplus}{v_0} \right) \right.$$
$$\left. - \text{erf}\left(\frac{v_{\text{min}} - v_\oplus}{v_0} \right) \right] \tag{33}$$

The total local dark matter density ρ_χ is generally taken to be 0.3 GeV/cm^3. The above expression for differential rate is for a monoatomic detector like Ge but it can be easily extended for a diatomic detector like NaI as well.

The measured response of the detector by the scattering of dark matter off detector nucleus is in fact a fraction of the actual recoil energy. Thus, the actual recoil energy E_R is quenched by a factor qn_X (different for different nucleus X) and we should express differential rate in Eq. (33) in terms of $E = qn_X E_R$.

Thus the differential rate in terms of the observed recoil energy E for a monoatomic detector like Ge detector can be expressed as

$$\frac{\Delta R}{\Delta E}(E) = \int_{E/qn_{Ge}}^{(E+\Delta E)/qn_{Ge}} \frac{dR_{Ge}}{dE_R}(E_R)\frac{dE_R}{\Delta E} \tag{34}$$

and for a diatomic detector like NaI, the above expression takes the form

$$\begin{aligned}
\frac{\Delta R}{\Delta E}(E) &= a_{Na}\int_{E/qn_{Na}}^{(E+\Delta E)/qn_{Na}} \frac{dR_{Na}}{dE_R}(E_R)\frac{dE_R}{\Delta E} \\
&+ a_I\int_{E/qn_I}^{(E+\Delta E)/qn_I} \frac{dR_I}{dE_R}(E_R)\frac{dE_R}{\Delta E}
\end{aligned} \tag{35}$$

where a_{Na} and a_I are the mass fractions of Na and I respectively in a NaI detector.

$$a_{Na} = \frac{m_{Na}}{m_{Na}+m_I} = 0.153 \quad a_I = \frac{m_I}{m_{Na}+m_I} = 0.847$$

The differential detection rates $\Delta R/\Delta E$ (/kg/day/keV) can thus be calculated for the case of a particular detector material.

There are certain ongoing experiments and proposed experiments for WIMP direct search. The target materials generally used are NaI, Ge, Si, Xe etc. NaI (100 kg) is used for DAMA experiment and near future LIBRA (Large sodium Iodine Bulk for RAre processes) experiment (250 kg of NaI) [28]. These set ups are at Gran Sasso tunnel in Italy. The DAMA collaboration claimed to have detected this annual modulation of WIMP through their direct WIMP detection experiments. Their analysis suggests possible presence of dark matter with mass around 50 GeV. This result is far below the range of LKP mass. The Cryogenic Dark Matter Search or CDMS detector employs low temperature Ge and Si as detector materials to detect WIMP's via their elastic scattering off these nuclei [29]. This is housed in a 10.6 m tunnel ($\sim$ 16 m.w.e) at Stanford Underground Facility beneath the University of Stanford. Although their direct search results are compatible with 3-σ allowed regions for DAMA analysis, it excludes DAMA results if standard WIMP interaction and a standard dark matter halo is assumed. The EDELWEISS dark matter search experiment which also uses cryogenic Ge detector at Frejus tunnel, 4800 m.w.e under French-Italian Alps observed no nuclear recoils in the fiducial volume [30]. This experiment excludes DAMA results at more than 99.8% C.L. The lower bound of recoil energy in this experiment was 20 keV. The Heidelberg Dark Matter Search (HDMS) uses in their inner detector, highly pure ^{73}Ge crystals [31] and with a very low energy threshold. They have made available their 26.5 kg day analysis. The recent low threshold experiment GENIUS (GErmenium in liquid NItrogen Underground Setup) [32] at Gran Sasso tunnel in Italy has started its operation. Although a project for $\beta\beta$-decay search, due to its very low threshold (and expected to be reduced futher) GENIUS is a potential detector for WIMP direct detection experiments and for detection of low energy solar neutrinos like pp-neutrinos or ^{7}Be neutrinos. In GENIUS experiment highly pure ^{76}Ge is used as detector material. For dark matter search, 100 kg. of the detector material is suspended in a tank of liquid nitrogen. The threshold for Germenium detectors is around 11 keV. But for GENIUS, this threshold will be reduced to 500 eV. The proposed XENON detector [33] consists of 1000 kg of ^{131}Xe with 4 keV threshold.

7 Discussions

The possible nature of the still unknown and overwhelming dark matter is discussed. Different theories predict different possibilities of dark matter candidates. Due to space constraints, the calculation of relic densities of such candidates could not be addressed. The theoretical calculation for direct detection rates in case of a detector material is also outlined. The experimental detection, if conclusively confirmed, will not only help us understand the nature and the particle constituents of dark matter, also it will open new vistas in understanding the fundamental laws of nature.

References

[1] C.L. Benett, M. Halpren *et al*, Ap. J. Supp. **148**, (2003) 1; N. Jarosik *et al* (WMAP Collaboration), arXiv: astro-ph/0603452.

[2] C. Clarkson, M. Cortes and B.A. Bassett, arXiv: astro-ph/0702670 and references therein.

[3] S. Banerjee, A. Bhattacharyya, S.K. Ghosh, S. Raha, B. Sinha and H. Toki, Mon. Not. Roy. Astron. Soc. **340**, (2003) 284; arXiv: astro-ph/0211560.

[4] E. Witten, Phys. Rev. **D30**, (1984) 272.

[5] K. Iso, H. Kodama and K. Sato, Phys. Lett. B **169**, (1986) 337.

[6] J. Alam, S. Raha and B. Sinha, Ap. J. **513**, (1999) 572.

[7] A. Bhattacharyya, J. Alam, S. Sarkar, P. Roy, B. Sinha, S. Raha and P. Bhattacharjee, Phys. Rev. D **61**, (2000) 083509.

[8] J.E. Kim, Phys. Rev. Lett., **43**, (1979) 103; M.A. Shifman, A.I. Vainshtein and V.I. Zakharov, Nucl. Phys. B **166**, (1980) 493.

[9] G. Raffelt, Phys. Rep. **198**, (1990) 1.

[10] K. Olive, arXiv: astro-ph/0301505.

[11] M. Drees, arXiv:hep-ph/9611409

[12] G. Jungman, M. Kamionkowski and K. Griest, Phys. Rep. **267**, (1996) 195.

[13] G. Gabadadze, *Lectures in ICTP Summer School on Astroparticle Physics and Cosmology*, Trieste, 2002.

[14] H.C. Cheng, K.L. Feng and K.T. Matchev, Phys. Rev. Lett. **89**, (2002) 211301.

[15] T. Appelquist, H.C. Cheng and B. Dobrescu, Phys. Rev. D **64**, (2001) 035002.

[16] K. Kong, K.T. Matchev, arXiv: hep-ph/0610057.

[17] H.-C. Cheng, K.T. Matchev and M. Schmaltz, Phys. Rev. D **66**, (2002) 036005.

[18] G. Servant and T.M.P. Tait, Nucl. Phys. **B360**, (2003) 391.

[19] G. Servant and T.M.P. Tait, New J. Phys. **4**, (2002) 99.

[20] D. Majumdar, Phys. Rev. D **67**, (2003) 095910.

[21] D. Majumdar, Mod. Phys. Lett. A **18**, (2003) 1705.

[22] R. Barbieri, L.J. Hall, V.S. Rychkov, Phys. Rev. D **74**, (2006) 015007.

[23] D. Majumdar and A. Ghosal, arXiv:hep-ph/0607067.

[24] R.H. Helm, Phys. Rev. **104**, (1956) 1466.

[25] J. Engel, Phys. Lett. B **264**, (1991) 114.

[26] P.J.T. Leonard, S. Tremaine, Astrophys. J. **353**, (1990) 486.

[27] C.S. Kochanek, Astrophys. J. **457**, (1996) 228.

[28] R. Bernabei *et al*, Phys. Lett. B **533**, (2000) 4.

[29] D. Abrams *et al*, CDMS collaboration, arXiv: astro-ph/0203500.

[30] A. Benoit *et al*, EDELWEISS collaboration, arXiv: astro-ph/0206271.

[31] H.V. Klapdor-Kleingrothaus *et al*, arXiv: hep-ph/0206151.

[32] H.V. Klapdor-Kleingrothaus, B. Majorovits in *York 2000, The identification of dark matter* (2000), arXiv: hep-ph/0103079; H.V. Klapdor-Kleingrothaus, Nucl. Phys. Proc. Suppl. **110**, (2002) 364.

[33] E. Aprile et al, arXiv: astro-ph/0207670.